Springer Monographs in Mathematics

David E. Edmunds
W. Desmond Evans

Hardy Operators, Function Spaces and Embeddings

With 6 Figures

David E. Edmunds
Department of Mathematics
Sussex University
Brighton BN1 9RF, United Kingdom
e-mail: d.e.edmunds@sussex.ac.uk

W. Desmond Evans
School of Mathematics
Cardiff University
Cardiff CF24 4YH, United Kingdom
e-mail: EvansWD@cardiff.ac.uk

Library of Congress Control Number: 2004108695

The cover figure is taken from a paper by L.E. Fraenkel and is Fig. 5.5 on page 232 of the text.

Mathematics Subject Classification (2000):
26D10, 26D15, 34L20, 35J05, 35P20, 45D05, 45P05, 46B50, 46E35, 47B06, 47B10

ISSN 1439-7382

ISBN 978-3-642-06027-4 ISBN 978-3-662-07731-3 (eBook)
DOI 10.1007/978-3-662-07731-3

springeronline.com

Originally published by Springer-Verlag Berlin Heidelberg New York in 2004
Softcover reprint of the hardcover 1st edition 2004

Typeset by the authors using a Springer LaTeX macro package
Production: LE-TeX Jelonek, Schmidt & Vöckler GbR, Leipzig
Cover design: Erich Kirchner, Heidelberg

Printed on acid-free paper SPIN: 10973073 41/3142YL - 5 4 3 2 1 0

Preface

Classical Sobolev spaces, based on Lebesgue spaces on an underlying domain with smooth boundary, are not only of considerable intrinsic interest but have for many years proved to be absolutely indispensable in the study of partial differential equations and variational problems. The embedding theorems and inequalities which feature so large in first courses on function spaces are key ingredients of the proofs of existence and regularity for elliptic boundary-value problems. There have been many developments of the basic theory since its inception, and of these we distinguish two which seem to us to be of particular interest:
(i) the consequences of working on space domains with irregular boundaries;
(ii) the replacement of Lebesgue spaces by more general Banach function spaces.

Both of these arise in response to demands imposed by concrete problems. For example, the ubiquitous nature of sets with fractal boundaries makes it unnecessary to give an extended justification of (i), while (ii) is very natural when faced with (degenerate) elliptic problems in which the coefficients of the differential operator satisfy more refined conditions than in classical situations. It is to be expected that these aspects of the theory will enjoy substantial further growth, but nevertheless we believe that the present state of affairs makes it desirable to have a connected account of those parts which seem to us to have reached a degree of maturity. This book is intended to do just that. Its main themes are Banach function spaces and spaces of Sobolev type based on them, especially when the space domain involved is a so-called generalised ridged domain; integral operators of Hardy type on intervals and on trees; and the distribution of the approximation numbers (singular numbers in the Hilbert space case) of embeddings of Sobolev spaces based on generalised ridged domains. A prerequisite for reading it is a good graduate course in real analysis.

Chapter 1 contains a variety of results and concepts which will be useful throughout the remainder of the book. It is for consultation, to be looked at when necessary. The next chapter is largely concerned with mappings T :

$L_p(a,b) \to L_q(a,b)$, where $1 \le p, q \le \infty$ and $-\infty < a < b \le \infty$, given by

$$(Tf)(x) = v(x) \int_a^x u(t)f(t)dt, \; x \in (a,b), \; f \in L_p(a,b).$$

Here u and v are given functions which satisfy certain integrability conditions. These operators are said to be of Hardy type, the original Hardy operator being given by taking $u = v = 1$ and $p = q$. They appear in a very natural way in connection with embeddings of Sobolev spaces based on the particular class of generalised ridged domains, as we shall see in Chapters 5 and 6; they also are of importance in some 'small ball' problems in probability theory, for which we refer to [157]. We give criteria for T to be bounded and determine its measure of non-compactness. This enables us to provide necessary and sufficient conditions for T to be compact. In the compact case, we furnish upper and lower bounds for the approximation numbers $a_n(T)$ of T which, when $p = q \in (1, \infty)$, lead to the interesting asymptotic result that

$$\lim_{n\to\infty} n a_n(T) = \frac{\gamma_p}{2} \int_a^b |u(t)v(t)| \, dt,$$

where $\gamma_p = \frac{1}{\pi} p^{1/p'} (p')^{1/p} \sin(\pi/p)$, $p' = p/(p-1)$, under appropriate conditions on u and v Further refinement of this is possible, giving remainder estimates for the approximation numbers. Finally, we show that many of these results can be taken over to the situation in which the interval (a,b) is replaced by a tree: this will be of crucial importance when we come in Chapters 5 and 6 to deal with embeddings of Sobolev spaces when the underlying subset of $\mathbb{R}^n$ is a generalised ridged domain. Amongst the byproducts of our analysis is the result (originally proved in [68] and [87]) that if (a,b) is a bounded interval in $\mathbb{R}$ and $1 < p < \infty$, then the approximation numbers $a_m(E_0)$ of the embedding E_0 of the Sobolev space $\mathring{W}_p^1(a,b)$ in $L_p(a,b)$ are given by

$$a_m(E_0) = \frac{\gamma_p}{2m}(b-a) \quad (m \in \mathbb{N}).$$

The precision of this owes much to the one-dimensional nature of the underlying domain (a,b), and is in marked contrast to known results (see, for example, [74]) concerning the approximation numbers of Sobolev embeddings of spaces based on open subsets of $\mathbb{R}^n$, which typically give sharp upper and lower bounds for these numbers but do not establish a genuine asymptotic behaviour.

In Chapter 3 we give an account of Banach function spaces and spaces of Sobolev type based on them. Our object here is to present, in a systematic way, some of the refinements of the classical Sobolev embedding theorems which have become known in the last ten years or so and which result from the use of scales of spaces which can be more finely tuned than the Lebesgue family. Originally, many of these results were proved for Sobolev spaces based on Lorentz-Zygmund or generalised Lorentz-Zygmund spaces rather than on

L_p, as in the classical situation. The strategy which we adopt, however, is to use Lorentz-Karamata spaces, which depend on the notion of slowly varying functions. By this means we are able to give a unified approach to the subject which is both more economical than a succession of ad hoc arguments used for particular circumstances and also helps to clarify the nature of the arguments which are deployed. As an illustration of the kind of result which we establish, consider the famous example involving the Sobolev space $W_n^1(\Omega)$, where Ω is a bounded domain in $\mathbb{R}^n$ $(n > 1)$ with smooth boundary. It is very well known that this space can be embedded in every $L_p(\Omega)$ space with $p \in (1, \infty)$, that it cannot be embedded in $L_\infty(\Omega)$, but that it can be embedded in an Orlicz space of exponential type, with Young function given by $\exp(t^{n/(n-1)}) - 1$. It turns out that if instead of using $L_n(\Omega)$ as the base for this Sobolev space we use an appropriate nearby Zygmund space, then the target space can be an Orlicz space of multiple exponential type.

Chapter 4 provides a discussion of Poincaré inequalities in the general setting of spaces $W(X, Y)$ of Sobolev type: here X and Y are Banach function spaces on a bounded domain Ω in $\mathbb{R}^n$, while $W(X, Y)$ is the set of all $f \in X$ with distributional gradient in Y. Connections are made between the Poincaré inequality and the measure of non-compactness of the embedding of $W(X, Y)$ in X, and numerous illustrative examples are given, including the cases in which X and Y are of Lebesgue or Orlicz type. The chapter contains a treatment of classical Sobolev and Poincaré inequalities under very weak conditions on Ω, such as that it should be a John domain. There is some discussion of the bewildering array of weak conditions on domains which can be found in the literature. To conclude we deal with the higher-dimensional analogue of the Hardy inequality handled in Chapter 2, again under quite weak conditions on the underlying space domain.

The generalised ridged domains which form the subject of Chapter 5 are certain domains in $\mathbb{R}^n$ which, roughly speaking, have a central axis, the so-called generalised ridge, which is the image in the domain of a tree under a Lipschitz map. This is a wide class of domains, and includes such diverse sets as horns, spirals, 'rooms and passages' and even snowflakes. There are two problems, in particular, addressed in this chapter: for Banach function spaces X, Y, Z defined on a generalised ridged domain Ω, what target space Z is permissible in an embedding $W(X, Y) \hookrightarrow Z$, and when is there a valid Poincaré inequality associated with the embedding $W(X, Y) \hookrightarrow X$? The second problem is shown to be related to the range of values of the measure of non-compactness of the embedding, a result originally due to Amick [6] when $X = Y = L_2$. Furthermore, it is proved that Poincaré-type inequalities yield the measure of non-compactness as a limit along a filter base of subsets of Ω, which makes precise the fact that any lack of compactness of the embedding is due to the singular nature of the set at the intersection of the boundary of Ω and the generalised ridge. The significance of generalised ridged domains stems from the fact that these problems can be reduced to corresponding ones on the associated trees, which, in certain cases, are simply intervals. This 'one-dimensionalisation' of the problems is a considerable advantage, and it is precisely this that makes tractable the problems which we study. Chapter 6 deals with the approximation numbers of Sobolev embeddings when the underlying space domain is a generalised ridged domain, and gives upper and lower estimates for these numbers. A point of interest is that we provide an L_p version of the Dirichlet-Neumann bracketing technique which is so familiar and effective in the L_2 theory of eigenvalues of elliptic operators.

Chapters are divided into sections, and some sections into subsections; the standard decimal classification is used. All chapters except the first contain a 'Notes and Remarks' section at their end, in which we provide supplementary information. Although we have not made a serious attempt to go into the history of the results given, we hope that the list of references will not be regarded as too exiguous and believe that the reader interested in historical matters will find it to be of some help. In addition to the bibliography, there is a glossary of terms and notation, together with author and subject indexes.

Brighton, Cardiff,
June 2004

David E. Edmunds,
W. Desmond Evans

Contents

Basic symbols

$B(x, r)$: open ball in $\mathbb{R}^n$, centre x and radius r.
$\mathbb{N}$: natural numbers.
$\mathbb{N}_0 = \mathbb{N} \cup \{0\}$.
$\mathbb{Z}$: integers.
$\mathbb{R}$: real numbers.
$\mathbb{C}$: complex numbers.
$A \subset B$: A contained in B, or possibly equal to B.
$A \subset\subset B$: the closure of A is compact and is contained in B.
Ω: open subset of $\mathbb{R}^n$; Ω is a domain if it is also connected.
$\partial\Omega$: boundary of Ω.
$\overline{\Omega}$: closure of Ω.
$\mu_n = |\cdot|_n$: n-dimensional Lebesgue measure.
$|\Omega|$: Lebesgue n-measure of $\Omega \subset \mathbb{R}^n$.
$d(x, F)$: distance of x from set F; also written as $d_F(x)$ and $d(x)$.
χ_E: characteristic function of E.
$D_i u = \partial u / \partial x_i$.
$D^\alpha u = \partial^{|\alpha|} u / \partial x_1^{\alpha_1} \cdots \partial x_n^{\alpha_n}$ if $\alpha = (\alpha_1, \cdots, \alpha_n)$, each $\alpha_j \in \mathbb{N}_0$, $|\alpha| = \alpha_1 + \cdots + \alpha_n$.
$\delta_+ = \max(\delta, 0)$.
$\|\cdot|X\|$: norm or quasi-norm on X.
$\ker(f)$: kernel of f.
$C^k(\Omega), k \in \mathbb{N}_0$: complex-valued functions f such that $D^\alpha f$ is continuous on Ω for $0 \leq |\alpha| \leq k$; denoted by $C(\Omega)$ for $k = 0$.
$C^\infty(\Omega) = \bigcap_{k=0}^\infty C^k(\Omega)$.
$C^\lambda(\overline{\Omega}), \lambda \in (0, 1]$: Hölder-continuous functions of exponent λ on $\overline{\Omega}$.
$C_0^k(\Omega)$: functions in $C^k(\Omega)$ with compact supports in Ω.
$C_0^\infty(\Omega) = \bigcap_{k=0}^\infty C_0^k(\Omega)$.
$L_p(\Omega)$: Lebesgue space of functions f with $|f|^p$ integrable on Ω if $0 < p < \infty$, $\operatorname{ess\,sup} u < \infty$ if $p = \infty$.
$L_{p,loc}(\Omega)$: functions in $L_p(K)$ for every compact subset K of Ω.
$l_p(\mathcal{I})$: space of sequences $\{x_i\}_{i \in \mathcal{I}}$, $x_i \in \mathbb{C}$, such that $\|\{x_i\}|l_p(\mathcal{I})\| < \infty$, where

$$\|\{x_i\}|l_p(\mathcal{I})\| = \begin{cases} \left(\sum_{i \in \mathcal{I}} |x_i|^p\right)^{1/p}, & \text{for } 1 \leq p < \infty, \\ \sup_{i \in \mathcal{I}} |x_i|, & \text{for } p = \infty. \end{cases}$$

$\omega_n = \pi^{n/2} / \Gamma(1 + \frac{n}{2})$: volume of unit ball in $\mathbb{R}^n$.
$c \lesssim d, d \gtrsim c$: c is bounded above by a multiple of d, the multiple being independent of any variables in c and d.
$c \approx d : c \lesssim d$ and $d \lesssim c$.
$X \hookrightarrow Y$: X is continuously embedded in Y.
$X \hookrightarrow\hookrightarrow Y$: X is compactly embedded in Y.
$f * g$: convolution of f and g.

1

Preliminaries

In this Chapter we collect some definitions and results which will be useful later in the book. Virtually no proofs are given, but we provide references to works where these matters are dealt with in a comprehensive way. All vector spaces which will be mentioned will be assumed to be over the complex field, unless otherwise stated.

1.1 Hausdorff and Minkowski dimensions

Definition 1.1.1. *Given any $s \geq 0$, $\varepsilon > 0$ and $E \subset \mathbb{R}^n$, we put*

$$H_\varepsilon^s(E) = \inf\left\{\sum_{j=1}^{\infty} \omega_s 2^{-s}(\text{diam } A_j)^s : E \subset \bigcup_{j=1}^{\infty} A_j, \text{diam } A_j < \varepsilon\right\},$$

where

$$\omega_s = \pi^{s/2}/\Gamma\left(\frac{s}{2}+1\right).$$

Note that H_ε^s is an outer measure on $\mathbb{R}^n$. Since $H_\varepsilon^s(E)$ is non-decreasing in ε, we define

$$H^s(E) = \lim_{\varepsilon \to 0} H_\varepsilon^s(E)$$

and call this the $s-$**dimensional Hausdorff outer measure of** E. The restriction of H^s to the $\sigma-$field of H^s-measurable sets (which can be shown to include the Borel sets) is called the $s-$**dimensional Hausdorff measure**

The reason for introducing H^s is to provide a means of distinguishing between various lower-dimensional subsets of $\mathbb{R}^n$. We summarise some of the more important properties of H^s as follows and refer to [90], [93] and [79] for proofs and further information:

(i) H^s is a Borel regular measure;

(ii) H^0 is counting measure;

(iii) H^n coincides with $n-$dimensional Lebesgue measure μ_n on $\mathbb{R}^n$;

(iv) if $s > n$, H^s is the zero measure on $\mathbb{R}^n$;
(v) $H^s(\lambda E) = \lambda^s H^s(E)$ for all $s \geq 0$, all $\lambda > 0$ and all $E \subset \mathbb{R}^n$;
(vi) if $f : \mathbb{R}^m \to \mathbb{R}^n$ is Lipschitz-continuous on $\mathbb{R}^m$ (that is, there exists $C > 0$ such that for all $x, y \in \mathbb{R}^m$, $|f(x) - f(y)| \leq C|x - y|$, where $|\cdot|$ denotes the corresponding Euclidean distance), then for all $E \subset \mathbb{R}^n$ and all $s \geq 0$,

$$H^s(f(E)) \leq C^s H^s(E).$$

To help us define the Hausdorff dimension of subsets of $\mathbb{R}^n$ the following Lemma will be very useful.

Lemma 1.1.2. *Let $E \subset \mathbb{R}^n$ and suppose that $0 \leq s < t < \infty$. If $H^s(E) < \infty$, then $H^t(E) = 0$; if $H^t(E) > 0$, then $H^s(E) = \infty$.*

Proof. First suppose that $H^s(E) < \infty$, and let $\delta > 0$. Then for some sets A_j ($j \in \mathbb{N}$) with diam $A_j \leq \delta$ and $E \subset \cup_{j=1}^{\infty} A_j$ we have

$$\sum_{j=1}^{\infty} \omega_s 2^{-s} (\text{diam } A_j)^s \leq H^s_\delta(E) + 1 \leq H^s(E) + 1.$$

Hence

$$H^t_\delta(E) \leq \sum_{j=1}^{\infty} \omega_t 2^{-t} (\text{diam } A_j)^t$$

$$= \frac{\omega_t}{\omega_s} 2^{s-t} \sum_{j=1}^{\infty} \omega_s 2^{-s} (\text{diam } A_j)^s (\text{diam } A_j)^{t-s}$$

$$\leq \frac{\omega_t}{\omega_s} 2^{s-t} \delta^{t-s} \{H^s(E) + 1\}.$$

Now let $\delta \to 0$: it follows that $H^t(E) = 0$. The rest is obvious. □

Definition 1.1.3. *The* ***Hausdorff dimension*** *of a subset E of $\mathbb{R}^n$ is defined to be*

$$d(E) = \inf \{s \in [0, \infty) : H^s(E) = 0\}.$$

Note that in view of Lemma 1.1.2,

$$H^s(E) = \begin{cases} 0 \text{ if } s > d(E), \\ \infty \text{ if } s < d(E). \end{cases}$$

Moreover, $0 \leq d(E) \leq n$; $d(E)$ need not be an integer, and even if $d(E) = k \in \mathbb{N}$, E need not be a "$k-$dimensional surface" in any reasonable sense.

Now we turn to the Minkowski dimension. Let E be a compact subset of $\mathbb{R}^n$, let $\varepsilon > 0$ and put

$$E_\varepsilon = \{x \in \mathbb{R}^n : d(x, E) < \varepsilon\}.$$

Definition 1.1.4. *Given any compact subset E of $\mathbb{R}^n$ and any $d \geq 0$,*

$$M^d(E) := \limsup_{\varepsilon \to 0+} \varepsilon^{-(n-d)} |E_\varepsilon|_n$$

*is the $d-$**dimensional upper Minkowski content** of E. Here $|E|_n$ is the Lebesgue n-measure of E. The **Minkowski dimension** of E is defined to be*

$$d_M(E) = \inf \left\{ d \geq 0 : M^d(E) = 0 \right\} = \sup \left\{ d \geq 0 : M^d(E) = \infty \right\}.$$

If $0 < M^{d_M(E)}(E) < \infty$ and

$$M^{d_M(E)}(E) = \lim_{\varepsilon \to 0+} \varepsilon^{-(n-d)} |E_\varepsilon|_n ,$$

*then E is said to be **Minkowski-measurable** and $M^{d_M(E)}(E)$ is called the **Minkowski measure** of E. It is known (see, for example, [153]) that if Ω is a non-empty open subset of $\mathbb{R}^n$ with boundary $\partial\Omega$, then $n-1 \leq d_M(\partial\Omega) \leq n$.*

1.2 The area and coarea formulae

These important formulae relate to functions $f : \mathbb{R}^m \to \mathbb{R}^n$, where $m, n \in \mathbb{N}$, which are **Lipschitz-continous** so that there is a constant $C > 0$ such that for all $x, y \in \mathbb{R}^m$,

$$|f(x) - f(y)| \leq C |x - y| .$$

The weaker notion of a **locally Lipschitz** function is also useful: by this we mean that for each compact set $K \subset \mathbb{R}^m$, there is a constant $C(K)$ such that for all $x, y \in K$,

$$|f(x) - f(y)| \leq C(K) |x - y| .$$

An important result for such functions is

Theorem 1.2.1. *(Rademacher's theorem) Let $f : \mathbb{R}^m \to \mathbb{R}^n$ be locally Lipschitz and let μ_m be Lebesgue $m-$measure on $\mathbb{R}^m$. Then f is differentiable μ_m-a.e. on $\mathbb{R}^m$.*

From this it is not difficult to prove

Corollary 1.2.2. *Let $f : \mathbb{R}^m \to \mathbb{R}^n$ be locally Lipschitz. Then its derivative $Df(x)$ is zero for μ_m-a.e. $x \in \ker(f)$.*

For these results see [79].

We shall also need the Jacobian of a Lipschitz map $f : \mathbb{R}^m \to \mathbb{R}^n$. For this we recall some basic facts concerning linear maps $L : \mathbb{R}^m \to \mathbb{R}^n$. Such a map is called **orthogonal** if $(Lx, Ly)_n = (x, y)_m$ for all $x, y \in \mathbb{R}^m$; it is **symmetric** if $m = n$ and $(x, Ly)_m = (Lx, y)_m$ for all $x, y \in \mathbb{R}^m$. Here $(\cdot, \cdot)_k$ denotes the inner product in $\mathbb{R}^k$. If $m \leq n$, there are a symmetric map $S : \mathbb{R}^m \to \mathbb{R}^m$ and an orthogonal map $O : \mathbb{R}^m \to \mathbb{R}^n$ such that $L = O \circ S$; if $m \geq n$, there are a symmetric map $S : \mathbb{R}^n \to \mathbb{R}^n$ and an orthogonal map $O : \mathbb{R}^n \to \mathbb{R}^m$ such that $L = S \circ O^*$, where O^* is the adjoint of O. In both cases we define the **Jacobian** of L to be $|\det S|$. For details of all this, and for a proof that the Jacobian is well-defined (that is, independent of the particular choices of O and S) we refer to [79].

Returning to our Lipschitz map $f : \mathbb{R}^m \to \mathbb{R}^n$, we note that by Rademacher's theorem, f is differentiable μ_m−a.e., so that its derivative $Df(x)$ exists and corresponds to a linear map from $\mathbb{R}^m$ to $\mathbb{R}^n$ for μ_m−a.e. $x \in \mathbb{R}^m$. The **Jacobian** of f at μ_m−a.e. $x \in \mathbb{R}^m$ is defined to be the Jacobian of this linear map and is denoted by $Jf(x)$.

After these preliminaries we can give the *area theorem*:

Theorem 1.2.3. *Let $m, n \in \mathbb{N}$, $m \leq n$, let $f : \mathbb{R}^m \to \mathbb{R}^n$ be Lipschitz-continuous and let A be a μ_m−measurable subset of $\mathbb{R}^m$. Then*

$$\int_A Jf dx = \int_{\mathbb{R}^n} H^0 \left(A \cap f^{-1}(y)\right) dH^m(y).$$

The corresponding result when $m \geq n$ is the *coarea theorem* :

Theorem 1.2.4. *Let $m, n \in \mathbb{N}$, $m > n$, let $f : \mathbb{R}^m \to \mathbb{R}^n$ be Lipschitz-continuous, let $g \in L_1(\mathbb{R}^m)$ and let A be a μ_m−measurable subset of $\mathbb{R}^m$. Then*

$$\int_A g(x) Jf(x) dx = \int_{\mathbb{R}^n} \int_{A \cap f^{-1}(y)} g(x) dH^{m-n}(x) dy.$$

For a proof of these important theorems we refer to [79] and [93].

From the special case $n = 1$, $m > 1$, $A = \mathbb{R}^m$ of the coarea theorem we have a result of particular interest.

Corollary 1.2.5. *Let $f : \mathbb{R}^m \to \mathbb{R}$ be Lipschitz-continuous and let $g \in L_1(\mathbb{R}^m)$. Then*

$$\int_{\mathbb{R}^m} g(x) \left|\nabla f(x)\right| dx = \int_0^\infty \int_{\{x \in \mathbb{R}^m : |f(x)| = t\}} g(x) dH^{m-1}(x) dt.$$

Proof. Just observe that $Jf = |\nabla f|$ and use Theorem 1.2.4. □

A more general version of the coarea formula will be useful. To explain this, let Ω be an open subset of $\mathbb{R}^n$ and let $f \in L_1(\Omega)$. We say that f is of **bounded variation** in Ω if its first-order distributional partial derivatives are signed Radon measures with finite total variation in Ω. The family of all functions of bounded variation on Ω is denoted by $BV(\Omega)$. If $u \in BV(\Omega)$, the distributional gradient Du of u is a vector-valued measure whose total variation $\|Du\|(\Omega)$ is a finite measure on Ω, and

$$\|Du\|(\Omega) = \sup\left\{ \int_{\Omega} u \operatorname{div} \phi \, dx : \phi \in C_0^\infty(\Omega, \mathbb{R}^n), |\phi(x)| \le 1 \text{ for all } x \in \Omega \right\}.$$

Given any $u \in BV(\Omega)$, the measure Du can be split into a part which is absolutely continuous with respect to Lebesgue measure, and a singular part. The density of the absolutely continuous part will be denoted by ∇u : thus if $u \in W_1^1(\Omega)$, $dDu = \nabla u dx$ and

$$\|Du\|(\Omega) = \int_{\Omega} |\nabla u| \, dx.$$

A set $E \subset \mathbb{R}^n$ is said to have **finite perimeter** if its characteristic function χ_E is in $BV(\mathbb{R}^n)$, in which case the **perimeter** of E is defined to be

$$P(E) = \|D\chi_E\|(\mathbb{R}^n).$$

It can be shown that sets with minimally smooth boundary, such as Lipschitz domains, have finite perimeter. The version of the coarea theorem which we shall need involves the perimeter of sets of the form

$$E_t := \{x \in \Omega : u(x) > t\}, \ t > 0, \ u \in BV(\Omega).$$

With this notation, the theorem reads as follows.

Theorem 1.2.6. *Let Ω be an open subset of $\mathbb{R}^n$ and let $u \in BV(\Omega)$. Then*

$$\|Du\|(\Omega) = \int_{\mathbb{R}} \|D\chi_{E_t}\|(\Omega) dt.$$

Moreover, if $u \in W_1^1(\mathbb{R}^n)$ (the Sobolev space consisting of functions which, together with their first-order distributional derivatives, are in $L_1(\mathbb{R}^n)$) and f is any Borel function on $\mathbb{R}^n$,

$$\int_{\mathbb{R}^n} f |\nabla u| \, dx = \int_{\mathbb{R}} \int_{\{u=t\}} f dH^{n-1}(x) dt.$$

For proofs of results of this nature we refer to the books of Giusti [100], Maz'ya [171] and Ziemer [231].

In conjunction with the coarea theorem we shall sometimes need the classical isoperimetric inequality:

Let E be a subset of $\mathbb{R}^n$ with finite $n-$measure $|E|_n$ and finite perimeter. Then

$$P(E) \ge n\omega_n^{1/n} |E|_n^{1-1/n}. \tag{1.2.1}$$

The books just mentioned may be consulted for details of this famous result. It implies that if E is a subset of $\mathbb{R}^n$ with finite $n-$measure and appropriate boundary, then

$$H^{n-1}(\partial E) \ge n\omega_n^{1/n} |E|_n^{1-1/n}. \tag{1.2.2}$$

Often it is applied in the situation where $E = \{x \in \mathbb{R}^n : |u(x)| > t\}$, $t > 0$, where u is a smooth function on $\mathbb{R}^n$ with compact support.

1.3 Approximation numbers

First, it may be helpful to give some information about quasi-normed spaces. A **quasi-norm** on a linear space X is a map $\|\cdot \mid X\| : X \to [0,\infty)$ which has the following three properties:
(i) $\|x \mid X\| = 0$ if, and only if, $x = 0$;
(ii) $\|\lambda x \mid X\| = |\lambda| \, \|x \mid X\|$ for all scalars λ and all $x \in X$;
(iii) there is a constant C such that for all $x, y \in X$;

$$\|x + y \mid X\| \leq C\left(\|x \mid X\| + \|y \mid X\|\right).$$

It is clear that $C \geq 1$. If it is possible to take $C = 1$, then (iii) is the familiar triangle inequality and $\|\cdot \mid X\|$ is a norm on X. A quasi-norm $\|\cdot \mid X\|$ defines a topology on X which is compatible with the linear structure of X : this topology has a basis of (not necessarily open) neighbourhoods of any point $x \in X$ given by the sets $\{y \in X : \|x - y \mid X\| < 1/n\}$, $n \in \mathbb{N}$. The pair $(X, \|\cdot \mid X\|)$ is said to be a **quasi-normed space** and is a special type of metrisable topological vector space. The notions of convergence and of Cauchy sequences are defined in the obvious way, and if every Cauchy sequence in X converges, to a point in X, then X is called a **quasi-Banach** space.

Let $p \in (0,1]$. By a $p-$**norm** on a linear space X is meant a map $\|\cdot \mid X\| : X \to [0,\infty)$ which has properties (i) and (ii) above and instead of (iii) satisfies (iii') $\|x + y \mid X\|^p \leq \|x \mid X\|^p + \|x \mid X\|^p$ for all $x, y \in X$.

Two quasi-norms or $p-$norms $\|\cdot \mid X\|_1$ and $\|\cdot \mid X\|_2$ on X are called **equivalent** if there is a constant $c \geq 1$ such that for all $x \in X$,

$$c^{-1} \|x \mid X\|_1 \leq \|x \mid X\|_2 \leq c \|x \mid X\|_1 .$$

It can be shown that if $\|\cdot \mid X\|_1$ is a quasi-norm on X, then there exist $p \in (0,1]$ and a $p-$norm $\|\cdot \mid X\|_2$ on X which is equivalent to $\|\cdot \mid X\|_1$; the connection between p and the constant C in (iii) is given by $C = 2^{\frac{1}{p}-1}$. Conversely, any $p-$norm is a quasi-norm with $C = 2^{\frac{1}{p}-1}$.

The standard examples of quasi-Banach spaces which are not Banach spaces are l_p and L_p, with $0 < p < 1$.

Let X, Y be quasi-Banach spaces and let $T : X \to Y$ be linear. As in the Banach space case, T is called **bounded** or **continuous** if

$$\|T\| := \sup\left\{\|Tx \mid Y\| : x \in X, \|x \mid X\| \leq 1\right\} < \infty.$$

Let X and Y be Banach spaces and let $\mathcal{B}(X,Y)$ be the space of all bounded linear maps from X to Y. If $T \in \mathcal{B}(X,Y)$ and $k \in \mathbb{N}$, the kth **approximation number** of T, denoted by $a_k(T)$, is defined by

$$a_k(T) = \inf \ \left\{\|T - L\| : L \in \mathcal{B}(X,Y), \ \text{rank}\ L < k\right\},$$

where rank $L = \dim\, L(X)$. The same definition can be used for the situation in which X and Y are quasi-Banach spaces.

It is easy to verify that if X, Y and Z are Banach spaces and $S,T \in \mathcal{B}(X,Y)$, $R \in \mathcal{B}(Y,Z)$, then
(i) $\|T\| = a_1(T) \geq a_2(T) \geq ... \geq 0$;
(ii) for all $k,l \in \mathbb{N}$,

$$a_{k+l-1}(S+T) \leq a_k(S) + a_l(T)$$

and

$$a_{k+l-1}(R \circ S) \leq a_k(R)a_l(S);$$

(iii) $a_k(T) = 0$ if, and only if, rank $T < k$;
(iv) if dim $X \geq n$ and $id : X \to X$ is the identity map, then $a_k(id) = 1$ for $k = 1, ..., n$.

With more effort (see [46], Prop. II.2.5), it can be shown that
(v) if T is compact, then $a_k(T) = a_k(T^*)$ for all $k \in \mathbb{N}$.

In view of (i) above, it is clear that

$$\alpha(T) := \lim_{k\to\infty} a_k(T)$$

exists. If $\alpha(T) = 0$, then T is the limit (in the operator norm sense) of a sequence of finite-dimensional maps and so is compact. However, if T is compact it does not follow that $\alpha(T) = 0$: this is a consequence of Per Enflo's work on the approximation problem (see [78]). Compactness of T *does* imply that the approximation numbers converge to zero if Y has the **bounded approximation property** (see [228]): we recall that this means that there is a constant C such that for every finite subset F of Y and every $\varepsilon > 0$, there is a bounded linear map $L : Y \to Y$ with finite rank such that $\|Ly - y \mid Y\| \leq \varepsilon$ for all $y \in F$, and $\|L\| \leq C$. This is so if, for example, Y is a Hilbert space or $Y = L_p(\Omega)$, where $1 \leq p < \infty$ and Ω is an open subset of $\mathbb{R}^n$; in both these cases (see [46], Corollary V.5.4),

$$\alpha(T) := \inf\{\|T-K\| : K \text{ is a compact linear map from } X \text{ to } Y\}.$$

An important property of the approximation numbers is their connection with eigenvalues, in a Hilbert space setting. Thus if H is a complex Hilbert space and T is a compact linear map from H to itself, then T^*T has a positive compact square root $|T|$, which accordingly has a sequence $\{\lambda_k(|T|)\}$ of positive eigenvalues, each repeated according to multiplicity and ordered so that

$$\lambda_1(|T|) \geq \lambda_2(|T|) \geq ... \geq 0.$$

If T has only a finite number of distinct positive eigenvalues and M is the sum of their multiplicities, we put $\lambda_k(|T|) = 0$ for all $k > M$. The eigenvalues $\lambda_k(|T|)$ of $|T|$ are called the **singular values** of T. It turns out (see, for example, Theorem II.5.10 of [46]) that for all $k \in \mathbb{N}$,

$$a_k(T) = \lambda_k(|T|).$$

In particular, if T is compact and positive (hence self-adjoint), then for all $k \in \mathbb{N}$,

$$a_k(T) = \lambda_k(T).$$

It is plain that for a compact map $T \in \mathcal{B}(X,Y)$ the approximation numbers may be thought of as providing a means of measuring 'how compact' it is, at least under some restrictions on Y. There are other sequences of numbers which perform the same function: here we single out the entropy numbers for special mention. Let X and Y be Banach spaces and let $U_X = \{x \in X : \|x \mid X\| \leq 1\}$. Given $T \in \mathcal{B}(X,Y)$ and $k \in \mathbb{N}$, the kth **entropy number** of T, denoted by $e_k(T)$, is defined by

$$e_k(T) = \inf \left\{ \varepsilon > 0 : T(U_X) \text{ can be covered by } 2^{k-1} \text{ balls in } Y \text{ of radius } \varepsilon \right\}.$$

It may be easily checked that properties (i) and (ii) above of the approximation numbers are also enjoyed by the entropy numbers. This is not so for (iii)-(v), however. Moreover,

$$\beta(T) := \lim_{k \to \infty} e_k(T)$$

is the **(ball) measure of non-compactness** of T; and T is compact if, and only if, $\beta(T) = 0$.

If T is a compact linear map from a Banach space X to itself, its spectrum, apart from the point 0, consists of eigenvalues of finite algebraic multiplicity: we let $\{\lambda_k(T)\}$ be the sequence of all non-zero eigenvalues of T, repeated according to algebraic multiplicity and ordered by decreasing modulus. If T has only a finite number of distinct eigenvalues and M is the sum of their algebraic multiplicities, then just as before we put $\lambda_k(T) = 0$ for all $k > M$. A most useful connection between the spectral properties of T and its geometrical characteristics as expressed by the entropy numbers is provided by Carl's inequality (see [28]):

$$|\lambda_k(T)| \leq \sqrt{2} e_k(T) \text{ for all } k \in \mathbb{N}.$$

For another proof of this and a more general inequality see [30]; an extension to quasi-Banach spaces is given in [74].

Two-sided estimates of the approximation numbers of embeddings between Sobolev spaces (and much more general spaces) are available. To illustrate this, let Ω be a bounded domain in $\mathbb{R}^n$ with smooth boundary, and for any $k \in \mathbb{N}$ and any $p \in (0, \infty]$ let $W_p^k(\Omega)$ be the Sobolev space of all functions u which, together with their distributional derivatives of all orders up to and including k, are in $L_p(\Omega)$. When endowed with the quasi-norm

$$\|u \mid W_p^k(\Omega)\| = \left(\sum_{|\alpha| \leq k} \|D^\alpha u \mid L_p(\Omega)\|^p \right)^{1/p}$$

(with the natural interpretation when $p = \infty$), this is a quasi-Banach space. Now suppose that

$$s_1, s_2 \in \mathbb{N};\ p_1, p_2 \in (0,\infty] \text{ and that } \delta^+ := s_1 - s_2 - n\left(\frac{1}{p_1} - \frac{1}{p_2}\right)_+ > 0.$$

Then $W_{p_1}^{s_1}(\Omega)$ is compactly embedded in $W_{p_2}^{s_2}(\Omega)$; denote the embedding map by id. It turns out that if in addition $0 < p_1 \leq p_2 \leq 2$, or $2 \leq p_1 \leq p_2 \leq \infty$, or $0 < p_2 \leq p_1 \leq \infty$, then

$$a_k(id) \approx k^{-\delta^+/n}.$$

The situation when p_1 and p_2 lie on opposite sides of 2, with $p_1 < p_2$, is more complicated, but it can be shown that if in addition to the hypothesis that $\delta^+ > 0$ we have $0 < p_1 < 2 < p_2 < \infty$ (or $1 < p_1 < 2 < p_2 = \infty$) and $\delta^+ < n/\min\{p_1', p_2\}$, then

$$a_k(id) \approx k^{-\frac{\delta^+}{2n}\min\{p_1', p_2\}}.$$

For these results we refer to [74], Chapter 3, and [26].

Additional results relating to the material in this section, and in particular concerning comparisons between approximation and entropy numbers, may be found in [29], [46], [74], [143], [198] and [199].

1.4 Inequalities

Here we give some inequalities which will be of help in the text. The first is of Minkowski type.

Theorem 1.4.1. *Let (S_1, μ_1) and (S_2, μ_2) be positive measure spaces and let K be a $\mu_1 \times \mu_2$–measurable function on $S_1 \times S_2$. Then if $1 \leq p < \infty$,*

$$\left\{\int_{S_1}\left[\int_{S_2} |K(s_1, s_2)|\, d\mu_1(s_1)\right]^p d\mu_2(s_2)\right\}^{1/p}$$
$$\leq \int_{S_1}\left\{\int_{S_2} |K(s_1, s_2)|^p\, d\mu_2(s_2)\right\}^{1/p} d\mu_1(s_1).$$

For this we refer to [44], Vol. 1, p. 530.

The next is Jensen's inequality.

Theorem 1.4.2. *Let (X, μ) be a finite measure space, let I be an interval in $\mathbb{R}$, let $\Phi : I \to \mathbb{R}$ be convex and suppose that $f \in L_1(X, \mu)$ is such that $f(X) \subset I$ and $\Phi \circ f \in L_1(X, \mu)$. Then*

$$\Phi\left(\frac{1}{\mu(X)}\int_X f\, d\mu\right) \leq \frac{1}{\mu(X)}\int_X (\Phi \circ f)\, d\mu.$$

We refer to [126], p.202, for this.

2

Hardy-type Operators

2.1 Introduction

In [118] Hardy proved the following celebrated inequality: let $1 < p < \infty$ and set $F(x) = \int_0^x f(t)dt$, where f is a non-negative measurable function on $(0, \infty)$. Then, if $\varepsilon < 1/p' = 1 - 1/p$,

$$\int_0^\infty F^p(x) x^{p(\varepsilon-1)} dx \leq C \int_0^\infty f^p(x) x^{\varepsilon p} dx \tag{2.1.1}$$

for some constant $C > 0$ independent of f. If $\varepsilon > 1/p'$, the inequality takes the form

$$\int_0^\infty G^p(x) x^{p(\varepsilon-1)} dx \leq C \int_0^\infty f^p(x) x^{\varepsilon p} dx \tag{2.1.2}$$

where $G(x) = \int_x^\infty f(t)dt$. The best possible constants C in (2.1.1) and (2.1.2) are equal and this common value was determined by Landau in [150] as

$$C = |\varepsilon - 1/p'|^{-p}. \tag{2.1.3}$$

The inequality (2.1.1) (and similarly (2.1.2)) can be interpreted in the following way. Let

$$h(x) = x^\varepsilon f(x), \quad (Hh)(x) := x^{\varepsilon-1} F(x) = x^{\varepsilon-1} \int_0^x t^{-\varepsilon} h(t) dt. \tag{2.1.4}$$

Then (2.1.1) expresses the fact that the operator $H : L_p(0, \infty) \to L_p(0, \infty)$ is bounded and (2.1.3) gives the value of its norm, $C = \|H\|^p$.

In this chapter we investigate properties of general operators T of the form

$$Tf(x) := v(x) \int_a^x u(t) f(t) dt \tag{2.1.5}$$

as mappings between L_p spaces. Such an operator is said to be of *Hardy type*; the special case $v = u = 1$ corresponds to the original Hardy operator. Initially we obtain criteria for T to be bounded as a map from $L_p(a,b)$ to $L_q(a,b)$, where $-\infty < a < b \leq \infty$ and $p,q \in [1,\infty]$, and determine its measure of non-compactness. In particular, this gives a necessary and sufficient condition for T to be compact. We then proceed to describe results in which upper and lower bounds are obtained for the approximation numbers of T and, in the special case $p = q$, an asymptotic formula. Results on the l_p and weak-l_p classes of the approximation numbers are subsequently established. In the course of the analysis, we determine the exact values of the approximation numbers of certain embeddings, such as the embedding of the familiar Sobolev space $\overset{\circ}{W}{}_p^1(a,b)$ in $L_p(a,b)$, when $1 < p < \infty$ and (a,b) is a bounded interval in $\mathbb{R}$.

Hardy-type operators play a fundamental role throughout analysis, and this has motivated a considerable amount of research into their properties when they act between a wide variety of function spaces. In Chapter 5 we shall show how Hardy-type operators defined on L_p and other spaces constructed on trees occur naturally in problems involving some domains in $\mathbb{R}^n$ with highly irregular boundaries. To prepare for our needs in Chapter 5, we present some of the basic results for operators on trees in Section 2.6 below.

2.2 Boundedness of T

We shall assume throughout this section that $p,q \in [1,\infty]$, unless the contrary is stated, and that u, v are prescribed real-valued functions such that for all $X \in (a,b)$,

$$u \in L_{p'}(a,X), \tag{2.2.1}$$

$$v \in L_q(X,b), \tag{2.2.2}$$

where $p' = p/(p-1)$. The Hardy-type operator T to be investigated is defined in (2.1.5). We denote by $\|\cdot\|_{p,I}$ the standard norm on $L_p(I)$ (or simply $\|\cdot\|_p$ when $I = (a,b)$), and denote $\|T|L_p(a,b) \to L_q(a,b)\|$ by $\|T\|$ when there is no chance of confusion.

Theorem 2.2.1. *Let $1 \leq p \leq q \leq \infty$, and suppose that (2.2.1) and (2.2.2) are satisfied. Then T in (2.1.5) is a bounded linear map of $L_p(a,b)$ into $L_q(a,b)$ if and only if*

$$A := \sup_{a<X<b} \{\|u\|_{p',(a,X)}\|v\|_{q,(X,b)}\} < \infty. \tag{2.2.3}$$

In this case we have

$$A \leq \|T\| \leq 4A. \tag{2.2.4}$$

Proof. Let

$$\alpha_X := \inf\{\|f\|_p : f \in L_p(a,b), \int_a^X |f(t)u(t)|dt = 1\}, \quad X \in (a,b). \tag{2.2.5}$$

We first show that

$$\alpha_X = \|u\|_{p',(a,X)}^{-1}. \tag{2.2.6}$$

By Hölder's inequality,

$$\int_a^X |f(t)u(t)|dt \leq \|f\|_p \|u\|_{p',(a,X)}$$

and hence $\alpha_X \geq \|u\|_{p',(a,X)}^{-1}$ for $1 \leq p \leq \infty$. If $p > 1$, the choice

$$f(x) = |u^{p'-1}(x)|\chi_{(a,X)}(x)\|u\|_{p',(a,X)}^{-p'},$$

where $\chi_{(a,X)}$ is the characteristic function of (a, X), gives

$$\int_a^X |f(t)u(t)|dt = 1$$

and thus, since $p(p'-1) = p'$,

$$\alpha_X \leq \|f\|_p = \|u\|_{p',(a,X)}^{-1}$$

whence (2.2.6) when $1 < p \leq \infty$. If $p = 1$, u is bounded a.e. on $[a, X]$ by (2.2.1). Given $\varepsilon > 0$, there exists a non-null set $S \subset [a, X]$ such that for all $x \in S$, $|u(x)| > (1+\varepsilon)^{-1}\|u\|_{\infty,(a,X)}$. Take $f(x) = \chi_S(x)[\int_S |u(t)|dt]^{-1}$. Then

$$\int_a^X |f(t)u(t)|dt = 1,$$

so that

$$\alpha_X \leq \|f\|_1 = \frac{|S|}{\int_S |u(t)|dt} \leq \frac{1+\varepsilon}{\|u\|_{\infty,(a,X)}}$$

and (2.2.6) again follows on allowing $\varepsilon \to 0$.

We are now able to prove easily that (2.2.3) implies that $T : L_p(a,b) \to L_q(a,b)$ is bounded. Let $q < \infty$: the proof for $q = \infty$ is similar. Define $\mathcal{I} = \mathbb{Z}$ when $u \notin L_{p'}(a,b)$ and $\mathcal{I} = \{k \in \mathbb{Z} : -\infty < k \leq M\}$ for some $M \in \mathbb{Z}$ when $u \in L_{p'}(a,b)$. For $f \in L_p(a,b)$ and $i \in \mathcal{I}$, let

$$X_i = \sup\{x \in (a,b) : \int_a^x |f(t)u(t)|dt = 2^i\}. \tag{2.2.7}$$

Then $\{X_i : i \in \mathcal{I}\}$ generates a partition of (a,b) and we have

$$\begin{aligned}
\|Tf\|_q^q &\leq \sum_{i\in\mathcal{I}} \int_{X_i}^{X_{i+1}} |v(x) \int_a^x |f(t)u(t)|dt|^q dx \\
&\leq \sum_{i\in\mathcal{I}} 2^{q(i+1)} \|v\|_{q,(X_i,b)}^q
\end{aligned}$$

$$\leq A^q \sum_{i\in\mathcal{I}} 2^{q(i+1)} \alpha^q_{X_i}$$

by (2.2.3) and (2.2.6). Since

$$\int_{X_{i-1}}^{X_i} |f(t)u(t)|dt = 2^i - 2^{i-1} = 2^{i-1}$$

we have

$$\alpha_{X_i} \leq 2^{1-i} \|f\chi_{[X_{i-1},X_i]}\|_p$$

and hence

$$\|Tf\|_q^q \leq (4A)^q \sum_{i\in\mathcal{I}} \|f\chi_{[X_{i-1},X_i]}\|_p^q$$
$$\leq (4A)^q \|f\|_p^q$$

as $p \leq q$.

To establish the necessity of (2.2.3), we choose, for a given $\varepsilon > 0$ and $X \in (a,b)$, an $f \in L_p(a,b)$ such that $uf \geq 0$, $\int_a^X f(t)u(t)dt = 1$ and $\|f\|_p \leq \alpha_X(1+\varepsilon)$. Then $|Tf(x)| \geq |v(x)|$ for $x \in [X,b)$, and if $T : L_p(a,b) \to L_q(a,b)$ is bounded, we have

$$\alpha_X(1+\varepsilon)\|T\| \geq \|T\|\|f\|_p \geq \|Tf\|_q \geq \|v\|_{q,(X,b)}.$$

This and (2.2.6) yield

$$(1+\varepsilon)\|T\| \geq \|u\|_{p',(a,X)}\|v\|_{q,(X,b)},$$

whence (2.2.3) and the first inequality in (2.2.4). The theorem is therefore proved. □

Remark 2.2.2. The proof of Theorem 2.2.1 is based on the method of Sawyer [205]. Unlike other methods which depend heavily on the use of the Hölder and Minkowski inequalities (see, for example, [171] and [191]), Sawyer's method is readily extended to the situation considered in Section 2.6 below when the interval (a,b) is replaced by a tree. However, the constant 4 multiplying A on the right-hand side of (2.2.4) is larger than necessary. Opic modified the earlier proofs in [31], [177], [21], [171], to obtain the upper bound

$$(1+q/p')^{1/q}(1+p'/q)^{1/p'}A$$

for $1 < p \leq q < \infty$ (see Comment 3.6 on page 27 in [191]). When $p = q$ this gives $p^{1/p}(p')^{1/p'}A$, a value also determined in the earlier works. This is best possible, for on choosing $a = 0, b = \infty, u(t) = t^{-\varepsilon}$ and $v(t) = t^{\varepsilon-1}$ with $\varepsilon < 1/p'$, we obtain

$$\{p^{1/p}(p')^{1/p'}A\}^p = (\frac{1}{p'} - \varepsilon)^{-p},$$

the value of the best possible constant in (2.1.1). However, when $1 < p < q < \infty, (a, b) = (0, \infty)$ and $u \notin L_{p'}(0, \infty)$, the best value for the upper bound in (2.2.4) achieved to date is that determined independently by Manakov in [163] and Read in [203], namely $k(p, q)A$, where

$$k(p, q) = \left[\frac{\Gamma(q/r)}{\Gamma(1 + 1/r)\Gamma([q-1]/r)} \right]^{r/q}, \quad r = \frac{q}{p} - 1. \tag{2.2.8}$$

This estimate is sharp for all u, v with a fixed A, in the sense that, for each $u \notin L_{p'}(0, \infty)$, there exists a function v such that $k(p, q)A$ is the best possible value for the upper bound in (2.2.4).

Remark 2.2.3. If $p = 1, q \in [1, \infty]$, or $q = \infty, p \in (1, \infty]$, then

$$\|T\| = A. \tag{2.2.9}$$

This is proved in [191], Lemma 5.4; see also [171].

Our form of the result for $p > q$ requires some additional terminology. Let $[a, b] = \cup_{i \in \mathcal{I}} \bar{B}_i$, where the $B_i = (c_i, c_{i+1})$ are non-empty and the index set $\mathcal{I}$ is arbitrary. Denote the set of all such decompositions of (a, b) by $\mathcal{C}$.

Theorem 2.2.4. *Let $1 \le q < p \le \infty, 1/s = 1/q - 1/p$, and suppose that (2.2.1) and (2.2.2) are satisfied. Then T in (2.1.5) is a bounded linear map of $L_p(a, b)$ into $L_q(a, b)$ if and only if*

$$B = \sup_{\mathcal{C}} \|\{\beta_i\}|l_s(\mathcal{I})\| < \infty, \tag{2.2.10}$$

where $\{\beta_i\}$ is the sequence

$$\beta_i = \|u\|_{p', B_{i-1}} \|v\|_{q, B_i}, \quad i \in \mathcal{I}, \tag{2.2.11}$$

corresponding to a decomposition $\{B_i\}_{i \in \mathcal{I}} \in \mathcal{C}$, and $l_s(\mathcal{I})$ denotes the usual sequence space. Moreover, if (2.2.10) is satisfied,

$$B \le \|T\| \le 4B. \tag{2.2.12}$$

Proof. Given $\{B_i\}_{i \in \mathcal{I}} \in \mathcal{C}$, let $B_i = [X_i, X_{i+1})$ and set

$$\alpha_i = \inf\{\|f\|_p : \text{supp} f \subset B_{i-1}, \int_{X_{i-1}}^{X_i} |f(t)u(t)| dt = 1\}. \tag{2.2.13}$$

It follows as in the proof of (2.2.6) that

$$\alpha_i = \|u\|_{p', B_{i-1}}^{-1}.$$

Also, for $f \in L_p(a, b)$ and $X_i, i \in \mathcal{I}$, defined as in (2.2.7), we have

$$\int_{X_{i-1}}^{X_i} |f(t)u(t)|dt = 2^{i-1}$$

and

$$\alpha_i \leq 2^{1-i}\|f\|_{p,B_{i-1}}.$$

Hence

$$\begin{aligned}
\|Tf\|_q^q &\leq \sum_{i\in\mathcal{I}} \int_{B_i} \left| v(x) \int_a^x |f(t)u(t)|dt \right|^q dx \\
&\leq \sum_{i\in\mathcal{I}} 2^{q(i+1)} \|v\|_{q,B_i}^q \\
&= \sum_{i\in\mathcal{I}} 2^{q(i+1)} \alpha_i^q \beta_i^q \\
&\leq 4^q \sum_{i\in\mathcal{I}} \beta_i^q \|f\|_{p,B_{i-1}}^q \\
&\leq 4^q \left(\sum_{i\in\mathcal{I}} \beta_i^s\right)^{q/s} \left(\sum_{i\in\mathcal{I}} \|f\|_{p,B_{i-1}}^p\right)^{q/p} \\
&\leq (4B)^q \|f\|_p^q
\end{aligned}$$

which proves that (2.2.10) is sufficient for T to be bounded, and $\|T\| \leq 4B$.

Suppose that $T : L_p(a,b) \to L_q(a,b)$ is bounded and let $\{B_i\}_{i\in\mathcal{I}} \in \mathcal{C}$. Given $\varepsilon > 0$ and $i \in I$, choose f_i such that $f_i \geq 0, \text{supp } f_i \subset B_{i-1}, \int_{B_{i-1}} f_i(t)u(t)dt = 1$ and $\|f_i\|_p \leq \alpha_i(1+\varepsilon)$. Set $g_i(x) = (a_i/\alpha_i)f_i(x)$, where $\{a_i\}_{i\in I}$ is an arbitrary sequence of non-negative numbers in $l_p(\mathcal{I})$. Then $g := \sum g_i \in L_p(a,b)$ since

$$\begin{aligned}
\|g\|_p^p &= \sum \|(a_i/\alpha_i)f_i\|_{p,B_{i-1}}^p \\
&\leq (1+\varepsilon)^p \sum a_i^p < \infty.
\end{aligned}$$

Furthermore, on B_i we have

$$\begin{aligned}
Tg_i(x) &= v(x) \int_{B_{i-1}} (a_i/\alpha_i)f_i(t)u(t)dt \\
&= (a_i/\alpha_i)v(x)
\end{aligned}$$

and so

$$\begin{aligned}
\|Tg\|_q^q &= \sum \|Tg\|_{q,B_i}^q \geq \sum \|Tg_i\|_{q,B_i}^q \\
&= \sum (a_i/\alpha_i)^q \|v\|_{q,B_i}^q = \sum a_i^q \beta_i^q.
\end{aligned}$$

Since T is assumed to be bounded, we deduce that

$$\sum a_i^q \beta_i^q \leq \|T\|^q \|g\|_p^q \leq (1+\varepsilon)^q \|T\|^q (\sum a_i^p)^{q/p}$$

and, with $b_i = a_i^q$, on letting $\varepsilon \to 0+$, we obtain

$$\sum b_i \beta_i^q \leq \|T\|^q \|\{b_i\}\|_{l_{p/q}}.$$

Since this holds for an arbitrary non-negative sequence $\{b_i\} \in l_{p/q}$ by choice of $\{a_i\}$, and $q/p + q/s = 1$ it follows that $\{\beta_i^q\} \in l_{s/q}$ and $\|\{\beta_i\}\|_{l_s} \leq \|T\|$. The theorem is therefore proved. □

Remark 2.2.5. It is of interest to note that in the notation of Theorem 2.2.4, the constant A in (2.2.3) can be written

$$A = \sup_{\mathcal{C}} \|\{\beta_i\}|l_\infty(\mathcal{I})\|. \tag{2.2.14}$$

To see this, denote the right-hand side of (2.2.14) provisionally by B. We first observe that $A \leq B$ since the intervals $(a, X), [X, b)$ constitute a decomposition of (a, b) for any $X \in (a, b)$. Moreover, given $\{B_i\}_{i \in I} \in \mathcal{C}$ with $B_i = [X_i, X_{i+1})$, we have

$$A \geq \sup_{i \in I} \|u\|_{p', B_{i-1}} \|v\|_{q, B_i} = \|\{\beta_i\}|l_\infty(\mathcal{I})\|$$

and hence $A \geq B$, thus establishing (2.2.14).

Remark 2.2.6. The case $q < p$ covered by Theorem 2.2.4 was originally proved by Maz'ya [171] who obtained the result in the following form: $T : L_p(a,b) \to L_q(a,b)$ is bounded if and only if

$$C := \left\{ \int_a^b \left[\left(\int_x^b v^q(t)dt \right)^{1/q} \left(\int_a^x |u^{p'}(t)|dt \right)^{1/q'} \right]^s u^{p'}(x)dx \right\}^{1/s} < \infty, \tag{2.2.15}$$

in which case

$$q^{1/q} \left(\frac{p'q}{s} \right)^{1/q'} C \leq \|T\| \leq q^{1/q} (p')^{1/q'} C. \tag{2.2.16}$$

2.3 Compactness of T

Let $\mathcal{K}_{p,q} = \mathcal{K}_{p,q}(a,b)$ denote the set of all compact linear maps of $L_p(a,b)$ to $L_q(a,b)$, and $\mathcal{F}_{p,q} = \mathcal{F}_{p,q}(a,b)$ the subset of $\mathcal{K}_{p,q}$ whose elements are of finite rank. When $p = q$ we shall write $\mathcal{K}_p = \mathcal{K}_p(a,b)$ and $\mathcal{F}_p = \mathcal{F}_p(a,b)$ for these sets. The measure of non-compactness of T studied in this section is the number $\alpha(T)$ introduced in Section 1.3, which we write equivalently as

$$\alpha(T) := \inf\{\|T - P\| : P \in \mathcal{F}_{p,q}(a,b)\},$$

where we continue to use the notation $\|T\| = \|T|L_p(a,b) \to L_q(a,b)\|$. As noted in Section 1.3, if $1 \le q < \infty$, $L_q(a,b)$ has the bounded approximation property and this implies that

$$\alpha(T) = \inf\{\|T - P\| : P \in \mathcal{K}_{p,q}(a,b)\}.$$

From [46], Proposition II.3.6 and Theorem II.3.9 , it follows that when $q = 2$, the number $\alpha(T)$ is the *ball measure of non-compactness* of Section 1.3. In general it represents a natural measure of non-compactness of T, and clearly T is compact if and only if $\alpha(T) = 0$.

Theorem 2.3.1. *Let $1 \le p \le q < \infty$ or $1 < p \le q = \infty$, and suppose that (2.2.1) and (2.2.2) are satisfied. Then, if T in (2.1.5) is a bounded map from $L_p(a,b)$ into $L_q(a,b)$, and*

$$A(c,d) := \sup_{c<X<d} \{\|u\|_{p',(c,X)}\|v\|_{q,(X,d)}\}, \tag{2.3.1}$$

we have

$$\max\{\lim_{\phi\to a+} A(a,\phi), \lim_{\theta\to b-} A(\theta,b)\} \le \alpha(T) \le 4\{\lim_{\phi\to a+} A(a,\phi) + \lim_{\theta\to b-} A(\theta,b)\}. \tag{2.3.2}$$

In particular, T is compact if and only if $\lim_{\phi\to a+} A(a,\phi) = \lim_{\theta\to b} A(\theta,b) = 0$.

Proof. Given any ϕ, θ such that $a < \phi < \theta < b$, define maps S_ϕ, T_θ and $P_{\phi,\theta}$ by

$$S_\phi f(x) := \chi_{(a,\phi)}(x)v(x)\int_a^x u(t)f(t)dt, \tag{2.3.3}$$

$$T_\theta f(x) := \chi_{(\theta,b)}(x)v(x)\int_a^x \chi_{(\theta,b)}(t)u(t)f(t)dt, \tag{2.3.4}$$

$$P_{\phi,\theta} f(x) := \chi_{(\phi,b)}(x)v(x)\int_a^x \chi_{(a,\theta)}(t)u(t)f(t)dt, \tag{2.3.5}$$

where χ_I is the characteristic function of I. Then, we have

$$T = S_\phi + T_\theta + P_{\phi,\theta}. \tag{2.3.6}$$

For all $f \in L_p(a,b)$,

$$\begin{aligned}\|P_{\phi,\theta} f\|_q &\le |\int_a^\theta u(t)f(t)dt| \|\chi_{(\phi,b)}v\|_q \\ &\le \|u\|_{p',(a,\theta)}\|v\|_{q,(\phi,b)}\|f\|_p.\end{aligned}$$

Hence $P_{\phi,\theta} : L_p(a,b) \to L_q(a,b)$ is bounded. In fact $P_{\phi,\theta} \in \mathcal{K}_{p,q}(a,b)$. To show this we first observe that on setting $P_{\phi,\theta} f(x) = \chi_{(\phi,b)}(x)v(x)G_{\phi,\theta} f(x)$,

$$|G_{\phi,\theta}f(x)| \leq \|u\|_{p',(a,\theta)}\|f\|_p$$

and

$$\begin{aligned}|G_{\phi,\theta}f(x) - G_{\phi,\theta}f(y)| &\leq |\int_y^x u(t)f(t)\chi_{(a,\theta)}(t)dt| \\ &\leq \Big[\int_y^x |u(t)\chi_{(a,\theta)}(t)|^{p'}dt\Big]^{1/p'}\|f\|_p.\end{aligned}$$

Hence the set $\{G_{\phi,\theta}f_n : \|f_n\|_p \leq 1, n \in \mathbb{N}\}$ is uniformly bounded and equicontinuous on (a,b), and so, by the Arzelà-Ascoli Theorem, it contains a subsequence which converges uniformly on (a,b). Since

$$\|P_{\phi,\theta}f\|_q = \|vG_{\phi,\theta}f\|_{q,(\phi,b)},$$

the corresponding subsequence of $\{P_{\phi,\theta}f_n : \|f_n\| \leq 1, n \in \mathbb{N}\}$ therefore converges in $L_q(a,b)$. Thus $P_{\phi,\theta}$ maps the closed unit ball in $L_p(a,b)$ into a relatively compact subset of $L_q(a,b)$ and is therefore compact, as asserted.

From (2.3.6), we now infer that

$$\alpha(T) = \alpha(S_\phi + T_\theta) \leq \|S_\phi + T_\theta\| \leq \|S_\phi\| + \|T_\theta\|. \tag{2.3.7}$$

Since

$$T_\theta f(x) = \begin{cases} 0 & \text{if } a < x \leq \theta, \\ v(x)\int_\theta^x u(t)f(t)dt & \text{if } \theta < x < b, \end{cases}$$

we have by Theorem 2.2.1, with (a,b) now replaced by (θ,b), that with obvious notation,

$$\|T_\theta\| \leq 4A(\theta,b).$$

Similarly, $\|S_\phi\| \leq 4A(a,\phi)$, and hence

$$\alpha(T) \leq 4[A(a,\phi) + A(\theta,b)].$$

Since ϕ, θ are arbitrary ($\phi < \theta$), we conclude that

$$\alpha(T) \leq 4\{\lim_{\phi\to a+} A(a,\phi) + \lim_{\theta\to b-} A(\theta,b)\}.$$

We first obtain the lower bound for $\alpha(T)$ in (2.3.2) for $1 < p \leq q < \infty$. Let $\lambda > \alpha(T)$. Then, there exists $P \in \mathcal{F}_{p,q}(a,b)$ such that for all $f \in L_p(a,b)$,

$$\|Tf - Pf\|_q \leq \lambda\|f\|_p.$$

We now claim that we may assume that there exist $X, Y \in (a,b)$ such that $[X,Y] \subset (a,b)$ and for all $f \in L_p(a,b)$, $\text{supp}Pf \subset [X,Y]$. For suppose that $P \in \mathcal{F}_{p,q}(a,b)$ has rank N and is therefore of the form

$$Pf = \sum_{i=1}^N c_i(f)u_i,$$

where the $u_i, i = 1, \cdots, N$, are linearly independent functions in $L_q(a,b)$ with unit norm. Since all norms on the finite-dimensional range of P are equivalent, there exists a positive constant K such that

$$\sum_{i=1}^{N} |c_i(f)| \leq K\|Pf\|_{q,(a,b)} \leq K\|P\|\|f\|_{p,(a,b)}.$$

Given $\varepsilon > 0$, choose $\phi_i \in C_0^\infty(a,b)$ such that for $i = 1, \cdots, N$,

$$\|u_i - \phi_i\|_{q,(a,b)} < \varepsilon/(K\|P\|),$$

and set $Rf := \sum_{i=1}^{N} c_i(f)\phi_i$. Then $R \in \mathcal{F}_{p,q}$ and

$$\|Pf - Rf\|_{q,(a,b)} \leq \sum_{i=1}^{N} |c_i(f)|\|u_i - \phi_i\|_{q,(a,b)} < \varepsilon\|f\|_{p,(a,b)}.$$

Furthermore, R is supported in a compact subset of (a,b) and so our claim is verified.

Hence

$$\left(\int_a^X + \int_Y^b\right)|Tf(x)|^q dx \leq \lambda^q \|f\|_p^q \tag{2.3.8}$$

for all $f \in L_p(a,b)$. Let $\phi_1, \phi_2, \theta_1, \theta_2$ be such that $a < \phi_1 < \phi_2 < X < Y < \theta_1 < \theta_2 < b$ and choose $f_1 = \chi_{(a,\phi_1)}|u|^{p'-1}\text{sgn } u$, $f_2 = \chi_{(\theta_1,\theta_2)}|u|^{p'-1}\text{sgn } u$. Then

$$\int_a^X |Tf_1(x)|^q dx \geq \int_{\phi_1}^{\phi_2} |v(x)|^q \left(\int_a^{\phi_1} |u(t)|^{p'} dt|\right)^q dx,$$

$$\int_Y^b |Tf_2(x)|^q dx \geq \left(\int_{\theta_1}^{\theta_2} |u(t)|^{p'} dt\right)^q \left(\int_{\theta_2}^b |v(x)|^q dx\right)$$

and

$$\|f_1\|_p^p = \int_a^{\phi_1} |u(t)|^{p'} dt, \quad \|f_2\|_p^p = \int_{\theta_1}^{\theta_2} |u(t)|^{p'} dt.$$

We infer from (2.3.8) that

$$\|u\|_{p',(a,\phi_1)}\|v\|_{q,(\phi_1,\phi_2)} \leq \lambda, \quad \|u\|_{p',(\theta_1,\theta_2)}\|v\|_{q,(\theta_2,b)} \leq \lambda$$

and these yield

$$\max\{\lim_{\phi_2 \to a+} A(a,\phi_2), \quad \lim_{\theta_1 \to b-} A(\theta_1, b)\} \leq \alpha(T).$$

When $p = 1$ and $q \in [1,\infty)$, we again suppose that $a < \phi_1 < \phi_2 < X < Y < \theta_1 < \theta_2 < b$, where X, Y satisfy (2.3.8), and now set

$$u_1 := \|u\|_{\infty,(a,\phi_1)}, \quad u_2 := \|u\|_{\infty,(\theta_1,\theta_2)}.$$

Given $\varepsilon \in (0, \min\{u_1, u_2\})$, let $I_\varepsilon^1, I_\varepsilon^2$ be subsets of $[a, \phi_1]$ and $[\theta_1, \theta_2]$ respectively, of positive measure, which are such that

$$u_i - \varepsilon \leq |u(t)| \leq u_i, \quad t \in I_\varepsilon^i, i = 1, 2.$$

Then, with $f_i = \chi_{I_\varepsilon^i} u_i \mathrm{sgn}\, u$, we have $\|f_i\|_1 = u_i |I_\varepsilon^i$ and

$$\begin{aligned}
\int_a^X |Tf_1(x)|^q dx &\geq \int_{\phi_1}^{\phi_2} \left(\left| v(x) \int_{I_\varepsilon^1} |u(t)| u_1 dt \right| \right)^q dx \\
&\geq [u_1(u_1 - \varepsilon)|I_\varepsilon^1|]^q \|v\|_{q,(\phi_1,\phi_2)}^q, \\
\int_Y^b |Tf_2(x)|^q dx &\geq \int_{\theta_2}^{b} \left(\left| v(x) \int_{I_\varepsilon^2} |u(t)| u_2 dt \right| \right)^q dx \\
&\geq [u_2(u_2 - \varepsilon)|I_\varepsilon^2|]^q \|v\|_{q,(\theta_2,b)}.
\end{aligned}$$

The lower bound in (2.3.2) now follows as for the case $1 < p \leq q < \infty$.

It remains to consider the range $1 < p \leq q = \infty$. The proof of the upper bound for $\alpha(T)$ in (2.3.2) already given continues to hold, but that for the lower bound requires $C_0^\infty(a,b)$ to be dense in $L_q(a,b)$ and is therefore not applicable when $q = \infty$. We proceed as follows. The boundedness of $T : L_p(a,b) \to L_\infty(a,b)$ implies the boundedness of the map $T^* : L_1(a,b) \to L_{p'}(a,b)$ defined by

$$\int_a^b (T^*g)(x) f(x) dx = \int_a^b g(x)(Tf)(x) dx$$

for all $f \in L_p(a,b)$ and $q \in L_1(a,b)$. It is readily verified that

$$(T^*g)(x) = u(x) \int_x^b v(t) g(t) dt, \quad a < x < b. \tag{2.3.9}$$

Moreover,

$$\begin{aligned}
\alpha(T^*) &= \inf\{\|T^* - S\| : S \in \mathcal{F}_{1,p'}(a,b)\} \\
&\leq \inf\{\|T^* - P^*\| : P \in \mathcal{F}_{p,\infty}(a,b)\} \\
&\leq \inf\{\|T - P\| : P \in \mathcal{F}_{p,\infty}(a,b)\} \\
&= \alpha(T).
\end{aligned} \tag{2.3.10}$$

In (2.3.9) set $x = -y, a = -b', b = -a', \hat{g}(t) = g(-t)$. Then $T^* = MRM$, where $(Mg)(x) = \hat{g}(x)$ and

$$(Rh)(y) = \hat{u}(y) \int_{a'}^y \hat{v}(t) h(t) dt, \quad a' < y < b'. \tag{2.3.11}$$

We can now use what has already been established, namely that (2.3.2) holds for R and hence, in particular,

$$\alpha(R) \geq \max\{\lim_{\Phi \to a'+} A'(a', \Phi), \lim_{\Theta \to b'-} A'(\Theta, b')\}, \tag{2.3.12}$$

where, with $c' = -d, d' = -c$,

$$\begin{aligned} A'(c', d') &= \sup_{c' < X < d'} \{\|\hat{v}\|_{\infty,(c',X)} \|\hat{u}\|_{p',(X,d')}\} \\ &= A(c, d). \end{aligned} \tag{2.3.13}$$

Since M is an isometry on $L_1(a, b)$ and $L_{p'}(a, b)$, we have that $\alpha(T^*) = \alpha(R)$, and hence the lower bound in (2.3.2) follows from (2.3.10) and (2.3.12). □

Remark 2.3.2. It is proved in [55], Remark (b) after Theorem 4 that for the values $p = 1, q = \infty$ omitted in Theorem 2.3.1, $T : L_1(a, b) \to L_\infty(a, b)$ is never compact.

Remark 2.3.3. The constant multiples 1,4 in the lower and upper bounds respectively for $\alpha(T)$ in (2.3.2) are those obtained from (2.2.4). If the improved upper bound given by (2.2.8) for $\|T\|$ is used we obtain

$$\begin{aligned} &\max\{\lim_{\phi \to a+} A(a, \phi), \lim_{\theta \to b-} A_{\theta,b}\} \leq \alpha(T) \\ &\leq k(p, q)\{\lim_{\phi \to a+} A(a, \phi) + \lim_{\theta \to b-} A_{\theta,b}\}. \end{aligned} \tag{2.3.14}$$

The case $q < p$ is very different from that of $p \leq q$.

Theorem 2.3.4. *Let $1 \leq q < p \leq \infty, 1/s = 1/q - 1/p$ and suppose that (2.2.1) and (2.2.2) are satisfied. Then T in (2.1.5) is a compact map of $L_p(a, b)$ into $L_q(a, b)$ if and only if it is bounded, and this is equivalent to $B < \infty$ in the notation of Theorem 2.2.4.*

Proof. It suffices to prove that if $T : L_p(a, b) \to L_q(a, b)$ is bounded, or equivalently $B < \infty$, then T is compact, that is $\alpha(T) = 0$. Let C_ϕ, C_θ denote the set of all partitions $\{B_i\}_{i \in \mathcal{I}}$ of $(a, \phi), (\theta, b)$ respectively, and set

$$B(a, \phi) := \sup_{C_\phi} \|\{\beta_i\}\|_{l_s(\mathcal{I})}, \quad B(\theta, b) := \sup_{C_\theta} \|\{\beta_i\}\|_{l_s(\mathcal{I})}. \tag{2.3.15}$$

where $\beta_i = \|u\|_{p', B_{i-1}} \|v\|_{q, B_i}, i \in \mathcal{I}$. The first part of the proof of Theorem 2.3.1 continues to be valid in this case and we obtain, if T is bounded,

$$\alpha(T) \leq 4\{\lim_{\phi \to a+} B(a, \phi) + \lim_{\theta \to b-} B(\theta, b)\}. \tag{2.3.16}$$

It remains to prove that if $B \equiv B(a, b) < \infty$,

$$\lim_{\phi \to a+} B(a, \phi) = \lim_{\theta \to b-} B(\theta, b) = 0.$$

Suppose the contrary, and, in particular, assume that $\lim_{\theta\to b-} B(\theta,b) \neq 0$. Then, there exists $\varepsilon > 0$ such that, for all $c \in (a,b)$ there exists $\theta > c$ for which $B(\theta,b) > \varepsilon$: thus, for some $\{B_i\}_{i\in\mathcal{I}} \in \mathcal{C}_\theta$,

$$\sum_{i\in\mathcal{I}} \|u\|^s_{p',B_{i-1}} \|v\|^s_{q,B_i} > \varepsilon^s.$$

Hence there exists a finite subset $\mathcal{I}_N$ of $\mathcal{I}$, where $N = N(\theta)$, such that

$$\sum_{i\in\mathcal{I}_N} \|u\|^s_{p',B_{i-1}} \|v\|^s_{q,B_i} > \varepsilon^s.$$

We now define sequences $\{c_i\}_{i=1}^\infty, \{\theta_i\}_{i=1}^\infty$ and intervals $\{B_i\}$ iteratively as follows. Given $c_i \in (a,b)$, let $\theta_i \in (c_i,b)$ and $\{B_j\}_{j\in\mathcal{I}_{N(\theta_i)}}$ be as defined above; define $c_{i+1} \in (a,b)$ to be such that $c_{i+1} > \sup\{x : x \in \cup_{j\in\mathcal{I}_{N(\theta_i)}} B_j\}$. It follows that

$$\sum_i \sum_{j\in\mathcal{I}_{N(\theta_i)}} \|u\|^s_{p',B_{j-1}} \|v\|^s_{q,B_j} = \infty,$$

which contradicts $B(a,b) < \infty$. The assumption $\lim_{\phi\to a+} B(a,\phi) \neq 0$ leads similarly to the same contradiction, and the theorem is proved. □

2.4 Approximation numbers of T

We recall from Section 1.3 that for $m \in \mathbb{N}$, the *mth* approximation number of a bounded linear map $T : L_p(a,b) \to L_q(a,b)$ is defined to be

$$a_m(T) = \inf\{\|T-P\| : P \in \mathcal{F}_{p,q}(a,b), \text{rank } P < m\}; \tag{2.4.1}$$

and T is compact if and only if $\lim_{m\to\infty} a_m(T) = 0$. Furthermore,

$$\|T\| \geq a_1(T) \geq a_2(T) \geq \cdots. \tag{2.4.2}$$

As noted in Section 1.3, in the case $p = q = 2$, the approximation numbers of T are its singular values, that is, the eigenvalues of $|T| = (T^*T)^{1/2}$, and these are the moduli of the eigenvalues of T if T is self-adjoint, or indeed normal. However, no such straightforward relation holds in other cases. In fact, the operator T in (2.1.5) has no non-zero eigenvalues if $u(x) \neq 0$ *a.e.* on (a,b). For if $Tf = \lambda f$ for some $\lambda \neq 0$ and $f \neq 0$, we have that $F(x) = \int_a^x u(t)f(t)dt$ satisfies

$$\int_a^x u(t)v(t)F(t)dt = \lambda F(x),$$

whence

$$F(x) = C\exp[\lambda^{-1}\int_a^x u(t)v(t)dt]$$

for some constant C. Since $F(a) = 0$, it follows that $F(x) = 0$ for all x and hence $f = 0$, contrary to assumption.

2.4.1 The Hardy operator on a finite interval

We consider the Hardy operator H_c defined as an operator from $L_p(a,b)$ to $L_q(a,b)$, where $1 \le p, q \le \infty$, by

$$(H_c f)(x) := \int_c^x f(t)dt, \quad c \in [a,b], \tag{2.4.3}$$

with (a,b) a finite interval. For $q < \infty$, it follows from Hölder's inequality that

$$\begin{aligned}\int_a^b \Big|\int_c^x f(t)dt\Big|^q dx &\le \|f\|_{p,(a,b)}^q \int_a^b |x-c|^{q/p'} dx \\ &\le (b-a)^{\frac{q}{p'}+1} \|f\|_{p,(a,b)}^q.\end{aligned}$$

Hence $H_c : L_p(a,b) \to L_q(a,b)$ is bounded with norm

$$\|H_c : L_p(a,b) \to L_q(a,b)\| \le (b-a)^{1-\frac{1}{p}+\frac{1}{q}}.$$

The same result holds for $q = \infty$. Also, we have

$$\begin{aligned}\|H_a : L_p(a,b) \to L_q(a,b)\| &= \|H_b : L_p(a,b) \to L_q(a,b)\| \\ &= (b-a)^{1-\frac{1}{p}+\frac{1}{q}} \|H_0 : L_p(0,1) \to L_q(0,1)\| \\ &= (b-a)^{1-\frac{1}{p}+\frac{1}{q}} \gamma_{p,q},\end{aligned} \tag{2.4.4}$$

say.

Remark 2.4.1. The constant $\gamma_{p,q}$ in (2.4.4) was evaluated in [206], but the case $p = q$ was proved earlier in [155]; see also [16], [43] and [68]. For $p, q \in [1, \infty]$,

$$\begin{aligned}\gamma_{p,q} &:= \|H_0 : L_p(0,1) \to L_q(0,1)\| \\ &= \frac{r^{\frac{1}{r}-1}(p')^{1/p} q^{1/q'}}{\mathbf{B}\left(\frac{1}{p'}, \frac{1}{q}\right)},\end{aligned} \tag{2.4.5}$$

where $\frac{1}{r} = 1 + \frac{1}{q} - \frac{1}{p}$, $\frac{1}{s'} = 1 - \frac{1}{s}$ and $\mathbf{B}$ is the classical beta function

$$\mathbf{B}(s,t) := \frac{\Gamma(s)\Gamma(t)}{\Gamma(s+t)}.$$

The extremal functions, i.e. the $f \in L_p(0,1)$ such that $\|H_0 f\|_{q,(0,1)} = \gamma_{p,q}\|f\|_{p,(0,1)}$, are also determined. In particular, when $p = q$,

$$\gamma_p \equiv \gamma_{p,p} = \frac{1}{\pi} p^{1/p'} (p')^{1/p} \sin(\pi/p),$$

and, when $p \in (1,\infty)$, an extremal is given by

$$f(x) = \frac{\pi_p}{2} \cos_p\left(\frac{\pi_p}{2}x\right)$$

where $\frac{\pi_p}{2} = \int_0^1 \frac{dt}{(1-t^p)^{1/p}}$ is the first positive zero of $\cos_p(x) := \frac{d}{dx}\sin_p(x)$ and $\sin_p$ is the odd, $2\pi_p$-periodic function defined by

$$\sin_p^{-1}(x) = \int_0^x \frac{dt}{(1-t^p)^{1/p}}.$$

When $p = 1$ or ∞, $\gamma_p = 1$. There are no extremals when $p = 1$, but all constants are extremals when $p = \infty$.

In Theorem 2.4.3 below, $E_0 = E_0((a,b)) : \mathring{W}_p^1(a,b) \to L_p(a,b), E_{\mathbb{C}} = E_{\mathbb{C}}((a,b)) : W^1_{p,\mathbb{C}}(a,b) \to L_{p,\mathbb{C}}(a,b)$ are the natural embeddings. Here, $\mathring{W}_p^1(a,b)$ is the completion of $C_0^\infty(a,b)$ with respect to the norm,

$$\|f|\mathring{W}_p^1(a,b)\| \equiv \|f\|_{\mathring{W}_p^1(a,b)} = \|f'\|_{p,(a,b)} \tag{2.4.6}$$

and $W^1_{p,\mathbb{C}}(a,b), L_{p,\mathbb{C}}(a,b)$ are the quotient spaces $W_p^1(a,b)/\mathbb{C}, L_p(a,b)/\mathbb{C}$ respectively, where $W_p^1(a,b)$ is the usual Sobolev space of locally absolutely continuous functions on (a,b) with norm

$$\|f|W_p^1(a,b)\| \equiv \|f\|_{W_p^1(a,b)} = \|f'\|_{p,(a,b)} + \|f\|_{p,(a,b)}.$$

The norms on $W^1_{p,\mathbb{C}}(a,b)$ and $L_{p,\mathbb{C}(a,b)}$ are defined to be

$$\begin{aligned}
\|[f]|W^1_{p,\mathbb{C}}(a,b)\| &\equiv \|[f]\|_{W^1_{p,\mathbb{C}}(a,b)} = \|f'\|_{p,(a,b)}, \\
\|[f]|L_{p,\mathbb{C}}(a,b)\| &\equiv \|[f]\|_{L_{p,\mathbb{C}}(a,b)} = \inf\{\|f-c\|_{p,(a,b)} : c \in \mathbb{C}\};
\end{aligned} \tag{2.4.7}$$

note that the elements of $W^1_{p,\mathbb{C}}(a,b)$ and $L_{p,\mathbb{C}}(a,b)$ are equivalence classes $[\cdot]$ of functions which differ by a constant.

We need the following lemma, in which $\|\cdot\|_{q,I,\mu}$ denotes the norm in $L_q(I,\mu)$, where in our applications, μ will either be Lebesgue measure or $d\mu = v^q(x)dx$.

Lemma 2.4.2. *If $1 < q \leq \infty$, then given any $f, e_0 \in L_q(I,\mu)$, with $e_0 \neq 0$, there is a unique scalar c_{f,e_0} such that*

$$\|f - c_{f,e_0}e_0\|_{q,I,\mu} = \inf_{c\in\mathbb{C}} \|f - ce_0\|_{q,I,\mu}.$$

Proof. Since $\|f - ce_0\|_{q,I,\mu}$ is continuous in c and tends to ∞ as $c \to \infty$, the existence of c_{f,e_0} is guaranteed by the local compactness of $\mathbb{C}$. For $1 < q < \infty$, the uniqueness follows from the uniform convexity of $L_q(I,\mu)$. For if there were two such scalars, $c_{f,e_0}, c'_{f,e_0}, c_{f,e_0} \neq c'_{f,e_0}$ say, then,

$$\|f - \frac{1}{2}(c_{f,e_0} + c'_{f,e_0})e_0\|_{q,I,\mu} = \|\frac{1}{2}(f - c_{f,e_0}e_0) + \frac{1}{2}(f - c'_{f,e_0}e_0)\|_{q,I,\mu}$$
$$< \inf_{c\in\mathbb{C}} \|f - ce_0\|_{q,I,\mu}$$

which is a contradiction. When $q = \infty$, we also get the contradiction

$$\|f - \frac{1}{2}(c_{f,e_0} + c'_{f,e_0})e_0\|_{\infty,I,\mu} < \|f - c_{f,e_0}e_0\|_{\infty,I,\mu}.$$

□

To simplify notation we write $\|E_0\|$ for $\|E_0((a,b))|\mathring{W}^1_p(a,b) \to L_p(a,b)\|$ and similarly for $\|E_{\mathbb{C}}\|$.

Theorem 2.4.3. *Let* $1 < p \leq \infty$. *If* $c = \frac{(a+b)}{2}$,

$$\|E_0\| = \|E_{\mathbb{C}}\| = \|H_c|L_p(a,b) \to L_p(a,b)\| = \frac{\gamma_p}{2}(b-a), \tag{2.4.8}$$

where $\gamma_p = \gamma_{p,p}$ *is given in (2.4.5).*

Proof. Given $\varepsilon > 0$, there exists $\phi \in L_p(c,b)$ such that

$$\begin{aligned}\|H_c\phi\|_{p,(c,b)} &\geq \big(\|H_c|L_p(c,b) \to L_p(c,b)\| - \varepsilon\big)\|\phi\|_{p,(c,b)} \\ &= \big(\frac{\gamma_p}{2}(b-a) - \varepsilon\big)\|\phi\|_{p,(c,b)}\end{aligned} \tag{2.4.9}$$

by (2.4.4). Define the function

$$\psi(t) = \begin{cases} \phi(2c-t), & a < t \leq c, \\ \phi(t), & c < t < b \end{cases}$$

which is symmetrical about the mid-point c. Then $\Psi(x) = (H_c\psi)(x)$ is anti-symmetrical about c and we have

$$\|\psi\|_{p,(a,b)} = 2^{1/p}\|\phi\|_{p,(c,b)},$$
$$\|\Psi\|_{p,(a,b)} = 2^{1/p}\|H_c\phi\|_{p,(c,b)}.$$

These equations and (2.4.9) yield

$$\|H_c\psi\|_{p,(a,b)} \geq \big(\frac{\gamma_p}{2}(b-a) - \varepsilon\big)\|\psi\|_{p,(a,b)} \tag{2.4.10}$$

and hence

$$\|H_c|L_p(a,b) \to L_p(a,b)\| \geq \frac{\gamma_p}{2}(b-a). \tag{2.4.11}$$

We also have for the above Ψ that

$$\|\Psi\|_{L_{p,\mathbb{C}}(a,b)} = \|\Psi\|_{p,(a,b)}. \tag{2.4.12}$$

To see this, first note from Lemma 2.4.2 that, for any $f \in L_p(a,b)$, there exists a unique constant c_f such that

$$\|f\|_{L_{p,\mathbb{C}}(a,b)} = \inf_{k\in\mathbb{C}} \|f-k\|_{p,(a,b)} = \|f-c_f\|_{p,(a,b)}.$$

If $k_0 = c_\Psi$, we have, with $d = \frac{(b-a)}{2}$, since $\Psi(c+t) = -\Psi(c-t)$ for $0 \le t \le d$,

$$\begin{aligned}
\|\Psi - k_0\|^p_{p,(a,b)} &= \int_0^d |\Psi(c+t) - k_0|^p dt + \int_0^d |\Psi(c-t) - k_0|^p dt \\
&= \int_0^d |-\Psi(c-t) - k_0|^p dt + \int_0^d |-\Psi(c+t) - k_0|^p dt \\
&= \|\Psi + k_0\|^p_{p,(a,b)},
\end{aligned}$$

whence $k_0 = 0$ and (2.4.12). It therefore follows from (2.4.10) that

$$\|\Psi\|_{L_{p,\mathbb{C}}(a,b)} \ge \left(\frac{\gamma_p}{2}(b-a) - \varepsilon\right)\|\Psi'\|_{p,(a,b)}$$

and so

$$\|E_\mathbb{C}\| \ge \frac{\gamma_p}{2}(b-a). \tag{2.4.13}$$

For $f \in W^1_p(a,b)$,

$$(H_c f')(x) = \int_c^x f'(t)dt = f(x) - f(c).$$

Hence

$$\|f\|_{L_{p,\mathbb{C}}(a,b)} \le \|H_c f'\|_{p,(a,b)} \le \|H_c|L_p(a,b) \to L_p(a,b)\|\|f'\|_{p,(a,b)}$$

gives

$$\|E_\mathbb{C}\| \le \|H_c|L_p(a,b) \to L_p(a,b)\|.$$

Furthermore, by (2.4.4),

$$\begin{aligned}
\|H_c f\|^p_{p,(a,b)} &= \|H_c f\|^p_{p,(a,c)} + \|H_c f\|^p_{p,(c,b)} \\
&\le \left(\frac{\gamma_p}{2}(b-a)\right)^p \|f\|^p_{p,(a,b)}
\end{aligned}$$

and so

$$\|E_\mathbb{C}\| \le \|H_c|L_p(a,b) \to L_p(a,b)\| \le \frac{\gamma_p}{2}(b-a).$$

This and (2.4.13) give (2.4.8) for $\|E_\mathbb{C}\|$ and $\|H_c\|$ when $1 < p < \infty$. The case $p = \infty$ follows in the same way.

The proof for E_0 is similar, with the above Ψ replaced by

$$\Psi(x) = \begin{cases} (H_a\psi)(x), & a \le x \le c, \\ -(H_b\psi)(x), & c < x \le b. \end{cases}$$

We have as in (2.4.13) that $\|E_0\| \geq \frac{\gamma_p}{2}(b-a)$. Moreover, for $f \in \overset{\circ}{W}{}_p^1(a,b)$,

$$f(x) = (H_a f')(x) = -(H_b f')(x)$$

and

$$\begin{aligned}\|f\|_{p,(a,b)}^p &= \|H_a f'\|_{p,(a,c)}^p + \|H_b f'\|_{p,(c,b)}^p \\ &\leq \left(\frac{\gamma_p}{2}(b-a)\right)^p \|f'\|_{p,(a,b)}^p \\ &= \left(\frac{\gamma_p}{2}(b-a)\right)^p \|f\|_{\overset{\circ}{W}{}_p^1(a,b)}^p\end{aligned}$$

on using (2.4.4). Therefore $\|E_0\| \leq \frac{\gamma_p}{2}(b-a)$ and the proof is complete. □

In the next theorem we use the following notation:

$$\begin{aligned}\nu_{\mathbb{C}}(\varepsilon,\Omega) &:= \max\{m : a_m(E_{\mathbb{C}}(\Omega)) \geq \varepsilon\}, \\ \mu_{\mathbb{C}}(\varepsilon,\Omega) &:= \max\{\dim S : \alpha(S) \leq 1/\varepsilon\},\end{aligned}$$

where $\Omega = (a,b), \varepsilon > 0$ is sufficiently small and for each linear subspace S of $W_{p,\mathbb{C}}^1(\Omega)$,

$$\alpha(S) = \sup_{f \in S} \frac{\|f'\|_{p,\Omega}}{\|f\|_{L_{p,\mathbb{C}}(\Omega)}}$$

and $a_m(E_{\mathbb{C}}(\Omega))$ is the mth approximation number of $E_{\mathbb{C}}(\Omega)$. Also, $\nu_0(\varepsilon,\Omega)$, $\mu_0(\varepsilon,\Omega)$ are the corresponding numbers when $W_{p,\mathbb{C}}^1(\Omega)$ is replaced by $\overset{\circ}{W}{}_p^1(\Omega)$, and thus $E_{\mathbb{C}}(\Omega)$ by $E_0(\Omega)$.

Theorem 2.4.4. *For $1 < p < \infty$ and $n \in \mathbb{N}$ we have*

$$a_n(E_0(\Omega)) = a_n(E_{\mathbb{C}}(\Omega)) = \frac{\gamma_p}{2n}|\Omega| \tag{2.4.14}$$

$$\mu_0\left(\frac{\gamma_p}{2n}|\Omega|,\Omega\right) = \nu_0\left(\frac{\gamma_p}{2n}|\Omega|,\Omega\right) \tag{2.4.15}$$

$$= \nu_{\mathbb{C}}\left(\frac{\gamma_p}{2n}|\Omega|,\Omega\right) = n. \tag{2.4.16}$$

Hence

$$\begin{aligned}\lim_{\varepsilon\to 0} \varepsilon\mu_0(\varepsilon,\Omega) &= \lim_{\varepsilon\to 0} \varepsilon\nu_0(\varepsilon,\Omega) \\ &= \lim_{\varepsilon\to 0} \varepsilon\nu_{\mathbb{C}}(\varepsilon,\Omega) = \frac{\gamma_p}{2}|\Omega|.\end{aligned} \tag{2.4.17}$$

Proof. Let $\Omega = \bigcup_{i=0}^{n} \Omega_i$, where $\Omega_0 = (a, a_1), \Omega_i = [a_i, a_{i+1}), i = 1, \cdots, n-1, \Omega_n = [a_n, b)$, and, with $l_n = \frac{(b-a)}{n}$,

$$\begin{aligned} a_1 &= a + \frac{l_n}{2} \\ a_i &= a_1 + (i-1)l_n, \quad i = 1, \cdots, n-1, \\ a_n &= b - \frac{l_n}{2}. \end{aligned}$$

Set

$$Pf(x) = \sum_{i=1}^{n-1} f(c_i)\chi_i(x),$$

where $c_i = \frac{a_{i+1}+a_i}{2}$ and χ_i is the characteristic function of Ω_i. Then, by (2.4.8), for $f \in C_0^1(\Omega)$ and with $c_0 = a, c_n = b$,

$$\begin{aligned} \|E_0(\Omega)f - Pf\|_{p,\Omega}^p &= \sum_{i=0}^{n} \|H_{c_i} f'\|_{p,\Omega_i}^p \\ &\le \{(\frac{\gamma_p}{2n})|\Omega|\|f'\|_{p,\Omega}\}^p, \end{aligned}$$

where $|\Omega| = (b-a)$. Hence, since $C_0^1(\Omega)$ is dense in $\overset{\circ}{W}{}_p^1(\Omega)$, we have

$$\|E_0 - P|\overset{\circ}{W}{}_p^1(\Omega) \to L^p(\Omega)\| \le \frac{\gamma_p}{2n}|\Omega|. \tag{2.4.18}$$

Since rank $P \le n-1$, it follows that

$$a_n(E_0(\Omega)) \le \frac{\gamma_p}{2n}|\Omega| \tag{2.4.19}$$

and so

$$\nu_0(\frac{\gamma_p}{2n}|\Omega|, \Omega) \le n. \tag{2.4.20}$$

Next, we choose the partition $\Omega = \bigcup_{i=1}^{n} \Omega_i$, where $|\Omega_i| = \frac{|\Omega|}{n}$ for $i = 1, \cdots, n$. If $\mathcal{M}$ is the subspace of $\overset{\circ}{W}{}_p^1(\Omega)$ spanned by $\{\phi_1, \cdots, \phi_n\}$, where, for $\varepsilon > 0$, the $\phi_i \in \overset{\circ}{W}{}_p^1(\Omega_i)$ are translations of one another, and

$$\frac{\|\phi_i'\|_{p,\Omega_i}}{\|E_0(\Omega)\phi_i\|_{p,\Omega_i}} \le \frac{1}{(\frac{\gamma_p|\Omega|}{2n} - \varepsilon)},$$

then $\dim \mathcal{M} = n$ and for non-zero $\phi \in \mathcal{M}$,

$$\frac{\|\phi'\|_{p,\Omega}}{\|E_0(\Omega)\phi\|_{p,\Omega}} \le \frac{1}{(\frac{\gamma_p|\Omega|}{2n} - \varepsilon)}.$$

It follows that $\mu_0(\frac{\gamma_p}{2n}|\Omega|, \Omega) \ge n$. We shall prove in Lemma 6.3.1 that

$$\mu_0(\varepsilon,\Omega) \le \nu_0(\varepsilon,\Omega).$$

This implies that

$$\mu_0(\frac{\gamma_p}{2n}|\Omega|,\Omega) = \nu_o(\frac{\gamma_p}{2n}|\Omega|,\Omega) = n$$

and therefore $a_n(E_0(\Omega)) = \frac{\alpha_p}{2n}|\Omega|$.

To prove the result for $E_{\mathbb{C}}(\Omega)$, we again choose the partition $\Omega = \bigcup_{i=1}^n \Omega_i$, where $|\Omega_i| = \frac{|\Omega|}{n}$ for $i = 1,\cdots,n$. For $g \in [f] \in W^1_{p,\mathbb{C}}(\Omega)$, set $Pg(x) = \Sigma_{i=1}^n g(c_i)\chi_i(x)$, where c_i is the mid-point of Ω_i. Then $P : [f] \to [Pf] \in L_{1,\mathbb{C}}(\Omega)$ and P has rank $\le n-1$ as is seen on writing $Pf = f(c_1)+\Sigma_{i=2}^n(f(c_i)-f(c_1))\chi_i$. Furthermore,

$$\begin{aligned}
\|[f]-P[f]\|^p_{L_{p,\mathbb{C}}(\Omega)} &= \|[f]-[Pf]\|^p_{L_{p,\mathbb{C}}(\Omega)} \\
&\le \|f-Pf\|^p_{p,\Omega} \\
&= \Sigma_{i=1}^n \|H_{c_i}f'\|^p_{p,\Omega_i} \\
&\le \{(\frac{\gamma_p}{2n})|\Omega|\|f'\|_{p,\Omega}\}^p.
\end{aligned}$$

Hence

$$\|E_{\mathbb{C}}(\Omega) - P|W^1_{p,\mathbb{C}}(\Omega) \to L_{p,\mathbb{C}}(\Omega)\| \le (\frac{\gamma_p}{2n})|\Omega|$$

and so, since rank $P \le n-1$,

$$a_n(E_{\mathbb{C}}(\Omega)) \le (\frac{\gamma_p}{2n})|\Omega|.$$

From Theorem 2.4.3, for $i = 1,\cdots,n$,

$$\begin{aligned}
\|E_{\mathbb{C}}(\Omega_i)|W^1_{p,\mathbb{C}}(\Omega_i) \to L_{p,\mathbb{C}}(\Omega_i)\| &= \sup\{\inf_{c\in\mathbb{C}} \|f-c\|_{p,\Omega_i}; \|f'\|_{p,\Omega_i} = 1\} \\
&= \frac{\gamma_p}{2n}|\Omega|.
\end{aligned}$$

Thus, if $\delta < \frac{\alpha_p}{2n}|\Omega|$, there exists $\psi_i \in W^1_p(\Omega_i)$ such that $\|\psi_i\|_{p,\Omega_i} > 0$ and

$$\|[\psi_i]\|_{L_{p,\mathbb{C}}(\Omega_i)} \equiv \{\inf_{c\in\mathbb{C}} \|f-c\|_{p,\Omega_i}\} > \delta\|\psi_i'\|_{p,\Omega_i}.$$

Define ψ_i' to be zero outside Ω_i and such that $\psi_i \in W^1_p(\Omega)$. Now let $P : W^1_{p,\mathbb{C}}(\Omega) \to L_{p,\mathbb{C}}(\Omega)$ have rank $< n$. It follows that there exist constants $\lambda_i, i = 1,\cdots,n$, not all zero, such that

$$[\psi] = \Sigma_{i=1}^n \lambda_i[\psi_i], \quad P[\psi] = 0.$$

Then

$$\|(E_{\mathbb{C}}(\Omega) - P)[\psi]\|^p_{L_{p,\mathbb{C}}(\Omega)} = \|[\psi]\|^p_{L_{p,\mathbb{C}}(\Omega)}$$

$$= \Sigma_{i=1}^{n} \|\lambda_i[\psi_i]\|_{L_{p,\mathbb{C}}(\Omega_i)}^p$$
$$\geq \delta^p \Sigma_{i=1}^{n} \|\lambda_i \psi_i'\|_{p,\Omega_i}^p$$
$$= \delta^p \|\psi'\|_{p,\Omega}^p.$$

Since P (of rank $< n$) and $\delta < \frac{\gamma_p}{2n}|\Omega|$ are arbitrary, we have proved that

$$a_n(E_{\mathbb{C}}(\Omega)) \geq \frac{\gamma_p}{2n}|\Omega|,$$

and hence the proof is complete. □

The formula (2.4.14) for the approximation numbers of the embeddings E_0 and $E_{\mathbb{C}}$ is of particular interest. It was originally proved independently in [68] and [87]; the special case $p = 2$ is more familiar as it follows from the relationship between singular values and eigenvalues of the Laplacian. It remains to be seen whether or not there is an extension to Sobolev spaces on higher-dimensional domains.

2.4.2 The general case: Preliminaries

Our first objective in this section is to obtain upper and lower bounds for the $a_m(T)$. To achieve these we require some technical lemmas for a quantity $\mathbb{A}(I)$ defined on an arbitrary subinterval I of (a, b) and dependent on the functions u and v. This quantity will be used to define a partition of (a, b) which lies at the heart of the method. The account is based on the papers [83] and [84], concerning trees, but we restrict attention to an interval and comment later, in Sections 2.6 and 2.7, on the analogous results for trees.

We shall use I to denote an interval.

Definition 2.4.5. *Let $I \subset (a, b)$ be an interval and $c \in (a, b)$. For $1 \leq p, q \leq \infty$, we define:*

$$\mathbb{A}(I) \equiv \mathbb{A}(I, u, v) := \begin{cases} \sup_{f \in L_p(I), f \neq 0} \inf_{\alpha \in \mathbb{C}} \frac{\|T_{c,I}f - \alpha v\|_{q,I}}{\|f\|_{p,I}} & \textit{if } \mu(I) > 0, \\ 0 & \textit{if } \mu(I) = 0, \end{cases}$$

where

$$T_{c,I}f(x) := v(x)\chi_I(x) \int_c^x f(t)u(t)\chi_I(t)dt,$$

and

$$\mu(I) := \begin{cases} \int_I |v(t)|^q dt, \ 1 \leq q < \infty, \\ \sup_I |v(t)|, \ q = \infty. \end{cases}$$

We allow $I = (a, x), x \leq b$ if, and only if, $v \in L_q(a, b)$; this is to be understood hereafter, even when not explicitly mentioned. In the case $I = (a, b)$ we write T_c instead of $T_{c,I}$. Clearly, $\mathbb{A}(a, b) < \infty$ if and only if $T_c : L_p(a, b) \to L_q(a, b)$ is bounded.

We can assume, without loss of generality, that $u \geq 0, v \geq 0$ on (a, b). This is apparent on writing $u = |u| \text{ sgn } u, v = |v| \text{ sgn } v$, where $\text{sgn } u = u/|u|$ when $u \neq 0$ and is 1 otherwise.

Lemma 2.4.6. *The number* $\mathbb{A}(I, u, v)$ *is independent of* $c \in (a, b)$.

Proof. Let S denote the canonical map of $L_q(I)$ into its quotient by the space of scalar multiples of v. Then

$$\mathbb{A}(I) = \|ST_{c,I}|L_p(I) \to L_{q,\mathbb{C}}(I)\|.$$

For $d \in I$,

$$T_{d,I}f(x) = T_{c,I}f(x) + v(x)\chi_I(x) \int_d^c f(t)u(t)\chi_I(t)dt.$$

Thus, $ST_{d,I} = ST_{c,I}$, whence the lemma. □

Lemma 2.4.7. *For all subintervals* $I = (c, d) \subset (a, b)$ *and* $1 \leq p, q \leq \infty$ *we have that*

$$\mathbb{A}(I) \leq \inf_{x \in (a,b)} \|T_{x,I}|L_p(I) \to L_q(I)\|. \tag{2.4.21}$$

If $1 < p \leq \infty$, *the operators* $T_{x,I}, T_{x,(c,x)}, T_{x,(x.d)}$ *from* $L_p(I)$ *to* $L_q(I)$ *are such that their norms* $\|T_{x,I}\|, \|T_{x,(c,x)}\|, \|T_{x,(x,d)}\|$ *are continuous in* $x \in (a, b)$ *and there exists* $e \in I$ *such that*

$$\|T_{e,(c,e)}\| = \|T_{e,(e,d)}\|. \tag{2.4.22}$$

If $1 < p \leq q \leq \infty$, *then for any* $x \in I$,

$$\|T_{x,I}\| = \max\left\{\|T_{x,(c,x)}\|, \|T_{x,(x,d)}\|\right\}, \tag{2.4.23}$$

and

$$\min_{x \in I} \|T_{x,I}\| = \|T_{e,I}\|, \tag{2.4.24}$$

where $e \in I$ *satisfies (2.4.22).*

Proof. From Lemma 2.4.6, we have for any $x \in (a, b)$,

$$\begin{aligned} \mathbb{A}(I) &\leq \sup\left\{\|T_{x,I}f\|_{q,I} : \|f\|_{p,I} = 1\right\} \\ &= \|T_{x,I}|L_p(I) \to L_q(I)\| \end{aligned}$$

whence (2.4.21).

To prove the continuity of $\|T_{x,(x,d)}\|$, we first note that for $z, y \in (a,b)$, $z < y$,

$$T_{z,(z,d)}f(x) - T_{y,(y,d)}f(x) = v(x)\chi_{(y,d)}(x)\int_z^y u(t)f(t)dt + v(x)\chi_{(z,y)}(x)\int_z^x u(t)f(t)dt.$$

Hence, on applying Hölder's inequality,

$$\|T_{z,(z,d)}f - T_{y,(y,d)}f\|_{q,I} \le \|v\|_{q,(y,d)}\|u\|_{p',(z,y)}\|f\|_{p,I} + \|v\|_{q,(z,y)}\|u\|_{p',(z,y)}\|f\|_{p,I}$$

and so

$$\big|\|T_{z,(z,d)}\| - \|T_{y,(y,d)}\|\big| \le \|T_{z,(z,d)} - T_{y,(y,d)}|L_p(I) \to L_q(I)\| \le 2\|u\|_{p',(z,y)}\|v\|_{q,(z,d)}$$

which yields the continuity of $\|T_{x,(x,d)}\|$ when $p > 1$. Similarly for $\|T_{x,(c,x)}\|$ and $\|T_{x,I}\|$.

If supp $f \subset (y,d)$ then for $z < y$,

$$T_{z,(z,d)}f(x) = T_{y,(y,d)}f(x).$$

Hence

$$\|T_{y,(y,d)}\| \le \|T_{z,(z,d)}\|$$

and similarly

$$\|T_{z,(c,z))}\| \le \|T_{y,(c,y)}\|.$$

The identity (2.4.22) follows from these inequalities and the continuity of the norms $\|T_{x,(c,x)}\|$, $\|T_{x,(x,d)}\|$.

Let $f \in L_p(I)$ and set $f_1 = f\chi_{(c,x)}$, $f_2 = f\chi_{(x,d)}$. Then

$$(T_{x,I}f)(t) = (T_{x,(c,x)}f_1)(t) + (T_{x,(x,d)}f_2)(t).$$

This implies that

$$\begin{aligned}\|T_{x,I}f\|_{q,I}^q &= \|T_{x,(c,x)}f_1\|_{q,I}^q + \|T_{x,(x,d)}f_2\|_{q,I}^q \\ &\le \max\left\{\|T_{x,(c,x)}\|^q, \|T_{x,(x,d)}\|^q\right\}\left(\|f_1\|_{p,I}^q + \|f_2\|_{p,I}^q\right) \\ &\le \max\left\{\|T_{x,(c,x)}\|^q, \|T_{x,(x,d)}\|^q\right\}\left(\|f_1\|_{p,I}^p + \|f_2\|_{p,I}^p\right)^{q/p}\end{aligned}$$

since $q/p \ge 1$. As $\|f_1\|_{p,I}^p + \|f_2\|_{p,I}^p = \|f\|_{p,I}^p$, it follows that

$$\|T_{x,I}\| \le \max\left\{\|T_{x,(c,x)}\|, \|T_{x,(x,d)}\|\right\}.$$

The reverse inequality is obvious and (2.4.23) is proved. If $e < y$, then, from (2.4.23) and the above analysis,

$$\|T_{e,I}\| = \|T_{e,(c,e)}\| \leq \|T_{y,(c,y)}\| \leq \|T_{y,I}\|,$$

whence (2.4.24). □

In the next lemma $\|\cdot\|_{q,I,\mu}$ denotes the norm in $L_q(I,\mu)$, and the result relates to Lemma 2.4.2 with $e \in L_q(I,\mu)$ fixed.

Lemma 2.4.8. *Let $1 < q \leq \infty$ and given $f \in L_q(I,\mu)$, let $c_f = c_{f,e_0}$ be the unique scalar such that $\|f - c_f e_0\|_{q,I,\mu} = \inf_{c \in \mathbb{C}} \|f - ce_0\|_{q,I,\mu}$ for $e_0 \neq 0, e_0 \in L_q(I,\mu)$. Then the map $f \mapsto c_f : L_q(I,\mu) \to \mathbb{C}$ is continuous .*

Proof. Suppose that $g_n \to f$. Since $\{c_{g_n}\}$ is bounded, we may suppose that $c_{g_n} \to c$, say. Then

$$\|g_n - c_f e_0\|_{q,I,\mu} \geq \|g_n - c_{g_n} e_0\|_{q,I,\mu}$$

and so

$$\|f - c_f e_0\|_{q,I,\mu} \geq \|f - ce_0\|_{q,I,\mu}$$

which gives $c = c_f$. □

Theorem 2.4.9. *If $I \subset (a,b)$ and $1 < p = q < \infty$,*

$$\begin{aligned} \mathbb{A}(I) &= \min_{x \in} \|T_{x,I}|L_p(I) \to L_p(I)\| \\ &= \|T_{e,I}|L_p(I) \to L_p(I)\|, \end{aligned} \tag{2.4.25}$$

where $e \in I$ satisfies (2.4.22).

Proof. Let $\alpha < \|T_{e,I}\|$ and set $T_{e,I} = vF$, that is,

$$Ff(t) = F_{e,I}f(t) := \chi_I(t) \int_e^t u(t)\chi_I(t)f(t)dt.$$

Then, by (2.4.22) and (2.4.23), it follows that, with $I = (c,d)$, there exist $f_i, i = 1,2$, supported in $(c,e),(e,d)$, respectively, such that $\|f_i\|_{p,I} = 1, \|T_{e,I}f_i\|_{p,I} > \alpha$ and f_1 positive, f_2 negative. The same is true of the signs of the corresponding values of c_{Ff_1}, c_{Ff_2} in Lemma 2.4.2 since we are assuming, without loss of generality, that u and v are non-negative. Hence by the continuity established in Lemma 2.4.8, there is a $\lambda \in [0,1]$ such that $c_{Fg} = 0$ for $g = \lambda f_1 + (1-\lambda)f_2$, and

$$\|T_{e,I}g\|_{p,I}^p = \lambda^p \|T_{e,I}f_1\|_{p,I}^p + (1-\lambda)^p \|T_{e,I}f_2\|_{p,I}^p$$

$$> \alpha^p \|g\|_{p,I}^p.$$

Then, by Lemma 2.4.2 with $e_0 = g$ and the measure μ of Definition 2.4.5,

$$\begin{aligned} \mathbb{A}(I) &\geq \inf_{c\in\mathbb{C}} \|(F-c)g\|_{p,I,\mu}/\|g\|_{p,I} \\ &= \|Fg\|_{p,I,\mu}\|/\|g\|_{p,I} \\ &= \|T_{e,I}g\|_{p,I}/\|g\|_{p,I} \\ &> \alpha. \end{aligned}$$

Since $\alpha < \|T_{e,I}\|$ is arbitrary, $\mathbb{A}(I) \geq \|T_{e,I}\|$ and the result follows from Lemma 2.4.7. □

Lemma 2.4.10. *Let $I \subset (a,b)$ and $v \in L_q(I), 1 \leq q \leq \infty$, with $\mu(I) \neq 0$ in Definition 2.4.5. Then there exists $w_I \in \{L_q(I,\mu)\}^*$ such that:*

$$w_I(1) = 1,$$

$$\|w_I|\{L_q(I,\mu)\}^*\| = \frac{1}{\|v\|_{q,I}},$$

and

$$\inf_{c\in\mathbb{C}} \|(\varphi - c)\|_{q,I,\mu} \leq \|\varphi - w_I(\varphi)\|_{q,I,\mu} \leq 2 \inf_{c\in\mathbb{C}} \|\varphi - c\|_{q,I,\mu} \tag{2.4.26}$$

for all $\varphi \in L_q(I,\mu)$. In the case $q = 2$,

$$\inf_{c\in\mathbb{C}} \|\varphi - c\|_{2,I,\mu} = \|\varphi - w_I(\varphi)\|_{2,I,\mu}, \tag{2.4.27}$$

where

$$w_I(\varphi) = \frac{1}{\mu(I)} \int_I \varphi d\mu.$$

Proof. Define the linear functional w on the constants in $L_q(I,\mu)$ by $w(c.1) = c$. Then $w(1) = 1$ and $\|w|\{L_q(I,\mu)\}^*\|^q = 1/\mu(I)$ for $1 \leq q < \infty$, with $\|w\| = 1$ when $q = \infty$. The existence of w_I follows by the Hahn-Banach theorem, and (2.4.26) is immediate. The case $q = 2$ follows from the decomposition of $L_2(I,\mu)$ into $\mathbb{C}$ and its orthogonal complement. □

Lemma 2.4.11. *Let $I \subset (a,b), c \in I$ and $1 < p = q < \infty$. Then*

$$\mathbb{A}(I) = \|T_{c,I} - v\varpi|L_p(I) \to L_p(I)\|,$$

where ϖ is the bounded linear functional

$$\varpi(f) = \int_c^e fu$$

and $e \in I$ is given in Lemma 2.4.7.

Proof. From

$$T_{e,I}f = T_{c,I}f - v\chi_I \int_c^e fu\chi_I dt$$

we have by Lemma 2.4.9 that

$$\|T_{c,I} - P|L_p(I) \to L_q(I)\| = \mathbb{A}(I,u,v),$$

where

$$Pf(x) = v(x)\chi_I(x) \int_c^e fu\chi_I dt,$$

and the lemma follows. □

The dependence of $\mathbb{A}(I,u,v)$ on u,v is established in the next two lemmas.

Lemma 2.4.12. *Let u,v be constant functions over the interval $I = (c,d)$ and $1 \le p,q \le \infty$. Then $\mathbb{A}(I,u,v) = |v||u||I|\mathbb{A}((0,1),1,1)$, and*

$$\mathbb{A}((0,1),1,1) = \|E_{\mathbb{C}}|W^1_{p,\mathbb{C}}(0,1) \to L_{p,\mathbb{C}}(0,1)\| = \frac{1}{2}\gamma_{p,q}.$$

Proof. We have

$$\begin{aligned}
\mathbb{A}(I,u,v) &= \sup_{\|f\|_{p,I}\le 1} \inf_{\alpha\in\mathbb{C}} \left\| v\left(\int_c^x f(t)u(t)dt - \alpha\right)\right\|_{q,I} \\
&= |v||u| \sup_{\|f\|_{p,I}\le 1} \inf_{\alpha\in\mathbb{C}} \left\| \int_c^x f(t)dt - \alpha\right\|_{q,I} \\
&= |v||u||I| \sup_{\|f\|_{p,(0,1)}\le 1} \inf_{\alpha\in\mathbb{C}} \left\| \int_0^x f(t)dt - \alpha\right\|_{q,(0,1)} \\
&= |v||u||I|\mathbb{A}((0,1),1,1).
\end{aligned}$$

The rest follows from the definition of $\mathbb{A}$. □

Recall that the value of $\gamma_{p,q}$ is given in (2.4.5). Further remarks on the value of $\gamma_{p,q}$ are made in the notes on Section 2.4 at the end of this chapter.

Lemma 2.4.13. *Let $1 \le p,q \le \infty$, $u_1,u_2 \in L_{p'}(I)$ and $v \in L_q(I)$. Then*

$$|\mathbb{A}(I,u_1,v) - \mathbb{A}(I,u_2,v)| \le \|v\|_{q,I}\|u_1 - u_2\|_{p',I}.$$

Proof. We have

$$\mathbb{A}(I,u_1,v) = \sup_{\|f\|_{p,I}=1} \inf_{\alpha\in\mathbb{C}} \| \int_c^x f(t)(u_1(t)-u_2(t)+u_2(t))dt - \alpha\|_{q,I,\mu}$$
$$\leq \sup_{\|f\|_{p,I}=1} \inf_{\alpha\in\mathbb{C}} \left[\| \int_c^x f(t)(u_1(t)-u_2(t))dt\|_{q,I,\mu} + \| \int_c^x f(t)u_2(t)dt - \alpha\|_{q,I,\mu} \right]$$
$$\leq \sup_{\|f\|_{p,I}=1} \inf_{\alpha\in\mathbb{C}} \left[\|v\|_{q,I}\|u_1-u_2\|_{p',I} + \| \int_c^x f(t)u_2(t)dt - \alpha\|_{q,I,\mu} \right]$$
$$\leq \|v\|_{q,I}\|u_1-u_2\|_{p',I} + \mathbb{A}(I,u_2,v).$$

The same holds with u_1, u_2 interchanged, whence the lemma. □

Lemma 2.4.14. *Let $1 \leq p,q \leq \infty, u \in L_{p'}(I)$, and $v_1, v_2 \in L_q(I)$. Then*

$$|\mathbb{A}(I,u,v_1) - \mathbb{A}(I,u,v_2)| \leq 2\|v_1-v_2\|_{q,I}\|u\|_{p',I}.$$

Proof. We have

$$\mathbb{A}(I,u,v_1) = \sup_{\|f\|_{p,I}=1} \inf_{\alpha\in\mathbb{C}} \|v_1(x)\left[\int_c^x f(t)u(t)dt - \alpha\right]\|_{q,I}$$
$$= \sup_{\|f\|_{p,I}=1} \inf_{|\alpha|\leq\|u\|_{p',I}\|f\|_{p,I}} \|v_1(x)\left[\int_c^x f(t)u(t)dt - \alpha\right]\|_{q,I}.$$

Since

$$\|v_1(x)[\int_c^x f(t)u(t)dt-\alpha]\|_{q,I} \leq \|v_1-v_2\|_{q,I}\|u\|_{p',I}\|f\|_p + \|(v_1-v_2)\alpha\|_{q,I}$$
$$+ \|v_2[\int_c^x f(t)u(t)dt-\alpha]\|_{q,I},$$

it follows that

$$\mathbb{A}(I,u,v_1) \leq 2\|v_1-v_2\|_{q,I}\|u\|_{p',I}$$
$$+ \sup_{\|f\|_{p,I}=1} \inf_{|\alpha|\leq\|u\|_{p',I}} \|v_2(x)\left[\int_c^x f(t)u(t)dt - \alpha\right]\|_{q,I}$$
$$= 2\|v_1-v_2\|_{q,I}\|u\|_{p',I} + A(I,u,v_2).$$

The same holds with v_1 and v_2 interchanged. □

Lemma 2.4.15. *Let $1 < p \leq \infty, 1 \leq q < \infty$, and suppose that $K_1, K_2, K_1 \subset K_2$, are compact subintervals of (a,b). Then*

$$|\mathbb{A}(K_1) - \mathbb{A}(K_2)| \to 0 \qquad \text{as } |K_2 \setminus K_1| \to 0.$$

Proof. We see that $\mathbb{A}(K_1) = \mathbb{A}(K_1 \cup K_2, u\chi_{K_1}, v\chi_{K_1})$ and $\mathbb{A}(K_2) = \mathbb{A}(K_1 \cup K_2, u\chi_{K_2}, v\chi_{K_2})$. The lemma then follows from Lemmas 2.4.13 and 2.4.14. □

Lemma 2.4.16. *Let* $1 \leq p \leq q \leq \infty$, *and, for* $x \in I = (c,d) \subset (a,b)$, *define*

$$J_1(c,x) := \sup_{c<s<x} \{\|u\|_{p',(s,x)} \|v\|_{q,(c,s)}\}, \tag{2.4.28}$$

$$J_2(x,d) := \sup_{x<s<d} \{\|u\|_{p',(x,s)} \|v\|_{q,(s,d)}\}. \tag{2.4.29}$$

Then

$$\begin{aligned} \max\{J_1(c,x), J_2(x,d)\} &\leq \|T_{x,I}|L_p(I) \to L_q(I)\| \\ &\leq 4\max\{J_1(c,x), J_2(x,d)\}, \end{aligned} \tag{2.4.30}$$

and if

$$J(I) := \inf_{x\in I} [\{\max\{J_1(c,x), J_2(x,d)\}], \tag{2.4.31}$$

we have

$$J(I) \leq \mathbb{A}(I) \leq 4J(I). \tag{2.4.32}$$

Proof. From Theorem 2.2.1,

$$J_2(x,d) \leq \|T_{x,(x,d))}|L_p(x,d) \to L_q(x,d)\| \leq 4J_2(x,d). \tag{2.4.33}$$

We also have

$$J_1(c,x) \leq \|T_{x,(c,x)}|L_p(c,x) \to L_q(c,x)\| \leq 4J_1(c,x). \tag{2.4.34}$$

For, on substituting $t = -y, c = -C, x = -X, \hat{f}(y) = -f(t)$, we obtain

$$\begin{aligned} \|T_{x,(c,x)}f\|_{q,(c,x)} &= \|\hat{v}(y)\int_X^y \hat{u}\hat{f}ds\|_{q,(X,C)} \\ &= \|\hat{T}_{X,(X,C)}\hat{f}\|_{q,(X,C)} \end{aligned}$$

say, where $\hat{T}_{X,(X,C)}$ has the same norm as $T_{x,(c,x)}$. Theorem 2.2.1 applies to $\hat{T}_X$ with the quantity in (2.2.3) given by

$$\begin{aligned} &\sup_{X<y<C} \{\|\hat{u}\|_{p',(X,y)} \|\hat{v}\|_{q,(y,C)}\} \\ &= \sup_{c<t<x} \{\|u\|_{p',(t,x)} \|v\|_{q,(c,t)}\} \\ &= J_1(c,x), \end{aligned}$$

whence (2.4.34). The inequalities (2.4.30) follow from (2.4.23) and (2.4.32) is a consequence of (2.4.21). □

Remark 2.4.17. The constants in (2.4.30) can be made sharper on applying the improvements to (2.2.4) noted in Remarks 2.2.2 and 2.2.3.

2.4.3 Estimates for $a_m(T)$, $1 < p \le q < \infty$

In this section we shall assume that

$$v \in L_q(a, b). \tag{2.4.35}$$

We shall denote $T_a, T_{a,I}$ by T, T_I respectively.

For any interval $I \subset (a, b)$ and $\varepsilon > 0$, define

$$\begin{aligned} N(I, \varepsilon) &\equiv N(I, \varepsilon, u, v) \\ &:= \min \left\{n : I = \bigcup_{i=1}^{n} I_i, \mathbb{A}(I_i) \le \varepsilon\right\}, \end{aligned} \tag{2.4.36}$$

where the I_i are non-overlapping subintervals of I. If $I \subset\subset (a, b)$ (i.e. the closure of I is a compact subset of (a, b)) then $N(I, \varepsilon) < \infty$ by the continuity of $\mathbb{A}$ established in Lemma 2.4.15. When $I = (a, b), N(I, \varepsilon) < \infty$ if

$$\lim_{x \to a} \mathbb{A}(a, x) = \lim_{x \to b} \mathbb{A}(x, b) = 0. \tag{2.4.37}$$

This is satisfied if and only if T is compact. For by Theorem 2.3.1 and the proof of Lemma 2.4.16, T is compact if and only if

$$\lim_{x \to a} J_1(a, x) = \lim_{x \to b} J_2(x, b) = 0; \tag{2.4.38}$$

note that $J_1(a, \cdot), J_2(\cdot, b)$ are respectively monotonic increasing and decreasing on (a, b). The equivalence of (2.4.37) and (2.4.38) follows from Lemma 2.4.7, (2.4.33) and (2.4.34). Hence, when $I = (a, b), N(I, \varepsilon) < \infty$ for all $\varepsilon > 0$ if and only if T is compact.

If $I = (c, d) \subset (a, b)$, and T is compact when $I = (a, b)$, we have that there exist intervals $I_i = (c_i, c_{i+1}), i = 1, 2, \cdots, N$, which are such that

$$c = c_1 < c_2 < \cdots < c_{N+1} = d,$$

with

$$\mathbb{A}(c_i, c_{i+1}) = \varepsilon, \; i = 1, 2, \cdots, N-1, \quad \mathbb{A}(c_N, c_{N+1}) \le \varepsilon. \tag{2.4.39}$$

This follows from the monotonicity and continuity of $\mathbb{A}$ which hold for the present range of values of p and q, namely, $1 < p \le q < \infty$.

We also define

$$\begin{aligned} M(I, \varepsilon) &:= M(I, \varepsilon, u, v) \\ &:= \max \{m : \text{there exist non-overlapping intervals} \\ &\qquad I_j \subset I, j = 1, 2, \cdots, m, \text{such that} \quad \mathbb{A}(I_j) \ge \varepsilon\}. \end{aligned}$$

Then, by (2.4.39),

$$M(I, \varepsilon) \ge N(I, \varepsilon) - 1. \tag{2.4.40}$$

Lemma 2.4.18. *If $N = N(I, \varepsilon) < \infty$, then $a_{N+1}(T_I) \leq \varepsilon$.*

Proof. Let $I_i = (c_i, c_{i+1}), i = 1, 2, \cdots, N$, and set $Pf = \sum_{i=1}^N P_i f$, where

$$P_i f(x) = \chi_{I_i}(x) v(x) \Big[\int_a^{c_i} u(t) f(t) dt + \varpi_i(f) \Big], \tag{2.4.41}$$

and ϖ_i, e_i, are as in Lemma 2.4.11 corresponding to I_i. Then rank $P \leq N$ and from Lemma 2.4.11, denoting T_{I_i} by T_i,

$$\begin{aligned} \|(T_I - P)f\|_{q,I}^q &= \sum_{i=1}^N \|T_I f - Pf\|_{q,I_i}^q \\ &= \sum_{i=1}^N \|T_i f - v\varpi_i(f)\|_{q,I_i}^q \\ &\leq \sum_{i=1}^N \mathbb{A}(I_i)^q \|f\|_{p,I_i}^q \\ &\leq \varepsilon^q \|f\|_{p,I}^q, \end{aligned}$$

whence the lemma. □

Lemma 2.4.19. *Let $\{I_i : i = 1, 2, \cdots, M\}$ be a set of non-overlapping subintervals of I such that $\mathbb{A}(I_i) \geq \varepsilon$ for $i = 1, \cdots, M$. Then $a_M(T_I) \geq \varepsilon M^{\frac{1}{q} - \frac{1}{p}}$.*

Proof. Let $\lambda \in (0, 1)$. From the definition of $\mathbb{A}(I_i)$ there exists $\phi_i \in L_p(I)$, with support in I_i, such that

$$\inf_{\alpha \in \mathbb{C}} \|T_I \phi_i - \alpha v\|_{q,I_i} \geq \lambda \varepsilon \|\phi_i\|_{p,I_i}.$$

Let $P : L_p(I) \to L_q(I)$ be bounded and linear with rank $P < M$. Then there exist constants $\lambda_1, \cdots, \lambda_M$, not all zero, such that

$$P\phi = 0, \quad \phi := \sum_{i=1}^M \lambda_i \phi_i.$$

Then, denoting summation over $\{i : \lambda_i \neq 0\}$ by $\sum'$ and setting $I_i = (c_i, c_{i+1})$, we have

$$\begin{aligned} \|T_I\phi - P\phi\|_{q,I}^q &= \|T_I \phi\|_{q,I}^q \\ &= \sum_{i=1}^M \|(T_I\phi)\chi_{I_i}\|_{q,I}^q \\ &= \sum{}' \|\chi_{I_i}(x) v(x) \Big(\int_{c_i}^x \lambda_i \phi_i(t) u(t) \chi_{I_i}(t) dt + \int_a^{c_i} \phi(t) u(t) dt \Big) \|_{q,I}^q \end{aligned}$$

$$= \sum{}' \| \left(T_I\phi_i(x) + v(x)\eta_i\lambda_i^{-1}\right) \lambda_i\|_{q,I_i}^q,$$

(where $\eta_i = \int_a^{c_i} u(t)\phi(t)dt$)

$$\begin{aligned}
&\geq \sum{}' \inf_{\alpha\in\mathbb{C}} \|T_I\phi_i - \alpha v\|_{q,I_i}^q |\lambda_i|^q \\
&\geq (\lambda\varepsilon)^q \sum{}' \|\phi_i\|_{p,I_i}^q |\lambda_i|^q \\
&\geq (\lambda\varepsilon)^q \Big(\sum{}' \|\lambda_i\phi_i\|_{p,I_i}^p\Big)^{q/p} M^{1-\frac{q}{p}} \\
&= (\lambda\varepsilon)^q \|\phi\|_{p,I}^q M^{1-\frac{q}{p}}.
\end{aligned}$$

Since P is an arbitrary bounded linear operator of rank less than M the result follows. □

Lemmas 2.4.18 and 2.4.19, and (2.4.40) combine to give

Theorem 2.4.20. *Let $1 < p \leq q < \infty$, and suppose that T_I is compact. Then given any $\varepsilon > 0$,*

$$a_{N+1}(T_I) \leq \varepsilon, \tag{2.4.42}$$

$$a_M(T_I) \geq M^{\frac{1}{q}-\frac{1}{p}}\varepsilon, \tag{2.4.43}$$

where $N = N(I,\varepsilon) < \infty$ and $M \geq N-1$.

Corollary 2.4.21. *Let $1 < p \leq q < \infty$, and suppose that*

$$u \in L_{p'}(I), v \in L_q(I).$$

Then given any $\varepsilon > 0, N = N(I,\varepsilon) < \infty$ and

$$a_{N+2}(T_I) \leq \|u\|_{p',I}\|v\|_{q,I}N^{-1}. \tag{2.4.44}$$

Proof. From (2.4.21) and (2.4.40), with $M = M(I,\varepsilon)$ and $N = N(I,\varepsilon)$, we have

$$\begin{aligned}
\varepsilon(N-1) \leq \varepsilon M &\leq \sum_{i=1}^M \mathbb{A}(I_i) \\
&\leq \sum_{i=1}^M \|T_{I_i}|L_p(I_i) \to L_q(I_i)\| \\
&\leq \sum_{i=1}^M \|v\|_{q,I_i}\|u\|_{p',I_i}
\end{aligned}$$

$$\le \|u\|_{p',I}\Big(\sum_{i=1}^{M}\|v\|_{q,I_i}^{p}\Big)^{1/p}$$
$$\le \|u\|_{p',I}\Big(\sum_{i=1}^{M}\|v\|_{q,I_i}^{q}\Big)^{1/q}$$
$$= \|u\|_{p',I}\|v\|_{q,I}.$$

Since (2.4.38) is clearly satisfied, T_I is compact, or equivalently $N(I,\varepsilon)<\infty$ for all $\varepsilon>0$, and (2.4.44) follows from (2.4.42). □

2.4.4 Estimates for $a_n(T)$ when $p=1$ or $q=\infty$

We first recall Remark 2.3.2, that $T: L_1(a,b)\to L_\infty(a,b)$ is never compact. When $p=1$ or $q=\infty$, the continuity of the function $\mathbb{A}(\cdot)$ in Definition 2.4.5 is not available. In this case we have instead of (2.4.40),

Lemma 2.4.22. *If $N(I,\varepsilon)<\infty$, then*

$$N(I,\varepsilon)\le 2M(I,\varepsilon)+1. \tag{2.4.45}$$

Proof. From the definition of $N=N(I,\varepsilon)$, there exist intervals $I_i, i=1,\cdots,N$, such that $\mathbb{A}(I_i)\le\varepsilon$, but $\mathbb{A}(I_i\cup I_j)>\varepsilon$ for $i\ne j$. Now set $J_1=I_1\cup I_2, J_2=I_3\cup I_4,\cdots,J_k=I_{N-1}\cup I_N$ if $N=2k$ and $J_{k+1}=I_N$ if $N=2k+1$. Thus $k\le M$ and the lemma is proved. □

On using Lemma 2.4.10 instead of Lemma 2.4.11, the proof of Lemma 2.4.18 remains valid with

$$P_if(x)=\chi_{I_i}(x)v(x)\Big[\int_a^{c_i}u(t)f(t)dt+w_{I_i}\Big(\int_{c_i}^{\cdot}u(t)f(t)dt\Big)\Big],$$

to give the estimate

$$a_{N+1}(T_I)\le\kappa_q\varepsilon$$

where $\kappa_q=2$ when $q\ne2$ and $\kappa_2=1$. Lemma 2.4.19 continues to hold and so we get

Theorem 2.4.23. *Let $1\le p\le q\le\infty$, $\varepsilon>0$ and $N=N(I,\varepsilon)<\infty$. Then*

$$a_{N+1}(T_I)\le\kappa_q\varepsilon \tag{2.4.46}$$

where $\kappa_q=2$ when $q\ne2$ and $\kappa_2=1$, and

$$a_{[N/2]-1}(T_I)\ge\varepsilon. \tag{2.4.47}$$

The measure of non-compactness $\alpha(T)$ of T_I (see Sections 1.3 and 2.3) satisfies

$$\alpha(T_I):=\lim_{n\to\infty}a_n(T_I)\approx\inf\{\varepsilon: N(I,\varepsilon,u,v)<\infty\}. \tag{2.4.48}$$

2.4.5 Approximation numbers of T when $1 \leq q < p \leq \infty$

In Theorem 2.3.4, we showed that when $1 \leq q < p \leq \infty$, $T : L_p(a,b) \to L_q(a.b)$ is compact if and only if it is bounded. The analysis that leads to Theorem 2.4.23 remains effective in this case: this was demonstrated by Achache in [1], using a partitioning function different from $\mathbb{A}$. The following result, which is slightly stronger than Theorem II.2.14 in [1], can be established.

Theorem 2.4.24. *Let $1 \leq q < p \leq \infty$ and suppose that (2.2.1), (2.2.2) and (2.2.10) are satisfied. Then $T : L_p(a,b) \to L_q(a,b)$ is compact, the number $N = N((a,b),\varepsilon)$ defined in (2.4.36) is finite for all $\varepsilon > 0$ and*

$$a_{N+2}(T) \leq (N+1)^{\frac{1}{q}-\frac{1}{p}}\varepsilon, \tag{2.4.49}$$

$$a_{N-1}(T) \geq \nu_q \varepsilon, \tag{2.4.50}$$

where $\nu_2 = 1$ and $\nu_q = 1/2$ for $q \neq 2$.

2.4.6 Asymptotic results for $p = q \in (1,\infty)$

Our account is based on [84] , but specialised to an interval. We shall remark on the general results in [84] for trees in Sections 2.6 and 2.7 below. We assume throughout that (2.2.1) and (2.4.35) are satisfied, that is, for all $X \in (a,b)$,

$$u \in L_{p'}(a,X), \quad v \in L_p(a,b). \tag{2.4.51}$$

The proofs of our main results can be adapted to allow for the assumption (2.2.2) only for v, but this adds to the analytic difficulties and distracts from the underlying ideas. However, in the problem on trees, (2.4.35) is crucial if we are to use the partitioning function $\mathbb{A}(\cdot)$, and benefit from the results of Section 2.4.1.

The first step in the proof of our main theorem is the following asymptotic result relative to a compact subinterval of (a,b). Given the number $N((a,b),\varepsilon)$ defined in (2.4.36) and $I \subset\subset (a,b)$, we denote by $N(I,\varepsilon)$ the number of the intervals $I_k = (c_k, c_{k+1})$ in (2.4.39) which lie inside I.

Theorem 2.4.25. *Let $p \in (1,\infty)$ and $I \subset\subset (a,b)$. Then*

$$\lim_{\varepsilon \to 0+} \varepsilon N(I,\varepsilon) = \frac{\gamma_p}{2} \int_I |u(t)v(t)|dt \tag{2.4.52}$$

where $\gamma_p = 2\mathbb{A}((0,1);1,1) = \frac{1}{\pi} p^{1/p'}(p')^{1/p} \sin(\frac{\pi}{p})$.

Proof. For each $\eta > 0$ there exist step functions u_η, v_η, on I such that

$$\|u - u_\eta\|_{p',I} < \eta, \quad \|v - v_\eta\|_{p,I} < \eta.$$

We may suppose that

$$u_\eta = \sum_{j=1}^{m} \xi_j \chi_{W(j)}, \quad v_\eta = \sum_{j=1}^{m} \eta_j \chi_{W(j)}, \tag{2.4.53}$$

where the $W(j)$ are closed subintervals of I with disjoint interiors and $I = \cup_{j=1}^{m} W(j)$.

Let $\{c_k : k = 1, \ldots, N\}$ be the sequence defined in (2.4.39) and, with $I_k(\varepsilon) = (c_k, c_{k+1})$, let $\{I_k(\varepsilon) : k = n_1, \ldots, n_2\}$ be the intervals which intersect I; hence

$$N(I,\varepsilon) \le n_2 - n_1 \le N(I,\varepsilon) + 2. \tag{2.4.54}$$

If $W(j)$ is contained in some $I_k(\varepsilon)$ for arbitrarily small values of ε, then $\mathbb{A}(W(j)) = 0$ since $\mathbb{A}(W(j)) \le \mathbb{A}(I_k(\varepsilon)) \le \varepsilon$. Let $W(j) = (c,d)$. Then, by (2.4.32),

$$J_1(c,e) := \sup_{c<s<e} \|u\|_{p',(s,e)} \|v\|_{p,(c,s)} = 0$$

and hence, either $\mu(W(j)) = 0$ and so $v = 0$ a.e. on $W(j)$, or else $uv = 0$ a.e. on (c,e). Similarly, $uv = 0$ a.e. on (e,d), and consequently, $uv = 0$ a.e. on $W(j)$.

We now define

$$\varepsilon_j = \inf\{\varepsilon > 0 : \text{there exists } k \text{ such that } I_k(\varepsilon) \subset W(j)\} \tag{2.4.55}$$

and put $\delta = \min\{\varepsilon_j : \varepsilon_j > 0\}$. Then, if $0 < \varepsilon < \delta$, it follows that $W(j) \not\subset I_k(\varepsilon)$ for all j and k. Also, with $\sum_j'$ denoting summation over those $j \in \{1, 2, \ldots, m\}$ for which $\int_{W(j)} |u(t)v(t)|dt \ne 0$, we have

$$\begin{aligned} \Big| \int_I |u(t)v(t)|dt &- \int_I \Big(\sum_j{}' \xi_j \eta_j \chi_{W(j)}(t)\Big)dt\Big| \\ &\le \Big| \int_I \{|u(t)v(t)| - |u_\eta(t)v_\eta(t)|\}dt\Big| \\ &\le \|u\|_{p',I}\|v - v_\eta\|_{p,I} + \|u - u_\eta\|_{p',I}\|v_\eta\|_{p,I}. \end{aligned} \tag{2.4.56}$$

However, $\int_{W(j)} |u(t)v(t)|dt \ne 0$ implies that $\varepsilon_j > 0$. Thus, if $0 < \varepsilon < \delta$, we see that, on using Lemma 2.4.12 and (2.4.55),

$$\begin{aligned} \frac{\gamma_p}{2}\xi_j\eta_j|W(j)| \le & \sum_{\{k: I_k(\varepsilon) \subset W(j)\}} \mathbb{A}(I_k(\varepsilon); \xi_j, \eta_j) \\ &+ \sum_{s=1}^{2} \mathbb{A}(I_{k(j;s)}(\varepsilon) \cap W(j); \xi_j, \eta_j), \end{aligned}$$

where $k(j;1), k(j;2)$ are such that the left- and right-hand end-points of $W(j)$ are respectively interior to $I_{k(j;1)}$ and $I_{k(j;2)}$; of course, one or both of the terms involving the $I_{k(j;s)}$ may be zero.

Hence, with the aid of Lemmas 2.4.13 and 2.4.14, we have

$$\begin{aligned}
\frac{\gamma_p}{2} \sum{}' \xi_j \eta_j |W(j)| &\leq \sum_{k=n_1}^{n_2} \mathbb{A}(I_k(\varepsilon); u_\eta, v_\eta) \\
&\quad + \sum_j \sum_{s=1}^{2} \mathbb{A}(I_{k(j;s)}(\varepsilon); u_\eta, v_\eta) \\
&\leq \sum_{k=n_1}^{n_2} \mathbb{A}(I_k(\varepsilon); u, v) + \sum_j \sum_{s=1}^{2} \mathbb{A}(I_{k(j;s)}(\varepsilon); u, v) \\
&\quad + \sum_{k=n_1}^{n_2} \big[\|u - u_\eta\|_{p', I_k(\varepsilon)} \|v_\eta\|_{p, I_k(\varepsilon)} \\
&\quad + 2\|u\|_{p', I_k(\varepsilon)} \|v - v_\eta\|_{p, I_k(\varepsilon)}\big] \\
&\quad + \sum_j \sum_{s=1}^{2} \big[\|u - u_\eta\|_{p', I_{k(j,s)}(\varepsilon)} \|v_\eta\|_{p, I_{k(j,s)}(\varepsilon)} \\
&\quad + 2\|u\|_{p', I_{k(j,s)}(\varepsilon)} \|v - v_\eta\|_{p, I_{k(j,s)}(\varepsilon)}\big] \\
&\leq (n_2 - n_1 + 2m)\varepsilon \\
&\quad + O\left(\|u - u_\eta\|_{p', I} \|v_\eta\|_{p, I} + \|u\|_{p', I} \|v - v_\eta\|_{p, I}\right) \\
&\leq (N(I, \varepsilon) + 2 + 2m)\varepsilon + O(\eta). \qquad (2.4.57)
\end{aligned}$$

From (2.4.56) and (2.4.57) we see that

$$\frac{\gamma_p}{2} \int_I |u(t)v(t)| dt \leq (N(I, \varepsilon) + 2 + 2m)\varepsilon + O(\eta).$$

Thus,

$$\frac{\gamma_p}{2} \int_I |u(t)v(t)| dt \leq \liminf_{\varepsilon \to 0+} \varepsilon N(I, \varepsilon) + O(\eta)$$

and so

$$\frac{\gamma_p}{2} \int_I |u(t)v(t)| dt \leq \liminf_{\varepsilon \to 0+} \varepsilon N(I, \varepsilon). \qquad (2.4.58)$$

Next, let $0 < \varepsilon < \delta$ and put $\mathcal{K} = \{k : \text{there exists } j \text{ such that } I_k(\varepsilon) \subset W(j)\}$. Then, $\sharp\mathcal{K} \geq N(I, \varepsilon) - 2m$ and, on using Lemmas 2.4.12, 2.4.13 and 2.4.14 again, we have

$$\begin{aligned}
(N(I, \varepsilon) - 2m)\,\varepsilon &\leq \sum_{k \in \mathcal{K}} \mathbb{A}(I_k(\varepsilon); u, v) \\
&\leq \sum_{k \in \mathcal{K}} \mathbb{A}(I_k(\varepsilon); u_\eta, v_\eta) + O(\eta)
\end{aligned}$$

$$\leq \frac{\gamma_p}{2} \sum_k |\xi_k \eta_k| |I_k(\varepsilon)| + O(\eta)$$
$$= \frac{\gamma_p}{2} \int_I |u_\eta(t) v_\eta(t)| dt + O(\eta)$$
$$= \frac{\gamma_p}{2} \int_I |u(t) v(t)| dt + O(\eta).$$

From this we infer that

$$\limsup_{\varepsilon \to 0+} \varepsilon N(I, \varepsilon) \leq \frac{\gamma_p}{2} \int |u(t) v(t)| dt$$

and hence the proof of the theorem is complete. □

Corollary 2.4.26. *Suppose that*

$$\liminf_{n \to \infty} n a_n(T) < \infty.$$

Then $uv \in L_1(a, b)$.

Proof. Let $N(\varepsilon) \equiv N((a, b), \varepsilon)$. Then, from Theorems 2.4.20 and 2.4.25,

$$\frac{\gamma_p}{2} \int_I |u(t) v(t)| dt = \lim_{\varepsilon \to 0+} \varepsilon N(I, \varepsilon)$$
$$\leq \liminf_{\varepsilon \to 0+} \varepsilon N(\varepsilon)$$
$$\leq \liminf_{n \to \infty} n a_n(T).$$

Since this is true for all $I \subset\subset (a, b)$, we infer that $uv \in L_1(a, b)$. □

To proceed, we need some more terminology. Set

$$U(x) = \int_a^x |u(t)|^{p'} dt, \quad x \in (a, b) \tag{2.4.59}$$

and define $\xi_k \in \mathbb{R}_+$ by

$$U(\xi_k) = 2^{kp'/p}. \tag{2.4.60}$$

Here k may be any integer if $u \notin L_{p'}(a, b)$, but if $u \in L_{p'}(a, b), 2^{k/p} \leq \|u\|_{p'}$; we shall refer to the range of possible values of k as being "admissible". For each admissible k we set

$$\sigma_k := \{ \int_{\xi_k}^{\xi_{k+1}} U^{p/p'}(t) |v(t)|^p dt \}^{1/p}, \quad Z_k = (\xi_k, \xi_{k+1}), \tag{2.4.61}$$

so that

$$2^k \int_{\xi_k}^{\xi_{k+1}} |v(t)|^p dt \le \sigma_k^p \le 2^{k+1} \int_{\xi_k}^{\xi_{k+1}} |v(t)|^p dt. \tag{2.4.62}$$

For non-admissible values of k we put $\sigma_k = 0$. The sequence $\{\sigma_k\}$ was used in [187] for the case $u = 1, p = 2$.

The following technical lemma plays a central role in the subsequent analysis.

Lemma 2.4.27. *Let $k_0, k_1, k_2 \in \mathbb{Z}$ with $k_0 \le k_1 \le k_2, k_0 < k_2$, and let $I_j (j = 0, 1, \cdots, l)$ be non-overlapping intervals in (a, b) with end-points a_j, b_j, such that $I_j \subset Z_{k_2} (j = 1, 2, \cdots, l), a_0 \in Z_{k_0}$ and $b_0 \in Z_{k_2}$; let $x_j \in I_j (j = 0, 1, \cdots, l)$ and $x_0 \in Z_{k_1}$. Then, if $\alpha \ge 1$,*

$$\begin{aligned} S &:= \sum_{j=0}^{l} \Big(\int_{a_j}^{x_j} |u(t)|^{p'} dt\Big)^{\alpha/p'} \Big(\int_{x_j}^{b_j} |v(t)|^p dt\Big)^{\alpha/p} \\ &\le 2^{\alpha/p}(2^{\alpha/p} + 1) \max_{k_1 \le n \le k_2} \sigma_n^\alpha. \end{aligned} \tag{2.4.63}$$

Proof. On using Jensen's inequality (see Theorem 19, p.28 in [119]) and Hölder's inequality,

$$\begin{aligned} S &\le \left(\int_{\xi_{k_0}}^{\xi_{k_1+1}} |u(t)|^{p'} dt\right)^{\alpha/p'} \left(\int_{\xi_{k_1}}^{\xi_{k_2+1}} |v(t)|^p dt\right)^{\alpha/p} \\ &+ \sum_{j=1}^{l} \left(\int_{I_j} |u(t)|^{p'} dt\right)^{\alpha/p'} \left(\int_{I_j} |v(t)|^p dt\right)^{\alpha/p} \\ &\le \left(2^{(k_1+1)p'/p} - 2^{k_0 p'/p}\right)^{\alpha/p'} \left(\sum_{n=k_1}^{k_2} \frac{\sigma_n^p}{2^n}\right)^{\alpha/p} \\ &+ \left\{\sum_{j=1}^{l} \left(\int_{I_j} |u(t)|^{p'} dt\right)^{1/p'} \left(\int_{I_j} |v(t)|^p dt\right)^{1/p}\right\}^{\alpha} \\ &\le 2^{(k_1+1)\alpha/p} \left(2^{1-k_1} \max_{k_1 \le n \le k_2} \sigma_n^p\right)^{\alpha/p} \\ &+ \left(\int_{\mathbf{Z}_{k_2}} |u(t)|^{p'} dt\right)^{\alpha/p'} \left(\int_{Z_{k_2}} |v(t)|^p dt\right)^{\alpha/p} \\ &\le 2^{2\alpha/p} \max_{k_1 \le n \le k_2} \sigma_n^\alpha + \frac{2^{(k_2+1)\alpha/p} \sigma_{k_2}^\alpha}{2 k_2^{\alpha/p}}. \end{aligned}$$

The result follows. □

Lemma 2.4.28. *The quantity A in (2.2.3), namely*

$$A = A(a,b) = \sup_{a<x<b} \left\{ \|u\|_{p',(a,x)} \|v\|_{p,(x,b)} \right\}$$

satisfies

$$A \leq 2^{1/p}(2^{1/p}+1) \sup_k \sigma_k \leq 2^{2/p}(2^{1/p}+1)A. \tag{2.4.64}$$

Proof. From Lemma 2.4.27 with $\alpha = 1$,

$$A \leq 2^{1/p}(2^{1/p}+1) \sup_k \sigma_k.$$

Also, by (2.4.61),

$$\begin{aligned}
\sigma_k^p &\leq 2^{k+1} \int_{\xi_k}^{\xi_{k+1}} |v(t)|^p dt \\
&\leq \frac{2^{k+1}}{U(\xi_k)^{p/p'}} \left\{ \left(\int_a^{\xi_k} |u(t)|^{p'} dt \right)^{1/p'} \left(\int_{\xi_k}^{\xi_{k+1}} |v(t)|^p dt \right)^{1/p} \right\}^p \\
&\leq 2A^p.
\end{aligned}$$

□

An immediate consequence of Theorem 2.2.1 and Lemma 2.4.28 is

Corollary 2.4.29. *The operator $T : L_p(a,b) \to L_p(a,b)$ is bounded if and only if $\{\sigma_k\} \in l_\infty(\mathbb{Z})$ and*

$$\|T\| \approx \|\{\sigma_k\}\|_{l_\infty(\mathbb{Z})}.$$

Corollary 2.4.30. *The operator $T : L_p(a.b) \to L_p(a,b)$ is compact if and only if $\lim_{|k|\to\infty} \sigma_k = 0$.*

Proof. On using Lemma 2.4.27 as in the proof of Lemma 2.4.28, we see that

$$\begin{aligned}
A(a,\xi_k) &\leq 2^{1/p}(2^{1/p}+1) \sup_{n\leq k} \sigma_n \\
A(\xi_k,b) &\leq 2^{1/p}(2^{1/p}+1) \sup_{n\geq k} \sigma_n.
\end{aligned}$$

The corollary now follows from Theorem 2.3.1. □

Theorem 2.4.31. *Let $1 < p < \infty$ and suppose that (2.4.51) is satisfied, i.e. $u \in L_{p'}(a, X)$ for all $X \in (a, b)$ and $v \in L_p(a, b)$. If $\{\sigma_n\} \in l_1(\mathbb{Z})$, then $T : L_p(a, b) \to L_p(a, b)$ is compact and*

$$\lim_{\varepsilon \to 0+} \varepsilon N(\varepsilon) = \frac{\gamma_p}{2} \int_a^b |u(t)v(t)|dt,$$

where $N(\varepsilon) = N((a, b), \varepsilon, u, v)$ and $\gamma_p = \gamma_{p,p}$ is given in (2.4.5).

Proof. Since $\{\sigma_n\} \in l_1(\mathbb{Z})$, it follows from Corollary 2.4.30 that T is compact and hence $N(\varepsilon) < \infty$ for all $\varepsilon > 0$. Denoting the number of the intervals $I_k = (c_k, c_{k+1})$ which lie inside $I \subset\subset (a, b)$ by $N(I, \varepsilon)$, we have as in Theorem 2.4.25,

$$\{N(\varepsilon) - N((\xi_l, \xi_m), \varepsilon) - 1\}\varepsilon = \sum_{\{k: I_k \not\subset (\xi_l, \xi_m)\}} \mathbb{A}(I_k).$$

Hence, by Lemmas 2.4.16 and 2.4.28,

$$\begin{aligned} 0 &\le \varepsilon\big\{N(\varepsilon) - N((\xi_l, \xi_m), \varepsilon)\big\} \\ &\le \varepsilon + K \sum_{\{k: I_k \not\subset (\xi_l, \xi_m)\}} J(I_k) \\ &\le \varepsilon + K \sum_{k<l, k>m} \sigma_k \end{aligned}$$

for some positive constant K. It follows from Theorem 2.4.25 that

$$\begin{aligned} \frac{\gamma_p}{2} \int_{\xi_l}^{\xi_m} |u(t)v(t)|dt &\le \liminf_{\varepsilon \to 0} \varepsilon N(\varepsilon) \\ &\le \limsup_{\varepsilon \to 0} \varepsilon N(\varepsilon) \\ &\le \frac{\gamma_p}{2} \int_{\xi_l}^{\xi_m} |u(t)v(t)|dt + K \sum_{k<l, k>m} \sigma_k. \end{aligned}$$

This yields the theorem. □

From Theorem 2.4.20 we therefore have

Corollary 2.4.32. *Let $1 < p < \infty$ and suppose that (2.4.51) is satisfied. Then if $\{\sigma_n\} \in l_1(\mathbb{Z}), T : L_p(a, b) \to L_p(a, b)$ is compact and*

$$\lim_{n \to \infty} n a_n(T) = \frac{\gamma_p}{2} \int_a^b |u(t)v(t)|dt, \qquad (2.4.65)$$

where $\gamma_p = \frac{1}{\pi} p^{1/p'} p'^{1/p} \sin(\pi/p)$.

In [151], the following remainder estimate (2.4.66) is derived when $1 < p < \infty$, extending an earlier result in [63] for $p = 2$.

Theorem 2.4.33. *let $u \in L_{p'}(a,b), v \in L_p(a,b)$, and suppose also that $u' \in L_{p'/(p'+1)} \cap C(a,b), v' \in L_{p/(p+1)}(a,b) \cap C(a,b)$. Then*

$$\begin{aligned} \limsup_{n\to\infty} n^{1/2} \left| \gamma_p \int_a^b |uv| dt - n a_n \right| &\leq \tfrac{3}{2} \gamma_p \int_a^b |uv| dt \\ &+ c(p) \left(\|u'\|_{p'/(p'+1),(a,b)} + \|v'\|_{p/(p+1),(a,b)} \right) \\ &\times \left(\|u\|_{p',(a,b)} + \|v\|_{p,(a,b)} \right), \end{aligned} \tag{2.4.66}$$

where $c(p)$ is a constant which depends only on p.

2.4.7 The cases $p = 1, \infty$

Analogues of Corollary 2.4.32 for the cases $p = 1, \infty$ are obtained in [83]. Substantial changes are necessary for the proof and, indeed, the form of the results is different. When $p = \infty$, the function

$$v_s(t) := \lim_{\varepsilon \to 0^+} \|v\|_{\infty,(t-\varepsilon,t+\varepsilon)} \tag{2.4.67}$$

features in the results, and the sequence corresponding to that in (2.4.61) is defined by

$$\sigma_k := \|uv\|_{\infty,(\xi_k,\xi_{k+1})}, \tag{2.4.68}$$

where

$$\int_a^{\xi_k} |u(t)| dt = 2^k. \tag{2.4.69}$$

Theorem 2.4.34. *Let $u \in L_1(a,X)$ for all $X \in (a,b)$ and $v \in L_\infty(a,b)$. Then*

1. *$T : L_\infty \to L_\infty$ is bounded if and only if $\{\sigma_k\} \in l_\infty(\mathbb{Z})$, and compact if and only if $\lim_{|k|\to 0} \sigma_k = 0$;*
2. *if $\{\sigma_k\} \in l_1(\mathbb{Z})$,*

$$\begin{aligned} (1/4) \int_a^b |u(t)| v_s(t) dt &\leq \liminf_{n\to\infty} a_n(T) \\ &\leq \limsup_{n\to\infty} a_n(T) \leq 2 \int_a^b |u(t)| v_s(t) dt, \end{aligned} \tag{2.4.70}$$

where v_s is defined in (2.4.67).

When $p = 1$, the function u_s (defined in the same way as v_s) makes its appearance, and we define the sequence $\{\sigma_k\}$ by

$$\sigma_k := \|uv\|_{\infty,(\xi_k,\xi_{k+1})}, \tag{2.4.71}$$

where

$$\int_{\xi_k}^{b} |v(t)|dt = 2^k. \tag{2.4.72}$$

Theorem 2.4.35. *Let $u \in L_\infty(a, X)$ for all $X \in (a,b)$ and $v \in L_1(a,b)$. Then*

1. *$T : L_1(a,b) \to L_1(a,b)$ is bounded if and only if $\{\sigma_k\} \in l_\infty(\mathbb{Z})$ and compact if and only if $\lim_{|k|\to\infty} \sigma_k = 0$;*
2. *if $\{\sigma_k\} \in l_1(\mathbb{Z})$,*

$$\begin{aligned}(1/4)\int_a^b u_s(t)|v(t)|dt &\le \liminf_{n\to\infty} na_n(T)\\ &\le \limsup_{n\to\infty} na_n(T) \le 2\int_a^b u_s(t)|v(t)|dt.\end{aligned} \tag{2.4.73}$$

The proofs of these results are given in [83].

2.5 l_α and $l_{\alpha,w}$ classes.

We denote by $l_\alpha(\Lambda)$, with $\Lambda = \mathbb{N}$ or $\mathbb{Z}$, the space of sequences $x = \{x_k\}_{k\in\Lambda}$ which are such that

$$\|x|l_\alpha(\Lambda)\| := \Big(\sum_{k\in\Lambda} |x_k|^\alpha\Big)^{1/\alpha} < \infty.$$

The *weak* $l_\alpha(\Lambda)$ space is denoted by $l_{\alpha,w}(\Lambda)$ ($l_{\alpha,\infty}(\Lambda)$ in the Lorentz scale); this is the space of sequences $x = \{x_k\}_{k\in\Lambda}$ which are such that

$$\|x|l_{\alpha,w}(\Lambda)\| := \sup_{t>0}\Big(t\big[\sharp\{k\in\Lambda : |x_k| > t\}\big]^{1/\alpha}\Big) < \infty.$$

We shall prove the following result for $T : L_p(a,b) \to L_p(a,b)$, where $p \in (1,\infty)$, and continue to assume throughout the section that (2.4.51) is satisfied.

Theorem 2.5.1. *Let $\alpha \in (1,\infty)$ and suppose that (2.4.51) is satisfied. Then*

1. *$\{a_k(T)\} \in l_\alpha(\mathbb{N})$ if, and only if, $\{\sigma_n\} \in l_\alpha(\mathbb{Z})$ and*

$$\|\{a_k(T)\}|l_\alpha(\mathbb{N})\| \approx \|\{\sigma_n\}|l_\alpha(\mathbb{Z})\|;$$

2. *$\{a_k(T)\} \in l_{\alpha,w}(\mathbb{N})$ if, and only if, $\{\sigma_n\} \in l_{\alpha,w}(\mathbb{Z})$ and*

$$\|\{a_k(T)\}|l_{\alpha,w}(\mathbb{N})\| \approx \|\{\sigma_n\}|l_{\alpha,w}(\mathbb{Z})\|.$$

The proof depends on the following two technical lemmas which enable us to compare the distribution functions of the two sequences.

Lemma 2.5.2. *Given any interval* $I = (c, d) \subset (a, b)$, *let*

$$A(I) = \sup_{X \in I} \{\|u\|_{p',(c,X)} \|v\|_{p,(X,d)}\},$$
$$A'(I) = \sup_{X \in I} \{\|u\|_{p',(X,b)} \|v\|_{p,(c,X)}\}.$$

Then, for all $k \in \mathbb{Z}$ *and* $Z_k = (\xi_k, \xi_{k+1})$ *defined in (2.4.61),*

$$A(\overline{Z}_k \cup \overline{Z}_{k+1}) \geq 2^{-1/p}(1 - 2^{-p'/p})^{1/p'} \sigma_{k+1},$$
$$A'(\overline{Z}_k \cup \overline{Z}_{k+1}) \geq 2^{+1/p}(1 - 2^{-p'/p})^{1/p'} \sigma_k.$$

Proof. From (2.4.59),

$$\begin{aligned} A(\overline{Z}_k \cup \overline{Z}_{k+1}) &\geq \|u\|_{p',Z_k} \|v\|_{p,Z_{k+1}} \\ &= \{U(\xi_{k+1}) - U(\xi_k)\}^{1/p'} \|v\|_{p,Z_{k+1}} \\ &\geq \{2^{(k+1)p'/p} - 2^{kp'/p}\}^{1/p'} 2^{-(k+2)/p} \sigma_{k+1} \end{aligned}$$

whence the result for A. The lower bound for A' is calculated in the same way. □

Lemma 2.5.3. *Let* $I = (c, d) \subset (a, b)$, *let* $\varepsilon > 0$ *and suppose that the set*

$$S(\varepsilon) := \{k \in \mathbb{Z} : Z_k \subset I, \sigma_k > 2^{1/p}(1 - 2^{-p'/p})^{-1/p'} \varepsilon\}$$

has at least 4 elements. Then the function $J(I)$ *defined in (2.4.31) satisfies* $J(I) > \varepsilon$, *and* $\mathbb{A}(I) > \varepsilon$.

Proof. Let $e \in (c, d)$. Since $S(\varepsilon)$ has at least 4 elements, at least one of the intervals $(c, e), (e, d)$ contains 2 members of $S(\varepsilon)$. If (c, e) has this property and $k_1 = \min\{k : k \in S(\varepsilon)\}$, then $Z_{k_1} \cup Z_{k_1+1} \subset (c, e)$ and, by Lemma 2.5.2,

$$\begin{aligned} J_1(c, e) &\geq A'(\overline{Z}_{k_1} \cup \overline{Z}_{k_1+1}) \\ &\geq 2^{1/p}(1 - 2^{-p'/p})^{1/p'} \sigma_k \\ &\geq \varepsilon 2^{2/p} > \varepsilon. \end{aligned}$$

A similar argument shows that if (e, d) contains 2 members of $S(\varepsilon)$ then $J_2(e, d) > \varepsilon$. Hence $J(I) > \varepsilon$, and $\mathbb{A}(\mathbb{I}) > \varepsilon$ follows from (2.4.32). □

Lemma 2.5.4. *Let* $\varepsilon > 0$ *and* $N(\varepsilon) = N((a,b), \varepsilon, u, v)$. *Then*

$$\sharp\{k \in \mathbb{Z} : \sigma_k > c\varepsilon\} \le 5N(\varepsilon),$$

where

$$c = 2^{1/p}(1 - 2^{-p'/p})^{-1/p'}. \tag{2.5.1}$$

Proof. Clearly, with $N = N(\varepsilon)$,

$$\sharp\{k \in \mathbb{Z} : c_i \in \bar{Z}_k \quad \text{for some } i, 1 \le i \le N\} \le 2N$$

and for every set not included in this set, $\overline{Z}_k \subset (c_i, c_{i+1})$ for some $i \in \{1, \cdots, N-1\}$. In any interval (c_i, c_{i+1}) there can be at most 3 intervals $\overline{Z}_k$ for which $\sigma_k > \varepsilon$ since otherwise $\mathbb{A}(c_i, c_{i+1}) > \varepsilon$ in view of Lemma 2.5.3, contrary to the definition of the sequence $\{c_i\}$ in (2.4.39). Hence

$$\sharp\{k \in \mathbb{Z} : \sigma_k > c\varepsilon\} \le 2N + 3N = 5N.$$

□

Lemma 2.5.5. *Let c be the constant in (2.5.1). Then, for all $t > 0$,*

$$\sharp\{k \in \mathbb{Z} : \sigma_k > t\} \le 5\sharp\{k \in \mathbb{N} : a_k(T) \ge t/c\} + 5.$$

Proof. From Lemma 2.4.19 and (2.4.40) we have that

$$\sharp\{k \in \mathbb{N} : a_k(T) \ge \varepsilon\} \ge N(\varepsilon) - 1.$$

Hence, by Lemma 2.5.4,

$$\begin{aligned}\sharp\{k \in \mathbb{Z} : \sigma_k > t\} &\le 5N(t/c)\\ &\le 5\sharp\{k \in \mathbb{N} : a_k \ge t/c\} + 5.\end{aligned}$$

□

The final preparatory lemma for the proof of Theorem 2.5.1 is

Lemma 2.5.6. *For all* $\alpha \in (1, \infty)$,

$$\|\{\sigma_k\}|l_\alpha(\mathbb{Z})\|^\alpha \le 5c^\alpha \|\{a_k(T)\}|l_\alpha(\mathbb{N})\|^\alpha + 5\|\{\sigma_k\}|l_\infty(\mathbb{Z})\|^\alpha,$$

where c is given by (2.5.1).

Proof. First observe that (see Proposition 3.2.5)

$$\|\{\sigma_k\}|l_\alpha(\mathbb{Z})\|^\alpha = \alpha \int_0^\infty t^{\alpha-1}\sharp\{k \in \mathbb{Z} : \sigma_k > t\}dt.$$

Put $S = \|\{\sigma_k\}\|_{l_\infty(\mathbb{Z})}$. Then Lemma 2.5.5 yields

$$\begin{aligned}\|\{\sigma_k\}|l_\alpha(\mathbb{Z})\|^\alpha \le 5\alpha &\int_0^\infty t^{\alpha-1}\sharp\{k \in \mathbb{N} : a_k(T) > t/c\}dt\\ &+ 5S^\alpha,\end{aligned}$$

whence the result. □

Proof of Theorem 2.5.1 (i) By Corollary 2.4.29, $\|T\| \approx \|\{\sigma_k\}|l_\infty(\mathbb{Z})\|$, and hence, since $\|T\| = a_1(T) \le \|\{a_k(T)\}|l_\alpha(\mathbb{N})\|$, it follows from Lemma 2.5.6 that

$$\|\{\sigma_k\}|l_\alpha(\mathbb{Z})\| \lesssim \|\{a_k(T)\}|l_\alpha(\mathbb{N})\|.$$

To obtain the reverse inequality, we group the intervals $I_k = [c_k, c_{k+1}]$ in (2.4.39) into families $\mathcal{F}_j, j = 1, 2, \cdots$, where each $\mathcal{F}_j$ consists of the maximal number of those intervals satisfying the hypothesis of Lemma 2.4.27 : they lie within (ξ_{k_0}, ξ_{k_2+1}) for some k_0, k_2 and the next interval I_k intersects Z_{k_2+1}. We infer from Lemmas 2.4.16 and 2.4.27 that there exists a constant $c > 0$ such that

$$\varepsilon \sharp \mathcal{F}_j \le c \max_{k_0 \le k \le k_2} \sigma_k =: c\sigma_{k_j},$$

say. Thus

$$\sharp \mathcal{F}_j \le [c\sigma_{k_j}/\varepsilon] =: n_j,$$

and

$$\begin{aligned} N(\varepsilon) &= \sum_j \# \mathcal{F}_j \le \sum_j \sum_{n=1}^{n_j} 1 \\ &= \sum_{n=1}^{\infty} \sum_{n_j \ge n} 1 \\ &= \sum_{n=1}^{\infty} \#\{j : c\sigma_j/\varepsilon \ge n\} \\ &\le \sum_{n=1}^{\infty} \#\{k : \sigma_k \ge n\varepsilon/c\}. \end{aligned} \tag{2.5.2}$$

Thus, if $\{\sigma_k\} \in l_\alpha$ for some $\alpha \in (1, \infty)$,

$$\begin{aligned} \alpha \int_0^\infty t^{\alpha-1} N(t) dt &\le \alpha \int_0^\infty \sum_{n=1}^{\infty} t^{\alpha-1} \#\{k : \sigma_k > nt/c\} dt \\ &= \alpha c_\alpha \int_0^\infty \sum_{n=1}^{\infty} n^{-\alpha} s^{\alpha-1} \#\{k : \sigma_k > s\} ds \\ &\lesssim \|\{\sigma_k\}|l_\alpha(\mathbb{Z})\|^\alpha. \end{aligned} \tag{2.5.3}$$

By (2.4.43),

$$\#\{k : a_k(T) > \varepsilon\} \le N(\varepsilon)$$

and so

$$\begin{aligned} \|\{a_k(T)\}|l_\alpha(\mathbb{N})\|^\alpha &= \alpha \int_0^\infty t^{\alpha-1} \#\{k : a_k(T) > t\} dt \\ &\le \alpha \int_0^\infty t^{\alpha-1} N(t) dt \end{aligned}$$

$$\lesssim \|\{\sigma_k\}|l_\alpha(\mathbb{Z})\|^\alpha$$

on using (2.5.3). Thus (i) is proved.

(ii) Suppose that $\{\sigma_k\} \in l_{\alpha,w}(\mathbb{Z})$. Then, from (2.5.2),

$$\begin{aligned} t^\alpha N(t) &\le \sum_{n=1}^{\infty} t^\alpha \#\{k : \sigma_k \ge n\varepsilon/c\} \\ &\le \sum_{n=1}^{\infty} \|\{\sigma_k\}|l_{\alpha,w}(\mathbb{Z})\|^\alpha (\frac{c}{n})^\alpha \\ &\lesssim \|\{\sigma_k\}|l_{\alpha,w}(\mathbb{Z})\|^\alpha. \end{aligned}$$

From (2.4.43)

$$\#\{k : a_k(T) > t\} \le N(t)$$

and this gives

$$\begin{aligned} \|\{a_k(T)\}|l_{\alpha,w}(\mathbb{N})\|^\alpha &\le \sup_{t>0} t^\alpha [N(t)] \\ &\lesssim \|\{\sigma_k\}|l_{\alpha,w}(\mathbb{Z})\|^\alpha. \end{aligned}$$

Finally, suppose that $\{a_k(T)\} \in l_{\alpha,w}(\mathbb{N})$. Lemma 2.5.4 and (2.4.43) give

$$\begin{aligned} \#\{k : \sigma_k > t\} &\le 5N(t/c) \\ &\le 5\left(\#\{k : a_k(T) \ge t/c\} + 1\right). \end{aligned}$$

Since

$$\#\{k : a_k(T) \ge t/c\} + 1 \ge N(t/c) \ge 1,$$

it follows that

$$\sup_{t>0} t^\alpha \#\{k : \sigma_k \ge t\} \lesssim \sup_{t>0} t^\alpha \#\{k : a_k(T) \ge t/c\}$$

and thus

$$\|\{\sigma_k\}|l_{\alpha,\omega}(\mathbb{Z})\| \lesssim \|\{a_k(T)\}|l_{\alpha,w}(\mathbb{N})\|.$$

The proof is therefore complete. □

Remark 2.5.7. The cases $p = 1, \infty$ are treated in [83]. Theorem 2.5.1 continues to hold when the σ_k are defined by (2.4.68) and (2.4.69) for $p = \infty$ and (2.4.71) and (2.4.72) for $p = 1$.

2.6 Hardy-type operators on trees

2.6.1 Analysis on trees

Let Γ be a *tree*, i.e. a connected graph without loops or cycles, where the edges are non-degenerate closed line segments whose end points are the vertices of the tree. We shall assume that Γ contains a finite or infinite number of vertices and that each vertex is of *finite degree*, i.e. only a finite number of edges emanate from each vertex. For every $x, y \in \Gamma$ there is a unique polygonal path in Γ which joins x and y. The distance between x and y is defined to be the length of this polygonal path, and in this way Γ is endowed with a metric topology.

Lemma 2.6.1. *Let $\tau(\Gamma)$ be the metric topology on Γ. Then*

1. *a set $A \subset \Gamma$ is compact if and only if it is closed and meets only a finite number of edges;*
2. *$\tau(\Gamma)$ is locally compact;*
3. *Γ is the union of a countable number of edges and vertices. Thus if Γ is endowed with the natural one-dimensional Lebesgue measure it is a σ-finite measure space.*

Proof. 1. Let A be compact and hence closed. Suppose A meets an infinite number of edges and choose a point t_k of A on each of these edges. A subsequence of $\{t_k\}$ converges to a point t lying on some edge of Γ. But this would imply that in each neighbourhood of t there exists an infinity of edges, contradicting the assumption that only a finite number of edges meet at a vertex.
Conversely, let A be closed and meet only a finite number of edges. Then the intersection of A with each edge is compact and A is the union of a finite number of compact sets. It is therefore compact.
2. Any point a on Γ lies on only a finite number of edges. Take a closed neighbourhood of a with diameter less than the distance from a to the nearest vertex different from a. This is compact by (1).
3. Let a be the mid-point of an edge. The set of finite sequences of vertices $x_1, \cdots, x_k$ which lie on the path joining a to x_k is uniquely determined by x_k. Since each vertex has finite degree the result follows.

□

For $a \in \Gamma$ we define $t \succeq_a x$(or equivalently $x \preceq_a t$) to mean that $x \in \Gamma$ lies on the path from a to $t \in \Gamma; t \succ_a x$ and $x \prec_a t$ have the obvious meaning. This is a partial ordering on Γ and the ordered graph so formed is referred to as the tree **rooted** at a and denoted by $\Gamma(a)$ when the root needs to be exhibited. If a is not a vertex we can make it one by replacing the edge on which it lies by two. In this way every rooted tree Γ is the unique finite union of subtrees which meet only at a.

The path joining two points $x, y \in \Gamma$ may be parameterised by $s(t) = \text{dist}(x, t)$ and for $g \in L_{1,loc}(\Gamma)$ we have, with $\langle x, y \rangle = \{t : x \preceq_a t \preceq_a y\}$,

$$\int_x^y g = \int_{\langle x,y \rangle} g(t)dt = \int_0^{\text{dist}(x,y)} g[(t(s)]ds.$$

The space $L_p(\Gamma)$ is defined in the natural way.

2.6.2 Boundedness of T

We shall use the following terminology in our analogues of Theorems 2.2.1 and 2.2.4 when the interval (a,b) is replaced by a tree Γ and T is the operator

$$Tf(x) := v(x)\int_a^x u(t)f(t)dt, \quad f \in L_p(\Gamma), \tag{2.6.1}$$

where Γ is rooted at a.

Definition 2.6.2. *Let K be a connected subset of $\Gamma = \Gamma(a)$ containing the root a , and denote the set of its boundary points by ∂K. A point $t \in \partial K$ is said to be* **maximal** *if every $x \succ_a t$ lies in $\Gamma \setminus K$. We shall denote by $\mathcal{I}_a(\Gamma)$ (or simply $\mathcal{I}_a$ when no confusion is likely) the set of all connected subsets K of Γ which contain a and all of whose boundary points are maximal.*

The assumptions on u, v in (2.6.1) which correspond to those in (2.2.1) and (2.2.2) are the following: for all $K \in \mathcal{I}_a(\Gamma)$ such that $K \subset\subset \Gamma$,

$$u \in L_{p'}(K), \quad v \in L_q(\Gamma \setminus K), \tag{2.6.2}$$

where $p, q \in [1, \infty]$.

Apart from the counterpart of (2.2.6), which is no longer valid for a general tree, the proof of Theorem 2.2.1 continues to hold (see [85]) and we have

Theorem 2.6.3. *Let $1 \le p \le q \le \infty$, and suppose that (2.6.2) is satisfied. For $K \in \mathcal{I}_a(\Gamma)$ define*

$$\alpha_K := \inf\{\|f\|_p : \int_a^X |u(t)f(t)|dt = 1 \quad \textit{for all} \quad X \in \partial K\}. \tag{2.6.3}$$

Then T in (2.6.1) is a bounded linear map from $L_p(\Gamma)$ into $L_q(\Gamma)$ if and only if

$$A := \sup_{K \in \mathcal{I}_a} \left\{ \frac{\|v\chi_{\Gamma\setminus K}\|_q}{\alpha_K} \right\} < \infty \tag{2.6.4}$$

in which case we have

$$A \le \|T\| \le 4A. \tag{2.6.5}$$

As in Theorem 2.2.4, we need some additional terminology for the case $p > q$. Let $B_i, i \in \mathcal{I}$, be non-empty disjoint subsets of Γ which are such that $\Gamma = \cup_{i\in\mathcal{I}} B_i$ and $B_i = K_{i+1} \setminus K_i$, where $K_i \in \mathcal{I}_a(\Gamma)$ and $K_i \subset K_{i+1}$. We shall denote the set of all such decompositions $\{B_i\}_{i\in\mathcal{I}}$ by $\mathcal{C}(\Gamma)$. The proof of Theorem 2.2.4 carries over to give

Theorem 2.6.4. *Let $1 \le q < p \le \infty$, $\frac{1}{s} = \frac{1}{q} - \frac{1}{p}$, and suppose that (2.6.2) is satisfied. For $\{B_i\}_{i\in\mathcal{I}} \in \mathcal{C}(\Gamma)$, define*

$$\alpha_i := \inf\{\|f\|_p : f \subset B_{i-1}, \int_a^X |u(t)f(t)|dt = 1 \quad for\ all\ X \in \partial K\} \quad (2.6.6)$$

and

$$\beta_i := \|v\chi_{B_i}\|_q/\alpha_i. \quad (2.6.7)$$

Then T in (2.6.1) is a bounded linear map of $L_p(\Gamma)$ into $L_q(\Gamma)$ if and only if

$$B := \sup_{C(\Gamma)} \|\{\beta_i\}|l_s(\mathcal{I})\| < \infty, \quad (2.6.8)$$

in which case

$$B \leq \|T\| \leq 4B. \quad (2.6.9)$$

Remark 2.6.5. In [85] algorithms are derived to assist in the calculation of the quantities in Theorems 2.6.3 and 2.6.4 and hence of $\|T\|$. Also comparisons with more manageable expressions are made. However, nothing can disguise the complexity of the problem, even for elementary trees with few edges.

2.7 Compactness of T and its approximation numbers

The methods presented in this chapter are shown in [84] to apply to the problem on a tree. However, the technicalities are naturally much more complicated and we give only a brief description of the changes introduced in [84] for the problem on a tree and state the asymptotic result.

With $U(x) := \int_a^x |u(t)|^{p'} dt, (x \in \Gamma)$, we define Z_k to be the closure of

$$\{x : x \in \Gamma, 2^{\frac{kp'}{p}} \leq U(x) < 2^{\frac{(k+1)p'}{p}}\}. \quad (2.7.1)$$

Here k may be any integer if $u \in L_{p',loc}(\Gamma) \setminus L_{p'}(\Gamma)$, while, if $u \in L_{p'}(\Gamma)$, $2^k \leq \|u\|^p_{p',\Gamma}$; we refer to these values of k as the *admissible* values.

We have that $Z_k = \bigcup_{i=1}^{n_k} Z_{k,i}$, where the $Z_{k,i}$ are the connected components of Z_k. Corresponding to each admissible k we set

$$\sigma^p_{k,i} := 2^k \mu(Z_{k,i}) \text{ for } i \in \{1, \ldots, n_k\} \quad (2.7.2)$$

and

$$\sigma^p_k := 2^k \mu(Z_k). \quad (2.7.3)$$

For non-admissible k we set $\sigma_k = 0$. We also set $\sigma_{k,i} = 0$ for $i \notin \{1, \ldots, n_k\}$.

Let

$$B_{k,i} := \#\partial Z_{k,i} - 1; \quad (2.7.4)$$

that is, $B_{k,i}$ is the number of boundary points of $Z_{k,i}$ excluding its root.

The analogue of Corollary 2.4.32 is

Theorem 2.7.1. *For $1 < p < \infty$ and $u \in L_{p'}(K), v \in L_p(\Gamma)$ for all $K \subset\subset \Gamma$, suppose that $B_{k,i}^{1/p'} \sigma_{k,i} \in l_i(\mathbb{Z} \times \mathbb{N})$. Then*

$$\lim_{n\to\infty} n a_n(T) = \gamma_p \int_\Gamma |u(t)v(t)|dt. \tag{2.7.5}$$

The proof of this and results analogous to those in Section 2.5 are given in [84].

In the case $p = 2$ and $u = 1$, a result similar to Theorem 2.7.1 is derived in [178], Theorem 4.1, when $v \in L_{p,loc}(\Gamma)$. The methods used in [178] use the underlying Hilbert space structure when $p = 2$ and are thus very different from the techniques here.

2.8 Notes

2.1. In [19], Bliss considered the inequality

$$\Big(\int_0^\infty |\int_0^x f(t)dt|^q x^{-1-q/p'} dx\Big)^{1/q} \le C\Big(\int_0^\infty |f(x)|^p dx\Big)^{1/p}$$

for $1 < p < q < \infty$ and proved that the best possible constant is

$$C = \Big(\frac{p' r^r}{q}\Big)^{1/q} \Big[\mathbf{B}\Big(\frac{1}{r}, \frac{q-1}{r}\Big)\Big]^{-r/q}, \quad r = q/p - 1,$$

where $\mathbf{B}$ is the beta function.

2.2. A comprehensive coverage of Hardy-type inequalities, including a full review of the contributions which had been published at the time of its publication in 1991, is given in [191]. This has been a topic of intensive study, especially since the work of Chisholm and Everitt [31] and Muckenhoupt [177] who established Theorem 2.2.1 in the cases $p = q = 2$ and $p = q \in (1, \infty)$ respectively: the finiteness of the quantity A in (2.2.3) is generally referred to as the Muckenhoupt condition. The first proof of Theorem 2.2.1 for $1 < p \le q < \infty$ was given by Bradley [21]. As noted in Remark 2.2.6, the case $q < p$ treated in Theorem 2.2.4 was first resolved by Maz'ya [171]. Since the publication of [191] the pace of activity has accelerated, and by now analogues to Theorems 2.2.1 and 2.2.4 are available for a wide variety of domain and target spaces of the operator T and also for more general operators of the form

$$Tf(x) := \int_a^x k(x, y) f(t)dt.$$

Examples may be found in [65], [145], [159], [160], [161] and [212].

If $0 < p < 1$, the *reverse* Hardy inequality holds, namely,

$$\int_0^\infty (x^{-1}\int_0^x |f(t)dt)^p dx \gtrsim \int_0^\infty |f(x)|dx.$$

This is the integral form of inequalities considered by Copson in [36], [37] for infinite series. General inequalities of the form

$$\Big(\int_a^b |v(x)\int_a^x u(t)f(t)dt|^q dx\Big)^{1/q} \gtrsim \Big(\int_a^b |f(x)|^p dx\Big)^{1/p},$$

for $0 < p < 1$ have been investigated by Beesack and Heinig [12] and Grosse-Erdmann [111]. Together they give a full characterisation for $0 < p < 1$ and $0 < q < \infty$.

2.3. The result in Theorem 2.3.1 concerning the compactness of T in the case $1 < p \leq q < \infty$ is proved in [191], Theorems 7.3 and 7.4. The equivalence of boundedness and compactness established in Theorem 2.3.4 when $1 < q < p < \infty$ is given in ([191], Theorem 7.5).

2.4. In [206] it is proved that (2.4.5) is in fact true for the range of values $q \in (0,\infty], p \in [1,\infty]$. For $p \in [1,\infty]$, Schmidt also proves the inequality

$$\int_0^1 \ln|z(t)| \leq \ln\big[1/G(\frac{p-1}{p})\big]\{\int_0^1 |\frac{dz}{dt}|^p dt\}^{1/p},$$

where

$$G(s) := e^s s^{-s}\Gamma(1+s).$$

The constant is shown to be sharp and the extremal functions are determined.

The proof of Theorem 2.4.4 is taken from that of Theorem 2.3 in [87]. However, the results are also established in [16], [43], [68] and [206].

In [17] the norm of the embedding operator

$$E_{M;p,q}(I) : W^1_{p,M}(I) \to L_{q,M}(I),$$

where

$$W^1_{p,M}(I) := \{f \in W^1_p(I) : f_I = 0\},$$
$$L_{q,M}(I) := \{f \in L_q(I) : f_I = 0\},$$

with $f_I = \frac{1}{|I|}\int_I f(t)dt$, is investigated for any $p, q \in (1,\infty)$. Analogues of these latter spaces defined on a domain $\Omega \subset \mathbb{R}^n$ will have a prominent role in Section 6.3; in Chapter 6 it is also important to know how these spaces are related to the quotient spaces $W^1_{p,\mathbb{C}}(\Omega) := W^1_p(\Omega)/\mathbb{C}, L_{q,\mathbb{C}}(\Omega) := L_q(\Omega)/\mathbb{C}$, and this is discussed in Section 6.2. It is proved in [17] that when the space $W^1_p(I)$ is *real* and $1 < q \leq 2p < \infty$, then with $1/r = 1 + 1/q - 1/p$,

$$\|E_{M;p,q}(I)|W^1_{p,M}(I) \to L_{q,M}(I)\| = \sup_{f\in W^1_p(I), f\neq 0} \frac{\|f - f_I\|_{q,I}}{\|f'\|_{p,I}}$$

$$= \frac{r^{(1/r)-1}p'^{1/p}q^{1/q'}}{\mathbf{B}(1/p', 1/q)}\left(\frac{|I|}{2}\right)^{1/r}$$
$$= \gamma_{p,q}\left(\frac{|I|}{2}\right)^{1/r},$$

where $\gamma_{p,q} := \|H_0 : L_p(0,1) \to L_q(0,1)\|$; see (2.4.5). When $p = q$, the assumption that $W_p^1(I)$ is real is not required in [17] and so

$$\|E_{M;p,p}(I)| = \|E_{\mathbb{C}}(I)\| = \gamma_p \frac{|I|}{2},$$

where $\gamma_p = \gamma_{p,p}$. Note that in Chapter 6, $E_{M;p,p}$ is denoted by E_M.

Estimates similar to those in Theorem 2.4.20 were first established in [47]. The main result in this section for the case $p = q \in (0, \infty)$ is Corollary 2.4.32. This asymptotic formula was proved in [48] for $p = q = 2$, using a similar technique but involving a partitioning function other than $\mathbb{A}$. However, it was also proved in [187] by a very different technique involving Hilbert space methods without counterparts when $p \neq 2$. Corollary 2.4.32 for any $p \in (1, \infty)$ was in fact first established as a special case of the result for a general tree in [84].

In [157] and [161] generalisations of (2.4.70), (2.4.73) and a similar result in [48] for $p = q \in (1, \infty)$, are established for T as a bounded operator from $L_p(0, \infty)$ to $L_q(0, \infty)$, under appropriate (and different) conditions on u and v. In both cases asymptotic estimates are obtained for the approximation numbers of T in terms of $\int |u(t)v(t)|^r dt$, where $\frac{1}{r} = 1 - \frac{1}{p} + \frac{1}{q}$.

It is also proved in [157] that for $I = (0, \infty)$ and $1 \le p, q \le \infty$, the entropy numbers $e_n(T)$ of T (see Section 1.3 and [46]) satisfy

$$\begin{aligned} c_1\|uv\|_{r,I} &\le \liminf_{n\to\infty} n e_n(T) \\ &\le \limsup_{n\to\infty} n e_n(T) \le c_2\|uv\|_{r,I}, \end{aligned} \tag{2.8.1}$$

where $1/r = 1 - 1/p + 1/q > 0$. These estimates are used to investigate the small ball behaviour of weighted Wiener processes in the $L_q(0, \infty)$-norm.

2.5. Results of the type in this section on the l_α and $l_{\alpha,\omega}$ classes of the approximation numbers were first obtained in the case $p = 2$ by Newman and Solomyak [187]. In [161], estimates are derived for the l_s and weak-l_s norms of the approximation numbers of the operator T of (2.1.5) as a map between $L_p(0, \infty)$ and $L_q(0, \infty)$. The $l_p, 1 < p < \infty$, classes of operators of the form

$$Sf(x) := \int_0^\infty \phi(\max\{x, y\}) f(y) dy$$

acting in $L_2(0, \infty)$ are considered in [4], the classification being given in terms of the function ϕ. The range of values $0 < p \le 1$ is also discussed.

3

Banach function spaces

3.1 Introduction

Classical function spaces, such as those of Lebesgue and Sobolev type, have played and continue to play a most important rôle in Analysis. With the passage of time, questions have naturally arisen which for their complete solution require scales of spaces more finely tuned than these famous predecessors. One of the ways of meeting this need is by means of Banach function spaces, which were introduced in 1955 by Luxemburg. They include not only the Lebesgue spaces mentioned above but also their more refined variants, the Lorentz, Zygmund, Lorentz-Zygmund (LZ) and generalised Lorentz-Zygmund (GLZ) spaces. There is much to be gained from the study and use of these spaces, and of spaces of Sobolev type based on them. For example, consider the critical case of the Sobolev embedding theorem (associated with Pohozaev, Trudinger and Yudovic, among others) in which the Sobolev space is based upon L_p and the target is an Orlicz space of exponential type. It turns out that if a Zygmund space $L_p(logL)_a$ is used as a base instead of L_p, then in certain cases the target can be an Orlicz space of multiple exponential type. This is but one of the consequences of the intensive study during the last decade of refinements of the Sobolev embedding theorems.

In this chapter we first provide some basic information about Banach function spaces and spaces of Sobolev type based on them. We then give an account of a number of the recently obtained refinements of the classical embedding theorems such as that referred to above. Although many of these refinements were originally derived in terms of LZ or GLZ spaces, we have chosen to present them using the more general notion of Lorentz-Karamata spaces, involving slowly varying functions. This has definite advantages: it enables us to unify the treatment of the different scales of spaces mentioned above, and makes it easier to lay bare the real nature of certain arguments. Moreover, the proofs are no harder in this setting than in more specialised contexts. The treatment is largely self-contained, but the proofs of some results have been omitted so as to maintain the momentum. The reader eager to form an impression of the embedding results may wish to go to Section 3.7, which contains numerous simple examples illustrating the way in which the classical Sobolev embedding theorems are extended. In these the typical domain space is one of Sobolev type based on a Zygmund space $L_p(logL)_a$ and the target space may be of Lebesgue, Lorentz, Orlicz or Hölder type, depending on the parameters.

3.1.1 Definitions

The general underlying setting is that of a $\sigma-$finite measure space (R, μ); μ is a (non-negative) measure and $R = \bigcup_{n=1}^{\infty} R_n$ for some sets R_n with $\mu(R_n) < \infty$ for all $n \in \mathbb{N}$. The family of all extended scalar-valued (real or complex) $\mu-$measurable functions is denoted by $\mathcal{M}(R, \mu)$; $\mathcal{M}_0(R, \mu)$ will stand for the subset of $\mathcal{M}(R, \mu)$ consisting of all those functions which are finite μ-*a.e.*; and $\mathcal{M}^+(R, \mu)$ (resp. $\mathcal{M}_0^+(R, \mu)$) will represent the subset of $\mathcal{M}(R, \mu)$ (resp. $\mathcal{M}_0(R, \mu)$) made up of all those functions which are non-negative μ-*a.e.* We shall write $\mathcal{M}(R), \mathcal{M}_0(R)$, etc., if there is no ambiguity. The characteristic function of a μ-measurable subset E of R will be denoted by χ_E. When R is a subset of $\mathbb{R}^n$ we shall take μ to be n-dimensional Lebesgue measure μ_n; we shall write μ instead of μ_1 if no ambiguity is possible.

Definition 3.1.1. *A* ***Banach function norm*** *on (R, μ) is a map ρ from $\mathcal{M}^+(R, \mu)$ to $[0, \infty]$ such that for all f, g, f_n $(n \in \mathbb{N})$ in $\mathcal{M}^+(R, \mu)$, all scalars $\lambda \geq 0$ and all μ-measurable subsets E of R, the following are true:*

(*P*1) *$\rho(f) = 0$ if, and only if, $f = 0$ μ-a.e., $\rho(\lambda f) = \lambda \rho(f)$, $\rho(f + g) \leq \rho(f) + \rho(g)$;*
(*P*2) *if $0 \leq g \leq f$ μ-a.e., then $\rho(g) \leq \rho(f)$;*

(*P*3) *if $0 \leq f_n \uparrow f$ μ-a.e., then $\rho(f_n) \uparrow \rho(f)$;*

(*P*4) *if $\mu(E) < \infty$, then $\rho(\chi_E) < \infty$;*

(*P*5) *if $\mu(E) < \infty$, then there is a constant $C = C(E, \rho) < \infty$ such that $\int_E f d\mu \leq C\rho(f)$.*

Given such a function norm ρ, the set $X = X(\rho)$ of all functions $f \in \mathcal{M}(R, \mu)$ (identifying functions equal μ-a.e.) such that $\rho(|f|) < \infty$ is called a ***Banach function space (BFS*** *for short), and we define*

$$\|f \mid X\| = \rho(|f|), \; f \in X.$$

With the natural linear space operations it is easy to check that X is a linear space and that $\|\cdot \mid X\|$ is a norm on it. In fact, $(X, \|\cdot \mid X\|)$ is a Banach space. To prove this it is convenient to establish the following result of Fatou type.

Lemma 3.1.2. *Let X be a Banach function space and suppose that $f_n \in X (n \in \mathbb{N}), \lim_{n\to\infty} f_n = f$ $\mu - a.e.$ and $\liminf_{n\to\infty} \|f_n \mid X\| < \infty$. Then $f \in X$ and*

$$\|f \mid X\| \leq \liminf_{n\to\infty} \|f_n \mid X\| .$$

Proof. Put $g_n(x) = \inf_{m \geq n} |f_m(x)| : 0 \leq |g_n(x)| \uparrow |f(x)|$ μ-*a.e.* Properties (P2) and (P3) now give

$$\rho(|f|) = \lim_{n\to\infty} \rho(g_n) \leq \lim_{n\to\infty} \left(\inf_{m\geq n} \rho(|f_m|) \right)$$

$$= \liminf_{n\to\infty} \|f_n \mid X\| < \infty.$$

Moreover, $f \in \mathcal{M}(R, \mu)$. Hence $f \in X$ and the result follows. □

Theorem 3.1.3. *Let X be a Banach function space. Then $(X, \|\cdot \mid X\|)$ is a Banach space.*

Proof. Let $(f_n)_{n\in\mathbb{N}}$ be a Cauchy sequence in X and let $(g_n)_{n\in\mathbb{N}}$ be a subsequence of it such that $h_n := g_n - g_{n-1}$ (with $g_0 = 0$) satisfies $\|h_n \mid X\| < 2^{-n}$ $(n \in \mathbb{N}, n > 1)$. For each $N > 1$, $g_N = \sum_{n=1}^N h_n$. Put $G = \sum_1^\infty |h_n|$, $G_N = \sum_1^N |h_n|$. Then $0 \leq G_N \uparrow G$,

$$\|G_N \mid X\| \leq \sum_1^N \|h_n \mid X\| \leq \sum_1^\infty \|h_n \mid X\| < \infty,$$

and it follows from (P3) that $G \in X$.

Now let E be a subset of X with $\mu(E) < \infty$. By (P5), given any $\varepsilon > 0$,

$$\mu\{x \in E : |G(x) - G_N(x)| > \varepsilon\} \leq \int_E \varepsilon^{-1} |G - G_N| \, d\mu$$

$$\leq \varepsilon^{-1} C(E, \rho) \rho(|G - G_N|)$$

$$\to 0$$

as $N \to \infty$. Hence $G_N \to G$ in measure on E, and so by Riesz's theorem (see [126], p.156) there is a subsequence of (G_N) which converges to G μ-*a.e.* on E; by diagonalisation we obtain a subsequence of (G_N) which converges to G μ-*a.e.* on R. Since $\sum_1^\infty |h_n|$ converges μ-*a.e.* on R, so does $\sum_1^\infty h_n$, to g, say; this means that $g_n \to g$ μ-*a.e.* Moreover,

$$\liminf_{N\to\infty} \|g_M - g_N \mid X\| \leq \liminf_{N\to\infty} \sum_{M+1}^N \|h_n \mid X\|$$

$$= \sum_{M+1}^{\infty} \|h_n \mid X\| < \sum_{M+1}^{\infty} 2^{-n} \to 0$$

as $M \to \infty$. Thus by Lemma 3.1.2, $g_M - g \in X$ and $\|g_M - g\| \to 0$ as $M \to \infty$. Hence $g \in X$ and the subsequence (g_n) of (f_n) converges to g. Since (f_n) is a Cauchy sequence, this implies that $f_n \to g$ as $n \to \infty$; the proof is complete. □

The prototype of the Banach function spaces is $L_p(R, \mu)$, with norm given by $\left(\int_R |f|^p \, d\mu\right)^{1/p}$ when $1 \le p < \infty$ and by $\operatorname{ess\,sup}_R |f(x)|$ when $p = \infty$; as usual, we shall denote this by $L_p(R, w)$ if R is a subset of $\mathbb{R}^n$ and $d\mu = w dx$ for some non-negative measurable function w. It is easy to check that these spaces have properties (P1)-(P5). Later on we shall see how other more complicated spaces fit into the Banach function space framework.

The following simple result will prove to be very useful.

Theorem 3.1.4. *Let X and Y be Banach function spaces over the same measure space, with $X \subset Y$. Then X is continuously embedded in Y.*

Proof. Suppose the theorem is false. Then there is a sequence (f_n) of elements of X with $\|f_n \mid X\| \le 1$ and $\|f_n \mid Y\| > n^3$ for each n; we may suppose that $f_n \ge 0$ for all n. Plainly $\sum n^{-2} f_n$ converges in X to a function $f \in X \subset Y$. Since $0 \le n^{-2} f_n \le f$ we must have $\|f \mid Y\| \ge n^{-2} \|f_n \mid Y\| > n$ for all n, which is impossible. □

Next we turn to the associate of a Banach function space.

Definition 3.1.5. *Let ρ be a Banach function norm on (R, μ). The **associate norm** ρ' is defined on $\mathcal{M}^+(R, \mu)$ by*

$$\rho'(g) = \sup\left\{\int_R fg d\mu : f \in \mathcal{M}^+(R, \mu), \rho(f) \le 1\right\}.$$

Theorem 3.1.6. *If ρ is a Banach function norm on (R, μ), then so is its associate norm ρ'.*

For a proof see [15], Theorem 1.2.2.

Definition 3.1.7. *Let X be the Banach function space determined by a Banach function norm ρ on (R, μ). The Banach function space determined by the associate norm ρ' is called the **associate space** of X and is denoted by X'.*

Notice that if $g \in X'$, its norm is given by

$$\|g \mid X'\| = \sup\left\{\int_R |fg|\, d\mu : f \in X, \|f \mid X\| \le 1\right\}. \qquad (3.1.1)$$

This leads naturally to

Theorem 3.1.8. ***(Hölder's inequality)*** *Let X be a Banach function space with associate space X'. Then for all $f \in X$ and all $g \in X'$,*

$$\int_R |fg|\, d\mu \le \|f \mid X\| \, \|g \mid X'\| .$$

Proof. If $\|f \mid X\| = 0$, then $f = 0$ μ-*a.e.* and the result is obvious. Suppose that $\|f \mid X\| > 0$. Then $\|f / \|f \mid X\| \mid X\| = 1$ and so by (3.1.1),

$$\int_R |fg|\, d\mu / \|f \mid X\| \le \|g \mid X'\| ,$$

as required. □

To complement Hölder's inequality there is

Lemma 3.1.9. *A function $g \in \mathcal{M}(R, \mu)$ belongs to X' if, and only if, $fg \in L_1(R, \mu)$ for all $f \in X$.*

Proof. Suppose that $fg \in L_1(R, \mu)$ for all $f \in X$ but that $\rho'(|g|) = \infty$. Then given any $n \in \mathbb{N}$, there exists a non-negative function f_n such that $\|f_n \mid X\| \le 1$ and $\int_R |f_n g|\, d\mu > n^3$. Since X is complete, the function $f := \sum_{n=1}^{\infty} n^{-2} f_n$ belongs to X. But for all $n \in \mathbb{N}$,

$$\int_R |fg|\, d\mu \ge n^{-2} \int_R |f_n g|\, d\mu > n.$$

Hence $\rho'(|g|) < \infty$; that is, $g \in X'$. The converse follows directly from Theorem 3.1.8. □

The next result indicates the difference between the associate space X' and the dual space X^*.

Theorem 3.1.10. *Let X be a Banach function space. Then $X'' := (X')' = X$ and $\|f \mid X\| = \|f \mid X''\|$ for all $f \in X$.*

For a proof of this result, which is due to Lorentz and Luxemburg, see [15], Theorem 1.2.7. Consideration of $L_\infty(R, \mu)$ shows that, in general, $X' \ne X^*$.

To discuss the circumstances under which $X' = X^*$ some new ideas are needed.

Definition 3.1.11. *Let X be a Banach function space. A function $f \in X$ is said to have* **absolutely continuous norm** *if $\lim_{n\to\infty} \|f\chi_{E_n} \mid X\| = 0$ whenever $\{E_n\}$ is a sequence of subsets of R such that $\chi_{E_n} \to 0$ μ-a.e. If every $f \in X$ has this property, X is said to have* **absolutely continuous norm***.*

The next lemma is of technical importance.

Lemma 3.1.12. *A function f in a Banach function space X has absolutely continuous norm if, and only if, $\|f\chi_{E_n} \mid X\| \downarrow 0$ whenever $\{E_n\}$ is a sequence of subsets of R such that $\chi_{E_n} \downarrow 0$ μ-a.e.*

Proof. Suppose that f has the given property for sequences $\{E_n\}$ of the type mentioned. Let $\{F_n\}$ be a sequence such that $\chi_{F_n} \to 0$ μ-a.e. Put $E_n = \bigcup_{m\geq n} F_m$ $(n \in \mathbb{N})$: $\chi_{E_n} \downarrow 0$ μ-a.e. and so $\|f\chi_{E_n} \mid X\| \downarrow 0$. But $F_n \subset E_n$ for all $n \in \mathbb{N}$. Thus $\|f\chi_{F_n} \mid X\| \to 0$, which means that f has absolutely continuous norm. The converse is trivial. □

When $1 \leq p < \infty$, it follows immediately from the dominated convergence theorem that $L_p(R,\mu)$ has absolutely continuous norm. This is not so, in general, for $L_\infty(R,\mu)$: see [15], p.15.

To give a more rounded picture of the notion of an absolutely continuous norm the following may be helpful.

Lemma 3.1.13. *Let f be an element of a Banach function space. If f has absolutely continuous norm, then given any $\varepsilon > 0$, there exists $\delta > 0$ such that $\|f\chi_E\| < \varepsilon$ whenever $\mu(E) < \delta$.*

Proof. If this were false, then there would be an $\varepsilon > 0$ and a sequence of sets $\{E_n\}$ such that $\mu(E_n) < 2^{-n}$ and $\|f\chi_{E_n}\| \geq \varepsilon$ for all $n \in \mathbb{N}$. Since

$$\mu\left(\bigcup_{n=m}^{\infty}\right) \leq \sum_{n=m}^{\infty} \mu(E_n) < 2^{-m+1},$$

it follows that $\chi_{E_n} \to 0$ μ-a.e., from which the absolute continuity enables us to see that $\|f\chi_{E_n}\| \to 0$: contradiction. □

Proposition 3.1.14. *Let f belong to a Banach function space X. Then f has absolutely continuous norm if, and only if, $\|f_n\| \downarrow 0$ whenever $\{f_n\}$ is a sequence of μ-measurable functions such that $|f| \geq f_n \downarrow 0$ μ- a.e.*

Proof. To establish sufficiency, take $f_n = f\chi_{E_n}$ and use Lemma 3.1.12. For necessity, suppose that f has absolutely continuous norm and let $\{f_n\}$ be a sequence with $|f| \geq f_n \downarrow 0$ μ-a.e. Write $R = \bigcup_{n=1}^{\infty} R_n$, where $R_n \subset R_{n+1}$ and $\mu(R_n) \in [0,\infty)$ for all $n \in \mathbb{N}$, put $S_n = R\backslash R_n$, and let $\varepsilon > 0$. Since $\chi_{S_n} \downarrow 0$ and f has absolutely continuous norm, it follows that for some $N \in \mathbb{N}$, $\|f\chi_{S_N}\| < \varepsilon$. Put $E_n = \{x \in R_N : f_n(x) > \varepsilon\}$ for all $n \in \mathbb{N}$. Since $f_n \downarrow 0$ μ-a.e. and $\mu(R_N) < \infty$, we must have

$$\lim_{n\to\infty} \mu(E_n) = \mu\left(\bigcap_{n=1}^{\infty} E_n\right) = 0.$$

Thus by Lemma 3.1.13, $\|f\chi_{E_n}\| < \varepsilon$ for $n \geq M$, say. Then

$$\begin{aligned}
\|f_n\| &\leq \|f\chi_{S_N}\| + \|f_n\chi_{R_N}\| \\
&\leq \|f_n\chi_{S_N}\| + \|f_n\chi_{E_n}\| + \left\|f_n\chi_{R_N\setminus E_n}\right\| \\
&\leq \|f\chi_{S_N}\| + \|f\chi_{E_n}\| + \varepsilon \left\|\chi_{R_N\setminus E_n}\right\| \\
&< \varepsilon + \varepsilon + \varepsilon \|\chi_{R_N}\|
\end{aligned}$$

if $n \geq \max(N, M)$. The result follows. □

The relationship between X' and X^* can now be explained in a very simple manner, using the notion of absolute continuity.

Theorem 3.1.15. *Let X be a Banach function space. Then X^* is canonically isometrically isomorphic to X' if, and only if, X has absolutely continuous norm. Moreover, X is reflexive if, and only if, both X and X' have absolutely continuous norms.*

For a proof of this see [15], Corollaries 1.4.3 and 1.4.4.

3.2 Rearrangements

As in the last section we suppose that μ is a non-negative measure and (R, μ) is a σ-finite measure space.

Definition 3.2.1. *Let $f \in \mathcal{M}_0(R,\mu)$. The* ***distribution function*** *of f is the map $\mu_f : [0,\infty) \to [0,\infty]$ defined by*

$$\mu_f(\lambda) = \mu\{x \in R : |f(x)| > \lambda\}$$

for every $\lambda \geq 0$.

The proposition below gives some basic properties of this function.

Proposition 3.2.2. *Let f, f_n $(n \in \mathbb{N})$ and g belong to $\mathcal{M}_0(R,\mu)$ and let $a \in \mathbb{R}, a \neq 0$. Then:*

(i) μ_f is decreasing and right-continuous;

(ii) if $\mu_f(\lambda) < \infty$ for all $\lambda > 0$, then
$\mu_f(\lambda-) = \mu\{x \in R : |f(x)| \geq \lambda\}$ for all $\lambda > 0$,
$\mu_f(\lambda-) - \mu_f(\lambda) = \mu\{x \in R : |f(x)| = \lambda\}$ for all $\lambda > 0$;

(iii) if $|g| \le |f|$ μ *a.e., then* $\mu_g \le \mu_f$;

(iv) $\mu_{af}(\lambda) = \mu_f(\lambda/|a|)$ *for all* $\lambda \ge 0$;

(v) $\mu_{f+g}(\lambda_1 + \lambda_2) \le \mu_f(\lambda_1) + \mu_g(\lambda_2)$ *for all* $\lambda_1, \lambda_2 \ge 0$;

(vi) if $|f| \le \liminf_{n\to\infty} |f_n|$ μ *-a.e., then* $\mu_f \le \liminf_{n\to\infty} \mu_{f_n}$;
thus $\mu_{f_n} \uparrow \mu_f$ *if* $|f_n| \uparrow |f|$ μ *-a.e.*

Proof. (i) It is obvious that μ_f is decreasing. For the right-continuity, let $\lambda_0 \ge 0$ and put $E(\lambda) = \{x \in R : |f(x)| > \lambda\}$ for $\lambda \ge 0$. Evidently $E(\lambda)$ increases as λ decreases and

$$E(\lambda_0) = \bigcup_{n=1}^{\infty} E(\lambda_0 + \frac{1}{n}).$$

By monotone convergence,

$$\mu_f(\lambda_0 + \frac{1}{n}) = \mu\left(E(\lambda_0 + \frac{1}{n})\right) \uparrow \mu\left(E(\lambda_0)\right) = \mu_f(\lambda_0).$$

(ii) Let $\lambda > 0$. Then

$$\begin{aligned} \mu\{x \in R : |f(x)| \ge \lambda\} &= \mu\left\{\bigcap_{n=1}^{\infty}\left\{x \in R : f(x) > \lambda - \frac{1}{n}\right\}\right\} \\ &= \lim_{n\to\infty} \mu\left(E(\lambda - \frac{1}{n})\right) = \mu_f(\lambda-). \end{aligned}$$

The rest of (ii) is plain.
(iii) and (iv) are obvious.
(v) Here we use the fact that

$$\begin{aligned} \{x \in R : |f(x) + g(x)| \ge \lambda_1 + \lambda_2\} \subset \{&x \in R : |f(x) + g(x)| \ge \lambda_1\} \\ &\cup \{x \in R : |f(x) + g(x)| \ge \lambda_2\}. \end{aligned}$$

(vi) Let $\lambda \ge 0$ and put $E = \{x \in R : |f(x)| > \lambda\}, E_n = \{x \in R : |f_n(x)| > \lambda\}$. Then $E \subset \cup_{m=1}^{\infty} \cap_{n>m} E_n$ and so

$$\begin{aligned} \mu\left(\cap_{n>m} E_n\right) &\le \inf_{n>m} \mu\left(E_n\right) \le \sup_m \inf_{n>m} \mu\left(E_n\right) \\ &= \liminf_{n\to\infty} \mu\left(E_n\right) \end{aligned}$$

for all $m \in \mathbb{N}$. Since $\cap_{n>m} E_n$ increases with m, monotone convergence shows that

$$\mu(E) \le \mu\left(\cup_{m=1}^{\infty} \cap_{n>m} E_n\right) = \lim_{m\to\infty} \mu\left(\cap_{n>m} E_n\right) \le \liminf_{n\to\infty} \mu\left(E_n\right).$$

The rest is clear. □

Definition 3.2.3. *Given any $f \in \mathcal{M}_0(R,\mu)$, the* ***non-increasing rearrangement*** *of f is the function $f^* : [0,\infty) \to [0,\infty]$ defined by*

$$f^*(t) = \inf\{\lambda \in [0,\infty) : \mu_f(\lambda) \le t\}$$

for all $t \ge 0$. Here the convention that $\inf \emptyset = \infty$ is used.

Note that since μ_f is decreasing,

$$f^*(t) = \sup\{\lambda \in \mathbb{R} : \mu_f(\lambda) > t\}, t \ge 0,$$

and so f^* is simply the distribution function of μ_f; the right-continuity of f^* implies that in the definition of f^* the infimum is really a minimum. Moreover, it is clear that if μ_f is continuous and strictly decreasing, then f^* is the inverse of μ_f on $[0,\infty)$. Further elementary properties of non-increasing rearrangements are summarised in the following Proposition.

Proposition 3.2.4. *Let f, f_n $(n \in \mathbb{N})$ and g belong to $\mathcal{M}_0(R,\mu)$ and let $a \in \mathbb{R}$. Then:*

(i) f^* is a non-negative, decreasing and right-continuous function on $[0,\infty)$;

(ii) $\{s \ge 0 : f^*(s) > t\} = [0, \mu_f(t))$ for all $t \ge 0$ with $\mu_f(t) > 0$;

(iii) $f^*(\mu_f(t)) \le t$ for all $t \ge 0$ with $\mu_f(t) < \infty$, $f^*(\mu_f(t)-) \ge t$ for all $t \in [0, ess \sup |f|)$,

$$[\mu_f(t), \mu_f(t-)) \subset \{s \ge 0 : f^*(s) = t\} \subset [\mu_f(t), \mu_f(t-)]$$

for all $t \ge 0$; $\mu_f(f^*(t)) \le t$ for all $t \ge 0$ with $f^*(t) < \infty$ and $\mu_f(f^*(t)-) \ge t$ for all $t \in [0, \mu(\text{supp } f)]$;

(iv) if $|g| \le |f|$ μ- a.e., then $g^* \le f^*$;

(v) $(af)^* = |a| f^*$;

(vi) $(f+g)^*(t_1+t_2) \le f^*(t_1) + g^*(t_2)$ for all $t_1, t_2 \ge 0$;

(vii) if $|f| \le \liminf_{n\to\infty} |f_n|$ μ-a.e., then $f^* \le \liminf_{n\to\infty} f_n^*$; thus

$$f_n^* \uparrow f^* \text{ if } |f_n| \uparrow |f| \ \mu\text{-a.e.};$$

(viii) for all $p \in (0,\infty)$, $(|f|^p)^* = (f^*)^p$;

(ix) if $\limsup_{n\to\infty} |f_n| \le |f|$ μ-a.e. and

$$|f_n| \le |g| \ \mu\text{-a.e. for all large enough } n,$$

$$g^*(t) < \infty \text{ for all } t > 0, \quad \lim_{t\to\infty} g^*(t) = 0,$$

then

$$\limsup_{n\to\infty} f_n^* \le f^*;$$

(x) if $|f| = \lim_{n\to\infty} |f_n|$ *μ-a.e., then*

$$f^* = \lim_{n\to\infty} f_n^*.$$

Proof. (i) Since f^* is itself a distribution function, the result follows from Proposition 3.2.2 (i).
(ii) If $f^*(s) > t$, then from the definition of f^* we see that $\mu_f(t) > s$. Conversely, if $s \in [0, \mu_f(t))$, then again from the definition of f^* it follows that $f^*(s) > t$.
(iii) Let $t \geq 0$ and suppose that $s = \mu_f(t) < \infty$. From the definition of f^* we have

$$f^*(\mu_f(t)) = f^*(s) = \inf\{\lambda \geq 0 : \mu_f(\lambda) \leq s\} \leq t,$$

which gives the first part.
For the second part, note that since f^* is non-increasing and

$$f^*(s) = \sup\{\lambda \geq 0 : \mu_f(\lambda) > s\},$$

$f^*(s-) = m(\{t \geq 0 : \mu_f(t) \geq s\})$ for all $s > 0$. Let $t \in [0, \text{ess sup } |f|)$. Then $f^*(\mu_f(t)-) = m(\{s \geq 0 : \mu_f(s) \geq \mu_f(t)\})$: recall that $\mu_f(t) > 0$ if, and only if, $t \in [0, \text{ess sup } |f|)$. But

$$\{s \geq 0 : \mu_f(s) \geq \mu_f(t)\} \supset [0, t]$$

since μ_f is decreasing. Thus $f^*(\mu_f(t)-) \geq t$.
To handle the third part, observe that if $t > 0$, then

$$\begin{aligned}\{s \geq 0 : f^*(s) = t\} &= \bigcap_{n=1}^{\infty} \left\{s \geq 0 : t\left(1 - \tfrac{1}{n}\right) < f^*(s) \leq t\right\} \\ &= \bigcap_{n=1}^{\infty} [\mu_f(t), \mu_f(t - \tfrac{1}{n})],\end{aligned}$$

by the first two parts.
For the next part, let $t \geq 0$ and suppose $\lambda = f^*(t) < \infty$. From the definition of f^* we see that there is a sequence $\{\lambda_n\}$ with $\lambda_n \downarrow \lambda$ and $\mu_f(\lambda_n) \leq t$ $(n \in \mathbb{N})$. By the right-continuity of μ_f,

$$\mu_f(f^*(t)) = \lim_{n\to\infty} \mu_f(\lambda_n) \leq t.$$

Finally, $\mu_f(f^*(t)-) = m(\{s \geq 0 : f^*(s) \geq f^*(t)\}) \supset [0, t]$. and so $\mu_f(f^*(t)-) \geq t$.
(iv),(v), (vii): These follow directly from (iii), (iv) and (vi) respectively of Proposition 3.2.2.
(vi) If $\lambda := f^*(t_1) + g^*(t_2) = \infty$ there is nothing to prove; we therefore suppose otherwise. Let $t = \mu_{f+g}(\lambda)$. Then

$$
\begin{aligned}
t &= \mu\left\{x \in R : |f(x)+g(x)| > f^*(t_1)+g^*(t_2)\right\} \\
&\leq \mu\left\{x \in R : |f(x)| > f^*(t_1)\right\} + \mu\left\{x \in R : |g(x)| > g^*(t_2)\right\} \\
&\leq \mu_f\left(f^*(t_1)\right) + \mu_f\left(g^*(t_2)\right) \\
&\leq t_1 + t_2,
\end{aligned}
$$

the final step following from (iii). Hence $t < \infty$. Thus

$$
\begin{aligned}
(f+g)^*(t_1+t_2) \leq (f+g)^*(t) &= (f+g)^*\left(\mu_{f+g}(\lambda)\right) \\
&\leq \lambda,
\end{aligned}
$$

where we have used (iii) again in the last step.
(viii) For all $\lambda \geq 0$, we see from (ii) that

$$
\mu_{|f|^p}(\lambda) = \mu_f\left(\lambda^{1/p}\right) = m(\{s \geq 0 : (f^*)^p(s) > \lambda\}).
$$

The result now follows from the definition of the non-increasing rearrangement.
(ix) When $\lambda \geq 0$ and $F \in \mathcal{M}(R,\mu)$, put

$$
E_\lambda(F) = \{x \in R : |F(x)| > \lambda\}.
$$

Since

$$
|f(x)| \geq \limsup_{n\to\infty} |f_n(x)| = \lim_{n\to\infty} h_n(x) \ \mu\text{-a.e.},
$$

where $h_n(x) = \sup_{m\geq n}|f_n(x)|$, it follows that $h_n \downarrow h \leq |f|$. Hence

$$
\mu_h(\lambda) \leq \mu_f(\lambda), \ \lambda \geq 0. \tag{3.2.1}
$$

Also

$$
E_\lambda(h) = \bigcap_{n=1}^{\infty} E_\lambda(h_n) = \bigcap_{n=1}^{\infty}\bigcup_{m\geq n} E_\lambda(f_m). \tag{3.2.2}
$$

The assumptions on f_n and g imply that

$$
\mu\left(\bigcup_{m\geq n} E_\lambda(f_m)\right) \leq \mu\left(E_\lambda(g)\right) = \mu_g(\lambda) < \infty.
$$

Together with (3.2.2) this gives

$$
\mu_h(\lambda) = \lim_{n\to\infty} \mu\left(\bigcup_{m\geq n} E_\lambda(f_m)\right).
$$

Since

$$\mu\left(\bigcup_{m\geq n} E_\lambda(f_m)\right) \geq \sup_{k\geq n} \mu\left(E_\lambda(f_k)\right) = \sup_{k\geq n} \mu_{f_k}(\lambda),$$

we see that

$$\mu_h(\lambda) \geq \lim_{n\to\infty} \sup_{k\geq n} \mu_{f_k}(\lambda) = \lim_{n\to\infty} \sup \mu_{f_n}(\lambda).$$

In view of (3.2.1), the result follows immediately.
(x) This is a simple consequence of (vii) and (ix). □

We can now see how L_p (quasi-)norms behave with regard to rearrangements.

Proposition 3.2.5. *Let $f \in \mathcal{M}_0(R,\mu)$. If $0 < p < \infty$, then*

$$\int_R |f|^p \, d\mu = p\int_0^\infty \lambda^{p-1}\mu_f(\lambda)d\lambda = \int_0^\infty f^*(t)^p dt.$$

Moreover,

$$\operatorname{ess\,sup}_{x\in R} |f(x)| = \inf\{\lambda : \mu_f(\lambda) = 0\} = f^*(0).$$

Proof. First suppose that $0 < p < \infty$ and that f is a non-negative simple function:

$$f(x) = \sum_{k=1}^n a_k \chi_{E_k}(x), \text{say},$$

where the E_k are pairwise disjoint subsets of R with finite measure, and $a_1 > a_2 > ... > a_n > 0$. Put

$$m_j = \sum_{i=1}^j \mu(E_i) \text{ for } j = 1, ..., n. \tag{3.2.3}$$

Then

$$\mu_f(\lambda) = \sum_{j=1}^n m_j \chi_{[a_{j+1},a_j)}(\lambda), \ \lambda \geq 0, \tag{3.2.4}$$

where $a_{n+1} = 0$. To see this, note that if $\lambda \geq a_1$, then plainly $\mu_f(\lambda) = 0$. If $a_{j+1} \leq \lambda < a_j$ for some $j \in \{1, ..., n\}$, then $\{x : f(x) > \lambda\} = \bigcup_{k=1}^j E_k$ and so $\mu_f(\lambda) = \sum_{k=1}^j \mu(E_k)$. Hence (3.2.4) follows; thus taking $m_0 = 0$,

$$f^*(t) = \sum_{j=1}^n a_j \chi_{[m_{j-1},m_j)}(t), \ t \geq 0. \tag{3.2.5}$$

Thus

$$\int_R |f|^p \, d\mu = \sum_{j=1}^n a_j^p \mu(E_j) = \sum_{j=1}^n a_j^p (m_j - m_{j-1}) = \int_0^\infty (f^*)^p \, dt. \qquad (3.2.6)$$

In the same way we see that

$$p \int_0^\infty \lambda^{p-1} \mu_f(\lambda) d\lambda = p \sum_{j=1}^n m_j \int_{a_{j+1}}^{a_j} \lambda^{p-1} d\lambda = \sum_{j=1}^n \left(a_j^p - a_{j+1}^p\right) m_j$$

$$= \sum_{j=1}^n a_j^p \mu(E_j) \qquad = \int_R |f|^p \, d\mu.$$

The desired result now follows for non-negative simple functions; for general functions it is merely necessary to use this together with Proposition 3.2.2 (vi), Proposition 3.2.4 (vii) and the monotone convergence theorem. The last part of the Proposition, corresponding to $p = \infty$, is plain. □

Definition 3.2.6. *Functions $f \in \mathcal{M}_0(R, \mu)$ and $g \in \mathcal{M}_0(S, \nu)$ are said to be **equimeasurable** if they have the same distribution function; that is, if $\mu_f(\lambda) = \mu_g(\lambda)$ for all $\lambda \geq 0$.*

Proposition 3.2.7. *Let $f \in \mathcal{M}_0(R, \mu)$. Then f and f^* are equimeasurable.*

Proof. First let f be a non-negative simple function,

$$f(x) = \sum_{k=1}^n a_k \chi_{E_k}(x),$$

say, where the E_k are pairwise disjoint subsets of R, each of finite measure, and $a_1 > a_2 > ... > a_n > 0$. As in the proof of Proposition 3.2.5,

$$\mu_f(\lambda) = \sum_{j=1}^n m_j \chi_{[a_{j+1}, a_j)}(\lambda), m_j = \sum_{i=1}^j \mu(E_i),$$

and

$$f^*(t) = \sum_{j=1}^n a_j \chi_{[m_{j-1}, m_j)}(t);$$

thus

$$\mu_{f^*}(\lambda) = \sum_{j=1}^n m_j \chi_{[a_{j+1}, a_j)}(\lambda) = \mu_f(\lambda).$$

For a general $f \in \mathcal{M}_0(R, \mu)$, let $\{f_n\}$ be a sequence of non-negative simple functions such that $f_n \uparrow |f|$; by Proposition 3.2.4 (vii), $f_n^* \uparrow f^*$. Since $\mu_{f_n}(\lambda) = \mu_{f_n^*}(\lambda)$ for all $n \in \mathbb{N}$ and all $\lambda \geq 0$, the result follows from Proposition 3.2.2 (vi). □

Corollary 3.2.8. *Let $\Phi : [0, \infty] \to [0, \infty]$ be continuous, strictly increasing and such that $\Phi(0) = 0$; let $f \in \mathcal{M}_0(R, \mu)$. Then*

$$\int_R \Phi(|f(x)|) \, d\mu(x) = \int_0^\infty \Phi(f^*(s)) \, ds.$$

Proof. For any $\lambda > 0$, since f and f^* are equimeasurable,

$$\mu_{\Phi(|f|)}(\lambda) = \mu\left\{x \in R : |f(x)| > \Phi^{-1}(\lambda)\right\}$$

$$= \mu_{f^*}\left(\Phi^{-1}(\lambda)\right) = \mu_{\Phi(f^*)}(\lambda).$$

Hence by the arguments used to prove Proposition 3.2.5,

$$\int_R \Phi(|f|)\, d\mu = \int_0^\infty \mu_{\Phi(|f|)}(\lambda)\, d\lambda = \int_0^\infty \mu_{\Phi(f^*)}(\lambda)\, d\lambda$$

$$= \int_0^\infty \Phi(f^*)\, ds.$$

□

This is a form of Cavalieri's principle, which states that two planar regions with the same cross-sectional lengths will have the same area.

We now turn to a most important inequality, due to Hardy and Littlewood, which lies at the heart of most theorems proved by means of rearrangements. It is convenient to begin with a Lemma.

Lemma 3.2.9. *Let f be a non-negative simple function on (R, μ) and let E be a μ-measurable subset of R. Then*

$$\int_E f d\mu \le \int_0^{\mu(E)} f^*(s) ds.$$

Proof. The function f may be represented in the form $f = \sum_{j=1}^n a_j \chi_{E_j}$ for some $a_j > 0$ $(j = 1, ..., n)$ and some sets E_j with $E_1 \subset E_2 \subset ... \subset E_n$. This follows from the standard representation

$$f = \sum_{j=1}^n b_j \chi_{F_j},$$

where the F_j are pairwise disjoint and $b_1 > b_2 > ... > b_n > 0$, on setting $a_j = b_j - b_{j+1}$ and $E_j = \bigcup_{i=1}^j F_i$. Then

$$f^* = \sum_{j=1}^n a_j \chi_{[0, \mu(E_j))}$$

and

$$\int_E f d\mu = \sum_{j=1}^n a_j \mu(E \cap E_j) \le \sum_{j=1}^n a_j \min(\mu(E), \mu(E_j))$$

$$= \sum_{j=1}^n a_j \int_0^{\mu(E)} \chi_{[0,\mu(E_j))}(s)\, ds = \int_0^{\mu(E)} f^*(s)\, ds.$$

□

The promised inequality due to Hardy and Littlewood is the following:

Theorem 3.2.10. *Let $f, g \in \mathcal{M}_0(R, \mu)$. Then*

$$\int_R |fg| \, d\mu \leq \int_0^\infty f^*(s) \, g^*(s) \, ds.$$

Proof. It is plainly enough to deal with the case in which f and g are non-negative: f^*, g^* depend on $|f|, |g|$ rather than on f, g. Now suppose that f and g are non-negative simple functions, with

$$f = \sum_{j=1}^m a_j \chi_{E_j}, \; E_1 \subset E_2 \subset ... \subset E_m, \; a_j > 0 \;\; (j = 1, ..., m).$$

Then

$$f^*(t) = \sum_{j=1}^m a_j \chi_{[0,\mu(E_j))}$$

and so, by Lemma 3.2.9,

$$\begin{aligned} \textstyle\int_R |fg| \, d\mu &= \textstyle\sum_{j=1}^m a_j \int_{E_j} g d\mu \leq \sum_{j=1}^m a_j \int_0^{\mu(E_j)} g^*(s) \, ds \\ &= \textstyle\int_0^\infty \sum_{j=1}^m a_j \chi_{[0,\mu(E_j))}(s) \, g^*(s) \, ds \\ &= \textstyle\int_0^\infty f^*(s) \, g^*(s) \, ds. \end{aligned}$$

The general result now follows from Proposition 3.2.4 (vii) and the monotone convergence theorem. □

Corollary 3.2.11. *Let $f, g, \widetilde{g} \in \mathcal{M}_0(R, \mu)$ and let $g, \widetilde{g}$ be equimeasurable. Then*

$$\int_R |f\widetilde{g}| \, d\mu \leq \int_0^\infty f^*(s) \, g^*(s) \, ds. \tag{3.2.7}$$

It turns out that it is desirable to discuss spaces (R, μ) in which the supremum, over all $\widetilde{g}$ equimeasurable with g, of the left-hand side of (3.2.7) coincides with the right-hand side. We address this matter next.

Definition 3.2.12. *The measure space (R, μ) is called* ***resonant*** *if for all $f \in \mathcal{M}_0(R, \mu)$,*

$$\int_0^\infty f^*(s) \, g^*(s) \, ds = \sup \int_R |f\widetilde{g}| \, d\mu,$$

where the supremum is taken over all functions $\widetilde{g}$ on R which are equimeasurable with g. If for each $f, g, \in \mathcal{M}_0(R, \mu)$ there is a function $\widetilde{g}$ on R which is equimeasurable with g and such that

$$\int_0^\infty f^*(s) \, g^*(s) \, ds = \int_R |f\widetilde{g}| \, d\mu,$$

the measure space (R, μ) is said to be ***strongly resonant****.*

A complete characterisation of resonant and strongly resonant spaces is given by the following theorem.

Theorem 3.2.13. *The measure space (R, μ) is resonant if, and only if, it is either*
(i) nonatomic,
or
(ii) completely atomic, with all atoms having equal measure.
It is strongly resonant if, and only if, it is resonant and $\mu(R) < \infty$.

For a proof of this see [15], Theorem 2.2.7 and Corollary 2.2.8.

Experience shows that it is often more convenient to work with the average of f^* than with f^* itself. We now turn to this average and some of its properties.

Definition 3.2.14. *Let $f \in \mathcal{M}_0(R, \mu)$. Then f^{**} is defined by*

$$f^{**}(t) = \frac{1}{t} \int_0^t f^*(s)\, ds, \ t > 0.$$

Proposition 3.2.15. *Let f, g, f_n $(n \in \mathbb{N})$ belong to $\mathcal{M}_0(R, \mu)$ and let a be any scalar. Then:*
(i) f^{**} is non-negative, decreasing and continuous;
(ii) $f^{**} = 0$ if, and only if, $f = 0$ μ-a.e.;
(iii) $f^* \leq f^{**}$;
(iv) if $|g| \leq |f|$ μ-a.e., then $g^{**} \leq f^{**}$;
(v) $(af)^{**} = |a| f^{**}$;
(vi) if $|f_n| \uparrow |f|$ μ-a.e., then $f_n^{**} \uparrow f^{**}$.

Proof. (i) It is plain that f^{**} is non-negative and continuous. Note, however, that $f^{**}(t)$ may be infinite for some $t > 0$; but if this is so, $f^{**}(t) = \infty$ for all $t > 0$, since f^* is decreasing. To show that f^{**} is decreasing, let $0 < t < s$. Then since f^* is decreasing,

$$\begin{aligned} f^{**}(s) &= s^{-1} \textstyle\int_0^s f^*(\tau)\, d\tau \leq s^{-1} \int_0^s f^*(t\tau/s)\, d\tau \\ &= t^{-1} \textstyle\int_0^t f^*(u)\, du = f^{**}(t). \end{aligned}$$

(ii) This follows immediately from the definition of f^{**} and Proposition 3.2.4.
(iii) Since f^* is decreasing,

$$f^*(t) = t^{-1} \int_0^t f^*(t)\, ds \leq t^{-1} \int_0^t f^*(s)\, ds = f^{**}(t), \ t > 0.$$

(iv)-(vi) These follow easily from the corresponding results in Proposition 3.2.4. □

Another useful property which f^{**} has is the following.

Theorem 3.2.16. *Let $f, g \in \mathcal{M}_0(R,\mu)$. Then for all $t > 0$,*

$$(f+g)^{**}(t) \leq f^{**}(t) + g^{**}(t).$$

For the proof of this we refer to [15], Theorem 2.3.4.

Note that f^* does not have this subadditivity property. To see this, take $R = \mathbb{R}$, $f = \chi_{[0,1)}$ and $g = \chi_{[1,2)}$.

Useful identifications of f^{**} can be made when the underlying measure space is resonant or strongly resonant.

Proposition 3.2.17. *Let $f \in \mathcal{M}_0(R,\mu)$ and suppose that $t > 0$ is in the range of μ. If (R,μ) is resonant, then*

$$f^{**}(t) = t^{-1} \sup \left\{ \int_E |f| \, d\mu : \mu(E) = t \right\};$$

if (R,μ) is strongly resonant, then there is a subset E of R, with $\mu(E) = t$, such that

$$f^{**}(t) = t^{-1} \int_E |f| \, d\mu.$$

Proof. Let F be a measurable subset of R with $\mu(F) = t$. Put $g = \chi_F$; thus $g^* = \chi_{[0,t)}$. Let $\widetilde{g} \in \mathcal{M}(R,\mu)$ and g be equimeasurable; then $|\widetilde{g}| = \chi_E$ μ-*a.e.* for some $E \subset R$ with $\mu(E) = \mu(F)$. The proposition now follows immediately from Definition 3.2.12. □

When R is an open subset of $\mathbb{R}^n$ and $\mu = \mu_n$, the n-dimensional Lebesgue measure on R, it is sometimes more convenient to work with the symmetric rearrangement of a function, which we now define. For the rest of this section G will be an open subset of $\mathbb{R}^n$.

Definition 3.2.18. *Let $f \in \mathcal{M}_0(G,\mu_n)$. The* ***symmetric rearrangement*** *of f is the function $f^\star$ defined by*

$$f^\star(x) = f^*(\omega_n |x|^n), \ x \in G^\star,$$

where $G^\star$ is the ball in $\mathbb{R}^n$ centred at 0 and with the same volume as G, if $\mu_n(G) < \infty$; $G^\star = \mathbb{R}^n$ otherwise.

It is clear that $f^\star$ is non-negative, radial (invariant under rotations about the origin) and radially nonincreasing: $f^\star(x)$ is nonincreasing as $|x|$ increases. Moreover, f and $f^\star$ are equimeasurable. From the Hardy-Littlewood inequality of Theorem 3.2.10 we immediately have

Theorem 3.2.19. *Let $f, g \in \mathcal{M}_0(G,\mu_n)$. Then*

$$\int_G |f(x)g(x)| \, dx \leq \int_{G^\star} f^\star(x) g^\star(x) dx.$$

We next give a form of the Pólya-Szegö principle which relates the behaviour of the gradients of $f^\bigstar$ and f. To do this it is convenient to establish the following Lemma.

Lemma 3.2.20. *Let f be a real-valued, Lipschitz-continuous function on $\mathbb{R}^n$ such that for all $t > 0$, $\mu_n(\{x \in \mathbb{R}^n : |f(x)| > t\}) < \infty$. Let $\Phi : [0,\infty) \to [0,\infty)$ be increasing and convex, with $\Phi(0) = 0$. Then*
(i) f^* is locally absolutely continuous on $(0,\infty)$;
(ii) for almost all $s > 0$,

$$\frac{d}{ds}\int_{G(s)} \Phi(|\nabla f(x)|)\,dx \geq \Phi\left(-n\omega_n^{1/n}s^{1-1/n}\frac{df^*}{ds}(s)\right), \tag{3.2.8}$$

where $G(s) = \{x \in \mathbb{R}^n : |f(x)| > f^*(s)\}$.

Proof. We first claim that

$$\int_{G(a,b)} |\nabla f(x)|\,dx \geq n\omega_n^{1/n}\left(f^*(a) - f^*(b)\right)a^{1-1/n} \tag{3.2.9}$$

and

$$\mu_n(G(a,b)) \leq b - a, \tag{3.2.10}$$

if $\mu_n(\text{supp } f) > b > a \geq 0$; here $G(a,b) = \{x \in \mathbb{R}^n : f^*(a) > |f(x)| > f^*(b)\}$. To establish this note that by the co-area formula (Corollary 1.2.5) and the isoperimetric theorem (1.2.2) in $\mathbb{R}^n$,

$$\int_{G(a,b)} |\nabla f(x)|\,dx = \int_{f^*(b)}^{f^*(a)} H^{n-1}(\{x \in \mathbb{R}^n : |f(x)| = t\})\,dt$$

$$\geq n\omega_n^{1/n}\int_{f^*(b)}^{f^*(a)} \left(\mu_n(\{x \in \mathbb{R}^n : |f(x)| \geq t\})\right)^{1-1/n} dt.$$

Since the integrand is monotone, this expression can be estimated from below by

$$n\omega_n^{1/n}\left(\mu_n(\{x \in \mathbb{R}^n : |f(x)| \geq f^*(a)\})\right)^{1-1/n}\left(f^*(a) - f^*(b)\right),$$

and as $\mu_n(\{x \in \mathbb{R}^n : |f(x)| \geq f^*(a)\}) \geq \mu_f(f^*(a)-) \geq a$, by Propositions 3.2.2 (ii) and 3.2.4 (iii), we obtain (3.2.9).

For (3.2.10), use Proposition 3.2.2 (ii) to obtain

$$\mu_n(G(a,b)) = \mu_f(f^*(b)) - \mu_f(f^*(a)-).$$

Proposition 3.2.4 (iii) now shows that $\mu_f(f^*(b)) \leq b$ and $\mu_f(f^*(a)-) \geq a$. Since supp $f^* = [0, \mu_n(\text{supp } f)]$, the local absolute continuity of f^* follows from (3.2.9) and (3.2.10).

We next assert that for almost all $s \geq 0$,

$$\frac{d}{ds}\int_{G(s)} |\nabla f(x)|\, dx \geq -n\omega_n^{1/n} s^{1-1/n} \frac{df^*}{ds}(s). \tag{3.2.11}$$

To see this, if $0 \leq s = \mu_n(\text{supp } f)$, the left-hand side of (3.2.11) equals

$$\lim_{h\to 0+} h^{-1} \int_{U(s,h)} |\nabla f(x)|\, dx,$$

where $U(s,h) = \{x \in \mathbb{R}^n : f^*(s) \geq |f(x)| > f^*(s+h)\}$. But by (3.2.9), this limit is bounded below by

$$n\omega_n^{1/n} s^{1-1/n} \lim_{h\to 0+} \frac{f^*(s) - f^*(s+h)}{h},$$

which gives the desired result. On the other hand, (3.2.11) is plain when $s \geq \mu_n(\text{supp } f)$ since the right-hand side is then zero.

Our arguments so far have shown that (3.2.8) holds for almost all $s \geq 0$ when $\Phi(t) = t$. To extend this to arbitrary functions Φ of the given type, we remark that either (i) $\frac{df^*}{ds}(s) = 0$, or (ii) f^* decreases strictly in some neighbourhood of s, or (iii) s belongs to some exceptional set of measure zero. There is nothing to prove in cases (i) and (iii); suppose that (ii) holds. Then we claim that $\mu_n(U(s,h)) = h$ if h is positive and sufficiently small. For $\mu_n(U(s,h)) = \mu_f(f^*(s+h)) - \mu_f(f^*(s))$. However, $\mu_f(t-) - \mu_f(t) = \mu(\{r \geq 0 : f^*(r) = t\})$, by Proposition 3.2.2 (ii); while

$$\mu_f(f^*(r)) \leq r \leq \mu_f(f^*(r)-) \text{ if } 0 \leq r < \mu_n(\text{supp } f),$$

by Proposition 3.2.4 (iii). Thus $\mu_f(f^*(r)) = r$ whenever r is close enough to s, and our claim is justified.

To complete the proof of (ii), we use Jensen's inequality (Theorem 1.4.2) to obtain

$$h^{-1}\int_{U(s,h)} \Phi(|\nabla f(x)|)\, dx \geq \Phi\left(h^{-1}\int_{U(s,h)} |\nabla f(x)|\, dx\right).$$

Hence

$$\frac{d}{ds}\int_{G(s)} \Phi(|\nabla f(x)|)\, dx \geq \Phi\left(\frac{d}{ds}\int_{G(s)} |\nabla f(x)|\, dx\right),$$

which together with (3.2.11) gives the result. □

Armed with this Lemma we can now give the promised result of Pólya-Szegö type.

Theorem 3.2.21. *Let $\Phi : [0,\infty) \to [0,\infty)$ be increasing and convex, with $\Phi(0) = 0$, and let f be a real-valued Lipschitz-continuous function on $\mathbb{R}^n$ such that for all $t > 0$, $\mu_n(\{x \in \mathbb{R}^n : |f(x)| > t\}) < \infty$. Then*

$$\int_{\mathbb{R}^n} \Phi\left(|\nabla f(x)|\right) dx \geq \int_0^\infty \Phi(-n\omega_n^{1-1/n} s^{1-1/n} \frac{df^*}{ds}(s))ds \tag{3.2.12}$$

and

$$\int_{\mathbb{R}^n} \Phi\left(|\nabla f(x)|\right) dx \geq \int_{\mathbb{R}^n} \Phi\left(\left|\nabla f^\star(x)\right|\right) dx. \tag{3.2.13}$$

Proof. Put $G(s) = \{x \in \mathbb{R}^n : |f(x)| > f^*(s)\}$. Then since

$$\int_{G(s)} \Phi\left(|\nabla f(x)|\right) dx \uparrow \int_{\mathbb{R}^n} \Phi\left(|\nabla f(x)|\right) dx$$

as $s \uparrow \infty$, we have

$$\int_{\mathbb{R}^n} \Phi\left(|\nabla f(x)|\right) dx \geq \int_0^\infty \left(\frac{d}{ds} \int_{G(s)} \Phi\left(|\nabla f(x)|\right) dx\right) ds.$$

By the definition of $f^\star$,

$$\int_{\mathbb{R}^n} \Phi\left(\left|\nabla f^\star(x)\right|\right) dx = \int_0^\infty \Phi(-n\omega_n^{1-1/n} s^{1-1/n} \frac{df^*}{ds}(s))ds.$$

The theorem now follows from Lemma 3.2.20. □

Another interesting consequence of Lemma 3.2.20 is

Theorem 3.2.22. *Let f be as in Theorem 3.2.21 and set $\mu_n(supp\ f) = V$. Then for almost all $s \in [0, V)$,*

$$f^*(0) - \frac{1}{n\omega_n^{1/n}} \int_0^s t^{-1+1/n} \left|\nabla f\right|^* (t)dt$$

$$\leq f^*(s) \leq \frac{1}{n\omega_n^{1/n}} \int_0^{V-s} (s+t)^{-1+1/n} \left|\nabla f\right|^* (t)dt,$$

with equality throughout if

$$f(x) = \begin{cases} 1 - |x|, & |x| < 1, \\ 0, & |x| \geq 1. \end{cases}$$

Proof. By Lemma 3.2.20, f^* is locally absolutely continuous and for almost all $t \geq 0$,

$$-n\omega_n^{1/n} \frac{df^*}{dt}(t) \leq t^{-1+1/n} \frac{d}{dt} \int_{G(t)} |\nabla f(x)|\, dx,$$

where $G(t) = \{x \in \mathbb{R}^n : |f(x)| > f^*(t)\}$. Since $\mathrm{supp} f^* = [0, V]$ we have for almost all $s \in [0, V)$,

$$n\omega_n^{1/n} f^*(s) \leq \int_s^V t^{-1+1/n} \left(\frac{d}{dt} \int_{G(t,s)} |\nabla f(x)| \, dx \right) dt,$$

where $G(t,s) = \{x \in \mathbb{R}^n : f^*(t) < |f(x)| \leq f^*(s)\}$. Using the fact that

$$\int_{G(t,s)} |\nabla f(x)| \, dx \leq \int_0^{\mu_n(G(t,s))} |\nabla f|^* (r) dr \leq \int_0^{t-s} |\nabla f|^* (r) dr,$$

we see on integrating by parts that

$$n\omega_n^{1/n} f^*(s) \leq V^{-1+1/n} \int_0^{V-s} |\nabla f|^* (t) dt$$

$$+ (1 - 1/n) \int_s^V t^{-2+1/n} \left(\int_0^{t-s} |\nabla f|^* (r) dr \right) dt$$

$$= \int_s^V t^{-1+1/n} |\nabla f|^* (t-s) dt.$$

The desired right-hand inequality follows. To obtain the left-hand inequality write

$$n\omega_n^{1/n} (f^*(0) - f^*(s)) = n\omega_n^{1/n} \int_0^s -\frac{df^*}{dt}(t) dt$$

$$\leq \int_0^s t^{-1+1/n} \left(\frac{d}{dt} \int_{G(t,s)} |\nabla f(x)| \, dx \right) dt$$

and proceed as above. The rest is clear. □

Remark 3.2.23. The hypothesis in these results that f is Lipschitz-continuous is not really necessary: it is enough to assume that $f \in L_{1,loc}(\mathbb{R}^n)$ and that the gradient of f (in the weak sense) is in $L_1(\mathbb{R}^n)$. The penalty for this is a little more sophistication in the proof. For example, the modifications needed in the proof of Lemma 3.2.20 are that we use the co-area formula in the form

$$I = \int_{G(a,b)} |\nabla f(x)| \, dx = \int_{f^*(b)}^{f^*(a)} P(\{x \in \mathbb{R}^n : |f(x)| > t\}) \, dt,$$

where P denotes the perimeter (see Section 1.2). Then by the isoperimetric theorem (1.2.1) we have

$$I \geq n\omega_n^{1/n} \int_{f^*(b)}^{f^*(a)} (\mu_n(\{x \in \mathbb{R}^n : |f(x)| > t\}))^{1-1/n} \, dt,$$

and the proof is back on course. From this the classical form of the Pólya-Szegö principle follows: if G is any bounded open subset of $\mathbb{R}^n$ and $1 \leq p \leq \infty$, then if $f \in \mathring{W}_p^1(G)$, the symmetric rearrangement $f^\star$ of f belongs to $\mathring{W}_p^1(G^\star)$ and

$$\left\|\nabla f^\star \mid L_p(G^\star)\right\| \leq \|\nabla f \mid L_p(G)\| .$$

Moreover, (3.2.13) holds for any weakly differentiable, compactly supported function f on $\mathbb{R}^n$ with $\int_{\mathbb{R}^n} \Phi(|\nabla f(x)|)\,dx < \infty$.

3.3 Rearrangement-invariant spaces

For the most part, we shall be concerned with Banach function spaces in which the function norm has the same value for all equimeasurable functions.

Definition 3.3.1. *Let ρ be a function norm over (R, μ). Then ρ is called* ***rearrangement-invariant*** *(r.i.) if $\rho(f) = \rho(g)$ whenever $f, g \in \mathcal{M}_0(R, \mu)$ are equimeasurable. The corresponding Banach function space $X = X(\rho)$ is in that case said to be* ***rearrangement-invariant***.

The Lebesgue spaces $L_p(R, \mu)$ $(1 \leq p < \infty)$ are rearrangement-invariant: this follows easily from Proposition 3.2.5.

For resonant spaces, the property of being rearrangement-invariant is inherited by the associate norm.

Proposition 3.3.2. *Let ρ be a rearrangement-invariant function norm over a resonant measure space (R, μ). Then the associate norm ρ' is also rearrangement-invariant,*

$$\rho'(g) = \sup\left\{\int_0^\infty f^*(s)g^*(s)ds : \rho(f) \leq 1\right\}, \quad g \in \mathcal{M}_0^+(R, \mu)$$

and

$$\rho'(g) = \sup\left\{\int_0^\infty f^*(s)g^*(s)ds : \rho'(g) \leq 1\right\}, \quad f \in \mathcal{M}_0^+(R, \mu).$$

Proof. Let $g \in \mathcal{M}_0^+(R, \mu)$. Then by Definition 3.1.5,

$$\rho'(g) = \sup\left\{\int_R fg d\mu : \rho(f) \leq 1\right\}.$$

If $\rho(f) \leq 1$ and $\widetilde{f} \in \mathcal{M}_0^+(R, \mu)$ is equimeasurable with f, then of course $\rho(\widetilde{f}) \leq 1$. Hence by Definition 3.2.12,

$$\begin{aligned}\rho'(g) &= \sup_{\rho(f)\leq 1} \sup\left\{\int_R \widetilde{f} g d\mu : \widetilde{f} \text{ is equimeasurable with } f\right\} \\ &= \sup_{\rho(f)\leq 1} \int_0^\infty f^*(s)g^*(s)ds.\end{aligned}$$

If g and h are equimeasurable, $g^* = h^*$, and so $\rho'(g) = \rho'(h)$: ρ' is rearrangement-invariant. By Theorem 3.1.10, $\rho'' = \rho'$; the formula for $\rho(f)$ now follows from the rearrangement-invariance of ρ' and the expression for $\rho'(g)$ which has just been established. □

Corollary 3.3.3. *(Hölder's inequality) Let ρ be an r.i. function norm over a resonant measure space (R, μ), and let $f, g \in \mathcal{M}_0(R, \mu)$. Then*

$$\left| \int_R fgd\mu \right| \leq \int_0^\infty f^*(s)g^*(s)ds \leq \rho(|f|)\, \rho'(|g|) = \|f \mid X\| \, \|g \mid X'\| .$$

Proof. Theorem 3.2.10 gives the first inequality; the second follows from Proposition 3.3.2. □

The next result, the Luxemburg representation theorem, shows that every r.i. function norm over a resonant measure space can be expressed in terms of an r.i. function norm over $\mathbb{R}^+$.

Theorem 3.3.4. *Let ρ be an r.i. function norm over a resonant measure space (R, μ) and let μ_1 be Lebesgue measure over $\mathbb{R}^+$. Then there is an r.i. function norm $\overline{\rho}$ over $(\mathbb{R}^+, \mu_1)$ such that for all $f \in \mathcal{M}_0^+(R, \mu)$,*

$$\rho(f) = \overline{\rho}(f^*);$$

$\overline{\rho}$ is uniquely determined by ρ if (R, μ) is non-atomic and $\mu(R) = \infty$. Moreover, the associate norm ρ' is given by

$$\rho'(g) = \overline{\rho}'(g^*) \quad \textit{for all } g \in \mathcal{M}_0^+(R, \mu).$$

For the proof of this important result see [15], Theorem 2.4.10 and the following discussion.

We now introduce a function, the *fundamental function* of an r.i. space, which plays a significant part in the theory, especially in connection with interpolation.

Definition 3.3.5. *Let X be an r.i. Banach function space over (R, μ). Given any t $(< \infty)$ in the range of μ, let $E \subset R$ be such that $\mu(E) = t$ and put $\phi_X(t) = \|\chi_E \mid X\|$. The function ϕ_X defined in this way is called the* ***fundamental function*** *of X.*

Note that ϕ_X is well-defined, for if $\mu(F) = \mu(E) = t$, then χ_F and χ_E are equimeasurable and so $\|\chi_F \mid X\| = \|\chi_E \mid X\|$ since X is an r.i. space. Observe also that if (R, μ) is non-atomic, then the range of μ is $[0, \mu(R)]$ and

$$\phi_{L_p}(t) = t^{1/p} \text{ if } 1 \leq p < \infty \;\; (0 \leq t < \mu(R)),$$

while

$$\phi_{L_\infty}(0) = 0, \; \phi_{L_\infty}(t) = 1 \;\; (0 < t < \mu(R)).$$

Some basic properties of the fundamental function are given in the following Theorem.

Theorem 3.3.6. *Let X be an r.i. Banach function space over a resonant measure space (R, μ). Then:*
(i) ϕ_X is increasing;
(ii) $\phi_X(t) = 0$ if, and only if, $t = 0$;
(iii) for each t $(< \infty)$ in the range of μ,

$$\phi_X(t)\phi_{X'}(t) = t,$$

where X' is the associate space of X;
(iv) $\phi_X(t)/t$ is decreasing;
(v) ϕ_X is continuous, except perhaps at $t = 0$.

Proof. (i) Let $0 < s < t < \infty$ and suppose $E, F \subset R$ are such that $\mu(E) = s, \mu(F) = t$. Then $\chi_E^* \leq \chi_F^*$ and so, by Proposition 3.3.2, $\rho(\chi_E) \leq \rho(\chi_F)$; that is, $\phi_X(s) \leq \phi_X(t)$.
(ii) This is clear.
(iii) By (ii) we may concentrate on the case in which $0 < t < \infty$ and t is in the range of μ. Let $E \subset R$ satisfy $\mu(E) = t$. By Corollary 3.3.3 (Hölder's inequality),

$$t = \int_E d\mu \leq \|\chi_E \mid X\| \, \|\chi_E \mid X'\| = \phi_X(t)\phi_{X'}(t).$$

The reverse inequality is established in [15], Theorem 2.5.2.
(iv) Since $\phi_X(t)/t = 1/\phi_{X'}(t)$, by (iii), the result follows from (i).
(v) As there is nothing to prove when (R, μ) is atomic, we assume that it is non-atomic. Then ϕ_X is increasing on $(0, \mu(R))$ and so any point of discontinuity $t > 0$ must be of jump type: this is impossible as it would contradict (iv). □

Note that ϕ_X may be discontinuous at 0, for this is the case, as we have seen, when $X = L_\infty(R, \mu)$ and (R, μ) is non-atomic.

Properties (i), (ii) and (iv) in Theorem 3.3.6 actually characterise those functions ϕ which can be fundamental functions of r.i. spaces. Before showing this we first give a name to functions having these properties.

Definition 3.3.7. *Let $\phi : [0, \infty) \to [0, \infty)$ be increasing on $(0, \infty)$, zero only at 0, and such that $\phi(t)/t$ is decreasing on $(0, \infty)$. Then ϕ is said to be* ***quasiconcave****.*

If $\phi : [0, \infty) \to [0, \infty)$ vanishes only at 0 and is concave and increasing, then ϕ is quasiconcave. To see this, note that if $0 \leq t_1 < t_2 < \infty$, then

$$\phi(t_1) \geq \frac{t_1}{t_2}\phi(t_2) + \frac{t_2 - t_1}{t_2}\phi(0),$$

so that $\phi(t_1)/t_1 \geq \phi(t_2)/t_2$. The function ϕ given by $\phi(t) = \max(1, t)$ if $t > 0$, $\phi(0) = 0$, is quasiconcave but not concave.

We have seen that the fundamental function of any r.i. space over $(\mathbb{R}^+, m)$ is quasiconcave. We now proceed in the opposite direction.

Definition 3.3.8. *Let ϕ be quasiconcave. By M_ϕ we shall mean the (Lorentz) space of all those $f \in \mathcal{M}_0(\mathbb{R}^+, \mu_1)$ such that*

$$\|f \mid M_\phi\| := \sup\{f^{**}(t)\phi(t) : 0 < t < \infty\} < \infty.$$

Proposition 3.3.9. *Given any quasiconcave function ϕ, the corresponding Lorentz space M_ϕ is an r.i. Banach function space when endowed with the norm $\|\cdot \mid M_\phi\|$, and its fundamental function is ϕ.*

Proof. We must first verify the Banach function norm properties (P1)-(P5) of Definition 3.1.1. With the help of Proposition 3.2.15 and Theorem 3.2.16, (P1)-(P3) follow immediately. To check (P4), let $E \subset \mathbb{R}^+$ be such that $\mu_1(E) = t < \infty$. Then $\chi_E^* = \chi_{[0,t)}$ and

$$\begin{aligned}\|\chi_E \mid M_\phi\| &= \sup\nolimits_{0<s<\infty}\{\chi_E^{**}(s)\phi(s)\} = \sup\nolimits_{0<s<\infty}\{\min(1, t/s)\,\phi(s)\}\\ &= \max\left\{\sup\nolimits_{0<s<t}\phi(s), t\sup\nolimits_{t\le s<\infty}\phi(s)/s\right\}\\ &= \phi(t),\end{aligned} \tag{3.3.1}$$

since ϕ increases and $\phi(s)/s$ decreases. Since $\phi(t) < \infty$, (P4) holds. As for (P5), let $f \in M_\phi$ and $E \subset \mathbb{R}^+$, with $\mu_1(E) = t \in (0, \infty)$. Then by Lemma 3.2.9,

$$\begin{aligned}\left|\int_E f(x)dx\right| &\le \int_0^t f^*(s)ds \le \frac{t}{\phi(t)} \sup_{0<s<\infty}\{f^{**}(s)\phi(s)\}\\ &= C(t)\,\|f \mid M_\phi\|.\end{aligned}$$

Since ϕ vanishes only at 0, $C(t) = t/\phi(t) < \infty$: (P5) follows. Thus M_ϕ is a Banach function space with function norm $\|\cdot \mid M_\phi\|$; it is r.i., since its norm is defined in terms of f^*. Its fundamental function is ϕ, by (3.3.1). □

We can now show that M_ϕ is the largest r.i. space with fundamental function ϕ.

Proposition 3.3.10. *Let X be an r.i. space over $(\mathbb{R}^+, m)$. Then X is continuously embedded in M_{ϕ_X}, and for all $f \in X$,*

$$\|f \mid M_{\phi_X}\| \le \|f \mid X\|.$$

Proof. Let $t > 0$ and $f \in X$. By Hölder's inequality and Theorem 3.3.6(iii),

$$\int_0^t f^*(s)ds \le \left\|\chi_{(0,t)} \mid X'\right\| \|f \mid X\| = \|f \mid X\|\,t/\phi(t).$$

Thus $f^{**}(t)\phi(t) \le \|f \mid X\|$ and the result follows. □

Turning to the question as to whether there is a smallest r.i. space with a given fundamental function, we remark that if ϕ is quasiconcave, it is dominated by the concave function $t \mapsto (1+t)\,\phi(1)$, since $\phi(t) \leq \phi(1) \max(1,t)$ if $t > 0$. The pointwise infimum of concave functions is concave, and so there is a smallest concave function which dominates ϕ.

Definition 3.3.11. *Let ϕ be quasiconcave. The smallest concave function ψ such that $\phi \leq \psi$ is called the **least concave majorant** of ϕ.*

Proposition 3.3.12. *Let $\widetilde{\phi}$ be the least concave majorant of a quasiconcave function ϕ. Then*

$$\frac{1}{2}\widetilde{\phi} \leq \phi \leq \widetilde{\phi}.$$

Proof. Let $x > 0$. Then for all $t \geq 0$, $\phi(t) \leq \left(1+\frac{t}{x}\right)\phi(x)$ and so $\widetilde{\phi}(t) \leq \left(1+\frac{t}{x}\right)\phi(x)$ for all $t \geq 0$. The result follows on taking $t = x$. □

Proposition 3.3.13. *Let X be an r.i. space over $(\mathbb{R}^+, \mu_1)$. Then there is an r.i. norm on X, equivalent to the original norm and with concave fundamental function.*

Proof. By Theorem 3.3.6, the fundamental function ϕ of X is quasiconcave; let $\widetilde{\phi}$ be the corresponding least concave majorant and $M_{\widetilde{\phi}}$ the Lorentz space associated with $\widetilde{\phi}$. For each $f \in \mathcal{M}_0^+(\mathbb{R}^+, \mu_1)$ put

$$\nu(f) = \max\left(\|f \mid X\|, \left\|f \mid M_{\widetilde{\phi}}\right\|\right).$$

Since X and $M_{\widetilde{\phi}}$ are r.i. spaces, ν is an r.i. function norm. Using Propositions 3.3.10 and 3.3.12 we see that

$$\|f \mid X\| \leq \nu(f) \leq \max\left(\|f \mid X\|, 2\left\|f \mid M_{\widetilde{\phi}}\right\|\right) \leq 2\,\|f \mid X\|.$$

Hence ν is a norm on X equivalent to $\|\cdot \mid X\|$. Since

$$\nu\left(\chi_{(0,t)}\right) = \max\left(\phi(t), \widetilde{\phi}(t)\right) = \widetilde{\phi}(t),$$

the Proposition follows. □

Definition 3.3.14. *Let X be an r.i. space over $(\mathbb{R}^+, \mu_1)$, and let it be renormed as in Proposition 3.3.13 so that its fundamental function ϕ_X is concave. The Lorentz space $M(X)$ is defined to be M_{ϕ_X} (see Definition 3.3.8), with norm*

$$\|f \mid M(X)\| = \sup_{0<t<\infty} \{f^{**}(t)\,\phi_X(t)\};$$

the Lorentz space $\Lambda(X)$ is

$$\left\{f \in \mathcal{M}_0^+(\mathbb{R}^+, \mu_1) : \|f \mid \Lambda(X)\| = \int_0^\infty f^*(s)\,d\phi_X(s) < \infty\right\},$$

and is equipped with the norm $\|\cdot \mid \Lambda(X)\|$.

It is possible to rewrite the expression for $\|f \mid \Lambda(X)\|$. By [144], Theorem 1.1, p.5,

$$\phi_X(s) = \int_0^s \Phi_X(t)\,dt \quad (s > 0)$$

for some non-negative, decreasing function Φ_X. Thus

$$\|f \mid \Lambda(X)\| = \|f \mid L_\infty\| \, \phi_X(0) + \int_0^\infty f^*(s)\,\Phi_X(s)\,ds.$$

Theorem 3.3.15. *Let X be an r.i. Banach function space over $(\mathbb{R}^+, \mu_1)$ and suppose X has been renormed so that its fundamental function ϕ_X is concave. Then $\Lambda(X)$ and $M(X)$ are r.i. Banach function spaces, each with ϕ_X as fundamental function. Moreover,*

$$\Lambda(X) \hookrightarrow X \hookrightarrow M(X),$$

each embedding having norm 1.

Proof. In view of Propositions 3.3.9 and 3.3.10, we need only deal with $\Lambda(X)$. Let $f, g \in \Lambda(X)$. By Theorem 3.2.16, $(f+g)^{**}(t) \le f^{**}(t) + g^{**}(t)$ for $t > 0$; that is,

$$\int_0^t (f+g)^*(s)ds \le \int_0^t f^*(s)ds + \int_0^t g^*(s)ds, \quad t > 0. \tag{3.3.2}$$

We claim that

$$\int_0^\infty (f+g)^*(s)\Phi_X(s)\,ds \le \int_0^\infty f^*(s)\Phi_X(s)\,ds + \int_0^\infty g^*(s)\Phi_X(s)\,ds. \tag{3.3.3}$$

To see this, note that Φ_X is non-negative and decreasing: the monotone convergence theorem shows that it is enough to prove (3.3.3) when Φ_X is replaced by a non-negative, decreasing step-function, η say,

$$\eta = \sum_{j=1}^n a_j \chi_{(0,t_j)},$$

where $a_j > 0$ and $0 < t_1 < \ldots < t_n$. But

$$\begin{aligned}
\int_0^\infty (f+g)^*(s)\eta(s)\,ds &= \sum_{j=1}^n a_j \int_0^{t_j} (f+g)^*(s)ds \\
&\le \sum_{j=1}^n a_j \int_0^{t_j} (f^* + g^*)\,ds \\
&= \int_0^\infty (f^* + g^*)\,\eta ds,
\end{aligned}$$

and (3.3.3) follows. Hence the triangle inequality holds in $\Lambda(X)$. That (P1)-(P3) hold is an immediate consequence of the corresponding properties of f^*. For (P4), let $E \subset \mathbb{R}^+$, $\mu(E) = t \in (0, \infty)$. Then

$$\|\chi_E \mid \Lambda(X)\| = \int_0^\infty \chi_{(0,t)}(s)\, d\Phi_X(s) = \phi_X(t);$$

hence (P4) holds and the fundamental function of $\Lambda(X)$ is ϕ_X.

Next we prove that for all $f \in \Lambda(X)$,

$$\|f \mid X\| \leq \|f \mid \Lambda(X)\|. \tag{3.3.4}$$

Both norms are r.i. and have the Fatou property (P3): thus it is enough to prove (3.3.4) when $f = f^*$ is a decreasing step-function,

$$f^* = \sum_{j=1}^n b_j \chi_{(0,t_j)},$$

where $b_j > 0$ and $0 < t_1 < ... < t_n$. In that case,

$$\begin{aligned} \|f \mid X\| \leq \textstyle\sum_{j=1}^n b_j \left\|\chi_{(0,t_j)} \mid X\right\| &= \textstyle\sum_{j=1}^n b_j \phi_X(t_j) \\ = \textstyle\int_0^\infty f^*(s)\, d\phi_X(s) &= \|f \mid \Lambda(X)\|. \end{aligned}$$

Property (P5) now follows from the fact that X has this property. □

We summarise the results just obtained as follows:

Corollary 3.3.16. *Let X be an r.i. space over $(\mathbb{R}^+, \mu_1)$ with concave fundamental function ϕ_X. Then the spaces $\Lambda(X), M(X)$ are respectively the smallest and the largest r.i. spaces with fundamental function ϕ_X.*

3.4 Examples

3.4.1 Lorentz, Lorentz-Zygmund and generalised Lorentz-Zygmund spaces

We saw in the last section that the Lebesgue spaces $L_p(R, \mu)$ are r.i. Banach function spaces. Useful though this scale of spaces is, there are times when a scale allowing more refined tuning is desirable, and the spaces which we now introduce have a significant part to play in this connection. We begin with Lorentz spaces.

Definition 3.4.1. *Let $p, q \in (0, \infty]$. The Lorentz space $L_{p,q} = L_{p,q}(R) = L_{p,q}(R, \mu)$ is the space of all $f \in \mathcal{M}_0(R, \mu)$ such that*

$$\|f \mid L_{p,q}\| := \begin{cases} \left(\int_0^\infty \left\{t^{1/p} f^*(t)\right\}^q \frac{dt}{t}\right)^{1/q} & \text{if } 0 < q < \infty, \\ \sup_{0<t<\infty} \left\{t^{1/p} f^*(t)\right\} & \text{if } q = \infty. \end{cases}$$

is finite.

We see from Proposition 3.2.5 that $L_{p,p}(R,\mu) = L_p(R,\mu)$ and that $\|f \mid L_{p,p}\| = \|f \mid L_p\|$ for all $f \in L_p(R,\mu)$. The space $L_{p,\infty}$ is often known as weak-L_p. It can be characterised in an alternative, and sometimes more convenient, manner when $p < \infty$.

Proposition 3.4.2. *Let $0 < p < \infty$. Then $f \in L_{p,\infty}(R,\mu)$ if, and only if,*

$$\sup_{\lambda>0} \lambda^p \mu\left(\{x \in R : |f(x)| > \lambda\}\right) < \infty.$$

Proof. Suppose that $f \in L_{p,\infty}(R,\mu)$. Then there exists $c > 0$ such that for all $t > 0$, $f^*(t) \le ct^{-1/p}$. Hence the distribution function μ_f of f satisfies

$$\inf\{\lambda \ge 0 : \mu_f(\lambda) \le t\} \le ct^{-1/p},$$

and as μ_f is right-continuous,

$$\mu_f\left(ct^{-1/p}\right) \le t, \ t > 0.$$

From this it is clear that $\mu_f(\lambda) \le (c/\lambda)^p \ (\lambda > 0)$, and the result follows.

Conversely, if there exists $C > 0$ such that for all $\lambda > 0$,

$$\mu_f(\lambda) \le C\lambda^{-p},$$

then if $C\lambda^{-p} \le t$, that is, $\lambda \ge (C/t)^{1/p}$, we have $\mu_f(\lambda) \le t$. Thus $f^*(t) \le (C/t)^{1/p}$ and $t^{1/p} f^*(t) \le C^{1/p}$ for all $\lambda > 0$. □

The dependence of the Lorentz spaces on the second index q is illustrated by the next result.

Proposition 3.4.3. *Let $0 < p \le \infty$ and $0 < q \le r \le \infty$. Then $L_{p,q}(R,\mu) \hookrightarrow L_{p,r}(R,\mu)$.*

Proof. If $p = \infty$, $L_{p,q} = \{0\}$ if $q < \infty$; hence we need only consider the case $p < \infty$ and $q < r$. Since f^* is decreasing, we have for all $t > 0$,

$$\begin{aligned} t^{1/p} f^*(t) &= \left\{\frac{q}{p}\int_0^t \left(s^{1/p} f^*(t)\right)^q \frac{ds}{s}\right\}^{1/q} \\ &\le \left\{\frac{q}{p}\int_0^t \left(s^{1/p} f^*(s)\right)^q \frac{ds}{s}\right\}^{1/q} \end{aligned}$$

$$\leq (q/p)^{1/q} \|f \mid L_{p,q}\| .$$

Hence

$$\|f \mid L_{p,\infty}\| \leq (q/p)^{1/q} \|f \mid L_{p,q}\| . \tag{3.4.1}$$

When $r < \infty$,

$$\begin{aligned}\|f \mid L_{p,r}\| &= \left\{ \int_0^\infty \left(t^{1/p} f^*(t) \right)^{r-q+q} \frac{dt}{t} \right\}^{1/r} \\ &\leq \|f \mid L_{p,\infty}\|^{1-q/r} \|f \mid L_{p,q}\|^{q/r} .\end{aligned}$$

Together with (3.4.1) this shows that

$$\|f \mid L_{p,r}\| \leq (q/p)^{(r-q)/(rq)} \|f \mid L_{p,q}\| .$$

□

Note that the Proposition asserts that for all $f \in L_{p,q}(R, \mu)$,

$$\|f \mid L_{p,r}\| \leq c \|f \mid L_{p,q}\|$$

and that the constant c may be taken to be $(q/p)^{(r-q)/(rq)}$, with the natural interpretation when $r = \infty$. That the target spaces in these embeddings may be genuinely bigger than the domain spaces is illustrated by the function f defined on $(0, 1)$ by $f(x) = 1/x$: when $R = (0, 1)$, $\mu = \mu_1$ and $0 < p \leq 1$, $f \in L_{p,\infty}(R, \mu) \backslash L_p(R, \mu)$.

As for the way in which the spaces depend upon the first index p, we have the following:

Proposition 3.4.4. *Let $0 < p < r \leq \infty$ and suppose that $q, s \in (0, \infty]$; assume also that $\mu(R) < \infty$. Then*

$$L_{r,s}(R, \mu) \hookrightarrow L_{p,q}(R, \mu);$$

that is, there is a constant c such that for all $f \in L_{r,s}(R, \mu)$,

$$\|f \mid L_{p,q}\| \leq c \|f \mid L_{r,s}\| .$$

The constant c may be taken as follows:

$$c = \begin{cases} (p/q)^{1/q} \mu(R)^{1/p} & \text{if } r = s = \infty, \\ \mu(R)^{1/p-1/r} \left[\frac{rp(s-q)}{sq(r-p)} \right]^{1/q-1/s} & \text{if } q < s, r < \infty, \\ (r/s)^{1/s-1/q} \max\left(1, \mu(R)^{1/p-1/r}\right) & \text{if } q \geq s, r < \infty. \end{cases}$$

When q or s is ∞, these expressions have to be interpreted in the natural way.

Proof. When $r = \infty$ we must have $s = \infty$ to avoid triviality of $L_{\infty,s}(R,\mu)$. Then

$$\|f \mid L_{p,q}\| \leq \|f \mid L_{\infty,\infty}\| \left(\int_0^{\mu(R)} t^{q/p-1} dt \right)^{1/q}$$
$$= \|f \mid L_{\infty,\infty}\| (p/q)^{1/q} \mu(R)^{1/p}$$

when $q < \infty$, with obvious modification when $q = \infty$.

Now suppose that $r < \infty$. If $s \leq q < \infty$, then

$$\|f \mid L_{p,q}\| \leq \max\left(1, \mu(R)^{1/p-1/r}\right) \left(\int_0^{\mu(R)} \left(t^{1/r} f^*(t)\right)^q \frac{dt}{t} \right)^{1/q}$$
$$\leq K (s/r)^{1/s-1/q} \|f \mid L_{r,s}\|,$$

where

$$K = \max\left(1, \mu(R)^{1/p-1/r}\right).$$

If $s = q = \infty$,

$$\|f \mid L_{p,\infty}\| \leq \sup_{t>0} K t^{1/r} f^*(t) = K \|f \mid L_{r,\infty}\|;$$

while if $s < q = \infty$,

$$\|f \mid L_{p,\infty}\| \leq K \|f \mid L_{r,\infty}\| \leq K (s/r)^{1/s} \|f \mid L_{r,s}\|,$$

by Proposition 3.4.3.

If $q < s < \infty$,

$$\|f \mid L_{p,q}\| = \left(\int_0^{\mu(R)} \left(t^{1/r} f^*(t)\right)^q t^{q(1/p-1/r)} \frac{dt}{t} \right)^{1/q}$$
$$\leq \left(\int_0^{\mu(R)} \left(t^{1/r} f^*(t)\right)^s \frac{dt}{t} \right)^{1/s} \left(\int_0^{\mu(R)} t^{q(1/p-1/r)(s/q)'} \frac{dt}{t} \right)^{\frac{1}{q(s/q)'}},$$

which gives the desired result. Finally, if $q < s = \infty$, the same splitting as in the last case gives

$$\|f \mid L_{p,q}\| \leq \|f \mid L_{r,\infty}\| \left(\int_0^{\mu(R)} t^{q(1/p-1/r)} \frac{dt}{t} \right)^{1/q}$$
$$= \|f \mid L_{r,\infty}\| \mu(R)^{1/p-1/r} \left(\frac{rp}{q(r-p)} \right)^{1/q}.$$

□

A difficulty with $\|\cdot \mid L_{p,q}\|$ is that it is not always a norm, even when $p, q > 1$. To overcome this, we simply replace f^* by f^{**} : the resulting quantity is then a norm for all $q \in [1, \infty]$ and all $p \in (1, \infty]$.

Definition 3.4.5. *Let $p, q \in (0, \infty]$ and for all $f \in \mathcal{M}_0(R, \mu)$ write*

$$\left\| f \mid L_{(p,q)} \right\| = \begin{cases} \left(\int_0^\infty \left\{ t^{1/p} f^{**}(t) \right\}^q \frac{dt}{t} \right)^{1/q} & \text{if } 0 < q < \infty, \\ \sup_{0<t<\infty} \left\{ t^{1/p} f^{**}(t) \right\} & \text{if } q = \infty. \end{cases}$$

Let $L_{(p,q)}(R, \mu)$ be the family of all f such that $\left\| f \mid L_{(p,q)} \right\| < \infty$.

A connection between $\|\cdot \mid L_{p,q}\|$ and $\left\|\cdot \mid L_{(p,q)}\right\|$ is given by the following result.

Lemma 3.4.6. *Let $p \in (1, \infty]$ and $q \in (0, \infty]$. Then for all $f \in \mathcal{M}_0(R, \mu)$,*

$$\|f \mid L_{p,q}\| \le \left\| f \mid L_{(p,q)} \right\| \le p' \|f \mid L_{p,q}\| .$$

Proof. Since $f^* \le f^{**}$ the first inequality is obvious. The second is immediate from Hardy's inequality (see [191], Theorems 5.9 and 5.10). □

Note that since $\|\cdot \mid L_{p,q}\|$ and $\left\|\cdot \mid L_{(p,q)}\right\|$ are equivalent, we may characterise $L_{p,q}$ as the set of all $f \in \mathcal{M}_0(R, \mu)$ for which $\left\| f \mid L_{(p,q)} \right\| < \infty$, provided that $p \in (1, \infty]$ and $q \in (0, \infty]$; without these restrictions on p and q all that can be claimed is that $L_{(p,q)} \hookrightarrow L_{p,q}$. In fact, we can say more about $L_{p,q}$, endowed with $\left\|\cdot \mid L_{(p,q)}\right\|$.

Theorem 3.4.7. *Suppose that $p \in (1, \infty)$ and $q \in [1, \infty]$, or that $p = q = \infty$. Then $\left(L_{p,q}, \left\|\cdot \mid L_{(p,q)}\right\|\right)$ is a rearrangement-invariant Banach function space.*

Proof. Since $(f+g)^{**} \le f^{**} + g^{**}$ (see Theorem 3.2.16), the triangle inequality for $\left\|\cdot \mid L_{(p,q)}\right\|$ follows from Minkowski's inequality, and consequently $\left\|\cdot \mid L_{(p,q)}\right\|$ is a norm on $L_{p,q}$. The remaining properties of an r.i. Banach function spaces are easy to check. □

We observe that if $1 < p < \infty$ and (R, μ) is strongly resonant, then a norm on $L_{p,\infty}(R)$ equivalent to $\| \cdot |L_{p,\infty}\|$ is given by

$$\sup \left\{ \mu(E)^{-1/p'} \int_E |f| d\mu \right\}, \quad f \in L_{p,\infty}(R),$$

where the supremum is taken over all $E \subset R$ with $\mu(E) < \infty$. For

$$\|f|L_{p,\infty}\| \approx \|f|L_{(p,\infty)}\| = \sup_{t>0} t^{1/p} f^{**}(t),$$

and the claim now follows from Proposition 3.2.17.

Note that if $1 < p < \infty$, then since the function ϕ defined by $\phi(t) = t^{1/p}$ is the fundamental function of L_p, then $L_{p,1}$ coincides with the Lorentz space Λ_ϕ (see Definition 3.3.14) and

$$\|f \mid L_{p,1}\| = \int_0^\infty t^{1/p} f^*(t) \frac{dt}{t} = p \|f \mid \Lambda_\phi\|;$$

moreover, $L_{p,\infty}$ coincides with the Lorentz space M_ϕ and

$$\left\|f \mid L_{(p,\infty)}\right\| = \sup_{t>0} t^{1/p} f^{**}(t) = \|f \mid M_\phi\|$$

(see Definition 3.3.14). This means that $L_{p,1}$ and $L_{p,\infty}$ are the smallest and the largest respectively of all r.i. spaces having the same fundamental function as L_p.

We also observe that if $1 < p < \infty, 1 \leq q \leq \infty$ and (R, μ) is a resonant measure space, then the associate space of $L_{p,q}(R, \mu)$ is, up to equivalence of norms, $L_{p',q'}(R, \mu)$. This will not be proved here as it is a special case of a result to be given in the next section.

To conclude these remarks on Lorentz spaces we observe that when $R = \Omega$, a measurable subset of $\mathbb{R}^n$, and μ is Lebesgue measure μ_n on $\mathbb{R}^n$, other equivalent quasi-norms on $L_{(p,q)}(\Omega)$ are known. Writing $|\Omega|$ instead of $\mu_n(\Omega)$ for shortness, one is given by

$$\left\| t^{1/r-1/q} \sup_{\tau \in (t, |\Omega|)} \tau^{1/p-1/r} f^{**}(\tau) \mid L_q(0, |\Omega|) \right\|$$

when $0 < q \leq \infty$ and either $0 < r \leq p \leq \infty$ or $0 < p < r < \infty$; another is

$$\left\| t^{1/r-1/q} \sup_{\tau \in (0,t)} \tau^{1/p-1/r} f^{**}(\tau) \mid L_q(0, |\Omega|) \right\|$$

when $p, q \in (0, \infty]$ and $-\infty < r < 0$; see [71]. Further results in this direction are contained in [189] and [190].

To introduce refinements of Lorentz spaces, the Lorentz-Zygmund and generalised Lorentz-Zygmund spaces, we need some efficient notation. Let $m \in \mathbb{N}$ and $\alpha \in \mathbb{R}^m$; define positive functions $l_0, ..., l_m$ on $(0, \infty)$ by

$$l_0(t) = \max\{t, 1/t\}, \; l_i(t) = 1 + \log l_{i-1}(t) \text{ for } i \in \{1, ..., m\}; \tag{3.4.2}$$

and for each $t \in (0, \infty)$ put

$$\vartheta_\alpha^m(t) = \prod_{i=1}^m l_i^{\alpha_i}(t). \tag{3.4.3}$$

Definition 3.4.8. *Let $p, q \in (0, \infty]$, $m \in \mathbb{N}$ and $\alpha \in \mathbb{R}^m$. The* **generalised Lorentz-Zygmund (GLZ)** *space $L_{p,q;\alpha}(R, \mu)$ is the set of all (equivalence classes of) functions $f \in \mathcal{M}_0(R, \mu)$ such that*

$$\|f \mid L_{p,q;\alpha}\| := \left\| t^{1/p-1/q} \vartheta_{\alpha}^{m}(t) f^{*}(t) \mid L_q(0,\infty) \right\|$$

is finite.

When convenient, we write

$$L_{p,q;\alpha_1,\ldots,\alpha_m}(R) \text{ or } L_{p,q}(\log L)_{\alpha_1} \ldots (\log\log \ldots \log L)_{\alpha_m}(R)$$

instead of $L_{p,q;\alpha}(R)$. When each $\alpha_j = 0$, the space $L_{p,q;\alpha_1,\ldots,\alpha_m}(R)$ coincides with the Lorentz space $L_{p,q}(R)$ introduced in Definition 3.4.1, and if in addition $p = q$, the space becomes $L_p(R)$. If $m = 1$, then $L_{p,q;\alpha_1}(R)$ is the Lorentz-Zygmund space introduced in [13] and which, when $p = q$, is the Zygmund space $L_p(\log L)_{\alpha_1}(R)$. The theory of GLZ spaces may be developed along lines similar to those given above for Lorentz spaces, but we shall not do this here as we shall discuss the more general Lorentz-Karamata spaces below.

3.4.2 Orlicz spaces

These are not only of intrinsic importance but also play a very significant part in the theory of embeddings of spaces of Sobolev type.

Definition 3.4.9. *A function $\Phi : [0,\infty) \to \mathbb{R}$ which is continuous, non-negative, strictly increasing and convex, will be called a* ***Young function*** *if*

$$\lim_{t\to 0+} \Phi(t)/t = \lim_{t\to\infty} t/\Phi(t) = 0.$$

It can be shown (see [144] and [202]) that

$$\Phi(t) = \int_0^t \phi(s)ds$$

for some non-decreasing, right-continuous function ϕ. The functions Φ_k ($k = 1,2,3,4,5$) given by $\Phi_1(t) = t^p$ $(1 < p < \infty)$, $\Phi_2(t) = t^p \log^q(2+t)$ $(p,q \in (1,\infty))$, $\Phi_3(t) = t\log^+ t$, $\Phi_4(t) = \exp(t^p) - 1$ $(1 < p < \infty)$, $\Phi_5(t) = \exp(\exp(t^a)) - e$ $(1 < a < \infty)$ are Young functions.

Function spaces may now be introduced in a natural way.

Definition 3.4.10. *Let Ω be a measurable subset of $\mathbb{R}^n$ and let Φ be a Young function. The* ***Orlicz class*** *$\widetilde{L}_{\Phi}(\Omega)$ is defined to be the set of all measurable functions f on Ω (identifying functions that are equal a.e.) such that*

$$\int_{\Omega} \Phi(|f(x)|)\,dx < \infty.$$

In general, $\widetilde{L}_{\Phi}(\Omega)$ is not a linear space. For example, if $n = 1, \Omega = (0,1)$ and $\Phi(t) = e^t - 1 - t$, then the function f given by $f(x) = -\frac{1}{2}\log x$ belongs to $\widetilde{L}_{\Phi}(\Omega)$, but $2f$ does not. However, there is the following

Proposition 3.4.11. *Let Φ be a Young function and let Ω be a measurable subset of $\mathbb{R}^n$. Then $\widetilde{L}_\Phi(\Omega)$ is convex and contains* 0.

Proof. Let $u, v \in \widetilde{L}_\Phi(\Omega)$ and let $\lambda \in (0,1)$. Since Φ is convex,

$$\Phi\left(\lambda\left|u(x)\right| + (1-\lambda)\left|v(x)\right|\right) \leq \lambda\Phi\left(\left|u(x)\right|\right) + (1-\lambda)\Phi\left(\left|v(x)\right|\right).$$

Hence by the monotonicity of Φ,

$$\int_\Omega \Phi\left(\left|\lambda u(x) + (1-\lambda)v(x)\right|\right)dx \leq \int_\Omega \Phi\left(\lambda\left|u(x)\right| + (1-\lambda)\left|v(x)\right|\right)dx < \infty.$$

Since $\Phi(0) = 0$ it is plain that $0 \in \widetilde{L}_\Phi(\Omega)$. □

It turns out that $\widetilde{L}_\Phi(\Omega)$ is a linear space if the Young function Φ does not grow too quickly. To handle this we introduce the following

Definition 3.4.12. *A Young function Φ is said to satisfy the Δ_2-condition if there are positive constants t_0 and c such that for all $t \geq t_0$,*

$$\Phi(2t) \leq c\Phi(t). \tag{3.4.4}$$

Evidently Young functions with polynomial growth satisfy the Δ_2-condition, while those with exponential growth do not. The importance of the Δ_2-condition is explained by the next result.

Theorem 3.4.13. *Let Φ be a Young function and let Ω be a measurable subset of $\mathbb{R}^n$. Then if $\mu_n(\Omega) < \infty$, $\widetilde{L}_\Phi(\Omega)$ is a linear space if, and only if, Φ satisfies the Δ_2-condition. If $\mu_n(\Omega) = \infty$ and Φ satisfies the Δ_2-condition with $t_0 =$ 0, then $\widetilde{L}_\Phi(\Omega)$ is a linear space.*

Proof. See [148], p.141. □

In view of Proposition 3.4.11, to generate a normed linear space from $\widetilde{L}_\Phi(\Omega)$ we think of the corresponding Minkowski functional. In detail, we proceed as follows.

Definition 3.4.14. *Let Φ be a Young function and let Ω be a measurable subset of $\mathbb{R}^n$. The* ***Luxemburg norm*** *ρ_Φ is defined by*

$$\rho_\Phi\left(\left|f\right|\right) = \inf\left\{\lambda > 0 : \int_\Omega \Phi\left(\left|f(x)\right|/\lambda\right)dx \leq 1\right\}, f \in \mathcal{M}_0\left(\mathbb{R}^n, \mu_n\right). \tag{3.4.5}$$

The infimum in (3.4.5) is actually a minimum, if it is positive: this follows immediately from the monotone convergence theorem and the continuity of Φ. To establish the properties of ρ_Φ a preliminary lemma is useful.

Lemma 3.4.15. *Let Φ be a Young function and let Ω be a measurable subset of $\mathbb{R}^n$. Then:*
(i) $f = 0$ a.e. on Ω if, and only if,

$$\int_\Omega \Phi(\mu|f(x)|)\,dx \leq 1 \text{ for all } \mu > 0;$$

(ii) if $\rho_\Phi(|f|) \leq 1$ (resp. > 1) then

$$\int_\Omega \Phi(|f(x)|)\,dx \leq \ (\text{resp. } \geq)\ \rho_\Phi(|f|);$$

thus

$$\rho_\Phi(|f|) \leq 1 \text{ if, and only if, } \int_\Omega \Phi(|f(x)|)\,dx \leq 1.$$

Proof. (i) Suppose that $\int_\Omega \Phi(\mu|f(x)|)\,dx \leq 1$ for all $\mu > 0$ but there exist $\varepsilon > 0$ and a set $E \subset \mathbb{R}^n$ with $\mu_n(E) > 0$ and $|f| \geq \varepsilon$ on E. Then

$$\int_\Omega \Phi(\mu|f(x)|)\,dx \geq \int_E \Phi(\mu\varepsilon)\,dx = \Phi(\mu\varepsilon)\,\mu_n(E)$$
$$\to \infty \text{ as } \mu \to \infty,$$

and we have a contradiction. The converse is obvious.

(ii) We may plainly suppose that $f \geq 0$. Let $\rho_\Phi(|f|) \leq 1$. If $\rho_\Phi(|f|) = 0$, then by (3.4.5) and (i), $f = 0$ a.e.; thus $\int_\Omega \Phi(f(x))\,dx = 0$. If $\rho_\Phi(|f|) > 0$, then

$$\int_\Omega \Phi(\mu_0 f(x))\,dx \leq 1,$$

where $\mu_0 = 1/\rho_\Phi(|f|) \geq 1$. Hence by the convexity of Φ,

$$\int_\Omega \Phi(f(x))\,dx = \int_\Omega \Phi\left(\mu_0^{-1}\mu_0 f(x)\right)dx \leq \mu_0^{-1}\int_\Omega \Phi(\mu_0 f(x))\,dx$$
$$\leq \rho_\Phi(f),$$

as required.

If $\rho_\Phi(f) > 1$, then $\int_\Omega \Phi(f(x)/\gamma)\,dx > 1$ whenever $1 < \gamma < \rho_\Phi(f)$. As Φ is convex,

$$1 < \int_\Omega \Phi(f(x)/\gamma)\,dx \leq \gamma^{-1}\int_\Omega \Phi(f(x))\,dx,$$

and so

$$\gamma \leq \int_\Omega \Phi(f(x))\,dx.$$

Since γ may be chosen arbitrarily close to $\rho_\Phi(f)$,

$$\rho_\Phi(f) \leq \int_\Omega \Phi(f(x))\,dx.$$

The rest is clear. □

Theorem 3.4.16. *Let Φ be a Young function and let Ω be a measurable subset of $\mathbb{R}^n$. The Luxemburg norm ρ_Φ defined by (3.4.5) is an r.i. Banach function norm.*

Proof. First we check properties (P1)-(P5) of Definition 3.1.1. We see immediately from Lemma 3.4.15 (i) that $\rho_\Phi(f) = 0$ if, and only if, $f = 0$ a.e. It is also quite clear that $\rho_\Phi(\lambda f) = \lambda\rho_\Phi(f)$ if $\lambda \geq 0$. As for the triangle inequality, suppose that $f, g \in \mathcal{M}_0^+(\Omega, \mu_n) \setminus \{0\}$ and that $\rho_\Phi(f) + \rho_\Phi(g) = \gamma < \infty$. Then $\alpha := \rho_\Phi(f)/\gamma > 0$, $\beta := \rho_\Phi(g)/\gamma > 0$ and $\alpha + \beta = 1$. By the definition of ρ_Φ,

$$\int_\Omega \Phi(f(x)/\rho_\Phi(f))\,dx, \int_\Omega \Phi(g(x)/\rho_\Phi(g))\,dx \leq 1.$$

Hence

$$\begin{aligned}\int_\Omega \Phi(\{f(x) + g(x)\}/\gamma)\,dx &= \int_\Omega \Phi\left(\frac{\alpha f(x)}{\rho_\Phi(f)} + \frac{\beta g(x)}{\rho_\Phi(g)}\right) dx \\ &\leq \alpha\int_\Omega \Phi\left(\frac{f(x)}{\rho_\Phi(f)}\right) dx + \beta\int_\Omega \Phi\left(\frac{g(x)}{\rho_\Phi(g)}\right) dx \\ &\leq \alpha + \beta = 1.\end{aligned}$$

Thus $\rho_\Phi(f+g) \leq \gamma = \rho_\Phi(f) + \rho_\Phi(g)$. Hence (P1) holds. For (P2), suppose that $f, g \in \mathcal{M}_0^+(\Omega, \mu_n)$, $0 \leq g \leq f$ a.e. and $0 < \rho_\Phi(f) < \infty$. Then

$$\int_\Omega \Phi(g(x)/\rho_\Phi(f))\,dx \leq \int_\Omega \Phi(f(x)/\rho_\Phi(f))\,dx \leq 1,$$

and so $\rho_\Phi(g) \leq \rho_\Phi(f)$.

To deal with (P3), let $0 \leq f_m \uparrow f$ μ_n–a.e. By (P2), the sequence $(\rho_\Phi(f_m))$ is increasing. Put $\alpha_m = \rho_\Phi(f_m)$, $\alpha = \sup_m \alpha_m$. Then $\rho_\Phi(f) \geq \alpha_m$ for all m, and so $\rho_\Phi(f) \geq \alpha$. Equality plainly holds here if $\alpha = 0$ or $\alpha = \infty$; suppose that $0 < \alpha_m < \alpha < \infty$ for all $m \in \mathbb{N}$. Then

$$\int_\Omega \Phi(f_m(x)/\alpha)\,dx \leq \int_\Omega \Phi(f_m(x)/\alpha_m)\,dx \leq 1,$$

from which we see, on using the monotone convergence theorem, that

$$\int_\Omega \Phi(f_m(x)/\alpha)\,dx \longrightarrow \int_\Omega \Phi(f(x)/\alpha)\,dx.$$

Hence $\int_\Omega \Phi(f(x)/\alpha)\,dx \leq 1$ and so $\rho_\Phi(f) \leq \alpha$. Thus $\rho_\Phi(f) = \alpha$ and (P3) follows.

Next, suppose that $E \subset \Omega$ and that $\mu_n(E) \in (0, \infty)$. Since $\Phi(0) = 0$, there exists $t > 0$ such that $\Phi(t) \leq 1/\mu_n(E)$. Thus

$$\int_{\Omega} \Phi\left(t\chi_E(x)\right) dx = \Phi(t)\mu_n(E) \le 1,$$

and so $\rho_\Phi(\chi_E) \le 1/\mu_n(E) < \infty$: (P4) holds.

To deal with (P5), let $E \subset \Omega$ be such that $\mu_n(E) \in (0,\infty)$. Let $f \in \mathcal{M}^+(\Omega,\mu_n)$ satisfy $0 < \rho_\Phi(f) < \infty$ and put $\lambda = 1/\rho_\Phi(f)$. Then by Jensen's inequality (Theorem 1.4.2),

$$\Phi\left(\frac{1}{\mu_n(E)}\int_E \lambda f(x)\,dx\right) \le \frac{1}{\mu_n(E)}\int_E \Phi(\lambda f(x))\,dx$$

$$\le \frac{1}{\mu_n(E)}\int_\Omega \Phi(\lambda f(x))\,dx \le \frac{1}{\mu_n(E)}.$$

Since Φ increases to infinity, there exists $c = c(\Phi,\mu_n(E))$ with

$$\frac{1}{\mu_n(E)}\int_E \lambda f(x)\,dx \le c;$$

thus $\int_E f(x)\,dx \le \mu_n(E)c/\lambda = \mu_n(E)c\rho_\Phi(f)$. This gives (P5). Hence ρ_Φ is a Banach function norm.

It remains to prove that ρ_Φ is rearrangement-invariant. To do this, it is enough to show that $\int_\Omega \Phi(f(x))\,dx = \int_\Omega \Phi(g(x))\,dx$ whenever f and g are equimeasurable. This can be reduced to the checking of the property when $g = f^*$: discussion of the case in which f is a simple function and use of the monotone convergence theorem give the result. □

Definition 3.4.17. *Let Φ be a Young function and let Ω be a measurable subset of $\mathbb{R}^n$. The* ***Orlicz space*** *$L_\Phi(\Omega)$ is the rearrangement-invariant Banach function space generated by ρ_Φ. Thus $L_\Phi(\Omega)$ is the space of all those measurable functions f on Ω such that $\rho_\Phi(|f|) < \infty$. The Luxemburg norm on $L_\Phi(\Omega)$ is given by*

$$\|f \mid L_\Phi(\Omega)\| = \rho_\Phi(|f|), \; f \in L_\Phi(\Omega).$$

Remark 3.4.18. Suppose that $1 < p < \infty$ and that $\Phi(t) = t^p$ for $t \ge 0$. Then it is easy to check that $L_\Phi(\Omega) = L_p(\Omega)$ and that $\rho_\Phi(|f|) = \left(\int_\Omega |f|^p\,dx\right)^{1/p}$ for all $f \in L_p(\Omega)$.

With associate and dual spaces in mind, we next introduce the complementary function of a Young function. This plays the same role in the theory of Orlicz spaces as the function $t \longmapsto t^{p'}$ does in L_p theory.

Definition 3.4.19. *Let Φ be a Young function. Its* ***complementary function*** *Ψ is defined by*

$$\Psi(t) = \sup\{st - \Phi(s) : s > 0\}, \; t \ge 0.$$

It is plain that Ψ is non-negative, continuous, strictly increasing and convex; it is also easy to see that

$$\Psi(t)/t = \sup\left\{t - t^{-1}\Phi(s) : s > 0\right\}$$

tends to 0 as $t \to 0$, and to ∞ as $t \to \infty$. Hence Ψ is a Young function, and we may refer to Φ and Ψ as complementary Young functions.

When $\Phi(t) = t^p/p$ with $1 < p < \infty$, it is easy to verify that its complementary function is given by $\Psi(t) = t^{p'}/p'$. The complementary function is usually difficult or impossible to determine precisely, but often inequalities can be obtained which serve just as well as exact functions.

The following inequality will be most useful.

Theorem 3.4.20. *(Young's inequality) Let Φ and Ψ be complementary Young functions. Then for all $s, t \geq 0$,*

$$st \leq \Phi(s) + \Psi(t),$$

with equality if, and only if, either $t = \phi(s)$ or $s = \psi(t)$, where (as indicated immediately after Definition 3.4.9) Φ and Ψ are the indefinite integrals of ϕ and ψ respectively.

For a proof of this we refer to [15], p.271.

The result is an extension of the familiar statement (also known as Young's inequality) that for all $s, t \geq 0$,

$$st \leq s^p/p + t^{p'}/p',$$

if $1 < p < \infty$. As an immediate consequence of Theorem 3.4.20 we have an analogue of the well-known result, arising from Hölder's inequality, that if $u \in L_p(\Omega)$ and $v \in L_{p'}(\Omega)$, then $uv \in L_1(\Omega)$.

Corollary 3.4.21. *Let Φ and Ψ be complementary Young functions, let Ω be a measurable subset of $\mathbb{R}^n$ and suppose that $u \in \widetilde{L}_\Phi(\Omega)$, $v \in \widetilde{L}_\Psi(\Omega)$. Then $uv \in L_1(\Omega)$ and*

$$\int_\Omega |u(x)v(x)|\, dx \leq \int_\Omega \Phi(|u(x)|)\, dx + \int_\Omega \Psi(|v(x)|)\, dx.$$

Now let Ψ be a Young function and let Ω be a measurable subset of $\mathbb{R}^n$. Given any $f \in (L_\Psi(\Omega))'$, by Definition 3.1.5,

$$\begin{aligned} \|f \mid (L_\Psi(\Omega))'\| &= \sup\left\{\int_\Omega |f(x)g(x)|\, dx : \int_\Omega \Psi(|g(x)|)\, dx \leq 1\right\} \\ &= \sup\left\{\int_\Omega |f(x)g(x)|\, dx : \|g \mid L_\Psi(\Omega)\| \leq 1\right\}, \end{aligned} \tag{3.4.6}$$

the last equality following from Lemma 3.4.15. This leads us to the next definition.

Definition 3.4.22. *Let Φ and Ψ be complementary Young functions and let Ω be a measurable subset of $\mathbb{R}^n$. Let $L^{\Phi}(\Omega)$ be the set of all measurable functions f on Ω such that*

$$\sup\left\{\int_{\Omega}|f(x)g(x)|\,dx : \int_{\Omega}\Psi(|g(x)|)\,dx \leq 1\right\} < \infty.$$

The ***Orlicz norm*** *on $L^{\Phi}(\Omega)$ is defined by*

$$\left\|f \mid L^{\Phi}(\Omega)\right\| = \left\|f \mid (L_{\Psi}(\Omega))'\right\|,$$

with (3.4.6) in mind.

Theorem 3.4.23. *Let Φ be a Young function and let Ω be a measurable subset of $\mathbb{R}^n$. Then $L_{\Phi}(\Omega) = L^{\Phi}(\Omega)$ and for all $f \in L_{\Phi}(\Omega)$,*

$$\|f \mid L_{\Phi}(\Omega)\| \leq \left\|f \mid L^{\Phi}(\Omega)\right\| \leq 2\,\|f \mid L_{\Phi}(\Omega)\|. \tag{3.4.7}$$

Proof. Let Ψ be the Young function complementary to Φ. Let $L_{\Phi}(\Omega)$, $f \neq 0$, and let $k = 1/\,\|f \mid L_{\Phi}(\Omega)\|$, so that

$$\int_{\Omega}\Phi(k\,|f(x)|)\,dx \leq 1.$$

Let g be such that $\int_{\Omega}\Psi(|g(x)|)\,dx \leq 1$. By Corollary 3.4.21,

$$\int_{\Omega}|kf(x)g(x)|\,dx \leq 2.$$

Hence

$$\int_{\Omega}|f(x)g(x)|\,dx \leq 2\,\|f \mid L_{\Phi}(\Omega)\|$$

and so

$$\left\|f \mid L^{\Phi}(\Omega)\right\| \leq 2\,\|f \mid L_{\Phi}(\Omega)\|.$$

To obtain the remaining inequality in (3.4.7) it is enough to prove it for non-negative simple functions, as once this is done, use of the Fatou property (possessed by the Banach function spaces $L^{\Phi}(\Omega)$ and $L_{\Phi}(\Omega)$) gives the general result. Let f be a non-negative simple function with $\left\|f \mid L^{\Phi}(\Omega)\right\| > 0$. All we have to do is to show that

$$\int_{\Omega}\Phi(\lambda\,|f(x)|)\,dx \leq 1,$$

where $\lambda = 1/\left\|f \mid L^{\Phi}(\Omega)\right\|$. Since f is simple, $\int_{\Omega}\Phi(\lambda\,|f(x)|)\,dx < \infty$. Now represent Φ in integral form: $\Phi(t) = \int_0^t \phi(s)ds$, say. The function g given by $g(x) = \phi(\lambda f(x))$ is simple. By Theorem 3.4.20,

$$\Psi(g(x)) + \Phi(\lambda f(x)) = \lambda f(x)g(x)$$

for all $x \in \Omega$. Hence

$$\int_{\Omega} \Phi(\lambda f(x))\, dx + \int_{\Omega} \Psi(g(x))\, dx = \int_{\Omega} \lambda f(x) g(x) dx, \tag{3.4.8}$$

and so

$$\int_{\Omega} \Psi(g(x))\, dx < \infty.$$

Since $L_{\Psi}(\Omega)$ is the associate space of $L^{\Phi}(\Omega)$, Hölder's inequality and Lemma 3.4.15 give

$$\begin{aligned}\int_{\Omega} \lambda f(x) g(x) dx &\le \|\lambda f \mid L^{\Phi}(\Omega)\| \, \|g \mid L_{\Psi}(\Omega)\| = g \mid L_{\Psi}(\Omega) \\ &\le \max\left\{1, \int_{\Omega} \Psi(g(x))\, dx\right\}.\end{aligned}$$

Thus from (3.4.8) we have

$$\int_{\Omega} \Phi(\lambda f(x))\, dx + \int_{\Omega} \Psi(g(x))\, dx \le 1 + \int_{\Omega} \Psi(g(x))\, dx,$$

and the result follows. □

Corollary 3.4.24. *Let Φ and Ψ be complementary Young functions and let Ω be a measurable subset of $\mathbb{R}^n$. The associate space of $(L_{\Phi}(\Omega), \|\cdot \mid L_{\Phi}(\Omega)\|)$ is $(L_{\Psi}(\Omega), \|\cdot \mid L^{\Psi}(\Omega)\|)$.*

We note that the fundamental function of the Orlicz space $L_{\Phi}(\Omega)$, endowed with the Luxemburg norm, is given by

$$\phi_{L_{\Phi}}(t) = 1/\Phi^{-1}(1/t), \; t \in (0, |\Omega|). \tag{3.4.9}$$

For this we refer to [144], (9.23).

We next turn to results which will enable us to link certain GLZ spaces with Orlicz spaces. In these, Ω will stand for a measurable subset of $\mathbb{R}^n$.

Lemma 3.4.25. *Let $t_0 \in (0, \infty)$, let $f \in \mathcal{M}_0(\Omega, \mu_n)$ and let Φ be a Young function. Then*

$$\int_{t_0}^{\infty} \Phi(f^*(t)/\lambda)\, dt \le 1 \text{ for all } \lambda > 0 \tag{3.4.10}$$

if, and only if, $f^*(t) = 0$ for all $t \ge t_0$.

Proof. Suppose that (3.4.10) holds but $f^*(t)$ is bounded away from 0, say $f^*(t) \ge \varepsilon > 0$, in some interval $(t_0, t_0 + a)$. Then for all $\lambda > 0$,

$$1 \ge \int_{t_0}^{t_0+a} \Phi(f^*(t)/\lambda)\, dt \ge \Phi(\varepsilon/\lambda) a,$$

which contradicts the fact that $\Phi(s) \uparrow \infty$ as $t \uparrow \infty$. Thus $f^*(t) = 0$ for all $t > t_0$ and also for $t = t_0$, as f^* is right-continuous. The converse is obvious. □

Theorem 3.4.26. *Let $0 < t_0 \leq \infty$ and $0 \leq L < \infty$; let $f \in \mathcal{M}_0(\Omega, \mu_n)$ and suppose that Φ is a Young function which satisfies the condition*

$$\int_0^{t_0} \Phi\left(\gamma \Phi^{-1}(1/t)\right) dt < \infty \text{ for some } \gamma > 0. \tag{3.4.11}$$

Then $f \in L_\Phi(\Omega)$ if, and only if,

$$\sup_{0<t<t_0} \frac{f^*(t)}{\Phi^{-1}(1/t)} < \infty \text{ and } \int_{t_0+L}^{\infty} \Phi(f^*(t)/\gamma_0)\, dt < \infty \quad \text{for some } \gamma_0 > 0. \tag{3.4.12}$$

In addition,

$$\|f \mid L_\Phi(\Omega)\| \approx \sup_{0<t<t_0} \frac{f^*(t)}{\Phi^{-1}(1/t)} + \inf\left\{\lambda > 0 : \int_{t_0+L}^{\infty} \Phi(f^*(t)/\lambda)\, dt \leq 1\right\}. \tag{3.4.13}$$

Proof. Let $f \in \mathcal{M}_0(\Omega, \mu_n)$, $f \neq 0$, and suppose that (3.4.12) holds. Put

$$\alpha = \inf\left\{\lambda > 0 : \int_{t_0+L}^{\infty} \Phi(f^*(t)/\lambda)\, dt \leq 1\right\};$$

note that the set over which the infimum is taken is non-empty since Φ is convex and zero at 0. First suppose that $\alpha > 0$. Then the infimum is attained: let $\lambda \downarrow \alpha$ in $\int_{t_0+L}^{\infty} \Phi(f^*(t)/\lambda)\, dt \leq 1$ and use the monotone convergence theorem. Set

$$K = \sup_{0<t<t_0} \frac{f^*(t)}{\Phi^{-1}(1/t)}, \quad \lambda = \frac{K}{\gamma} + \alpha.$$

Then

$$\begin{aligned}
\int_0^{\infty} \Phi\left(\frac{f^*(t)}{\lambda}\right) dt &\leq \int_0^{t_0} \Phi\left(\frac{f^*(t)}{\lambda}\right) dt + L\Phi\left(\frac{f^*(t_0)}{\lambda}\right) \\
&\quad + \int_{t_0+L}^{\infty} \Phi\left(\frac{f^*(t)}{\lambda}\right) dt \\
&\leq \left(\frac{L}{t_0} + 1\right) \int_0^{t_0} \Phi\left(\frac{f^*(t)}{\lambda}\right) dt + \int_{t_0+L}^{\infty} \Phi\left(\frac{f^*(t)}{\lambda}\right) dt \\
&\leq \left(\frac{L}{t_0} + 1\right) \int_0^{t_0} \Phi\left(\gamma \Phi^{-1}\left(\frac{1}{t}\right)\right) dt \\
&\quad + \int_{t_0+L}^{\infty} \Phi\left(\frac{f^*(t)}{\lambda}\right) dt.
\end{aligned} \tag{3.4.14}$$

If $t_0 < \infty$, it now follows that

$$\int_0^\infty \Phi\left(\frac{f^*(t)}{\lambda}\right) dt \leq \left(\frac{L}{t_0}+1\right) \int_0^{t_0} \Phi\left(\gamma \Phi^{-1}\left(\frac{1}{t}\right)\right) dt$$

$$+ \int_{t_0+L}^\infty \Phi\left(\frac{f^*(t)}{\alpha}\right) dt$$

$$\leq \left(\frac{L}{t_0}+1\right) \int_0^{t_0} \Phi\left(\gamma \Phi^{-1}\left(\frac{1}{t}\right)\right) dt + 1 < \infty,$$

which shows that $f \in L_\Phi(\Omega)$. Moreover, since Φ is convex and $\Phi(0) = 0$, there is a positive constant c such that

$$\int_0^\infty \Phi\left(\frac{f^*(t)}{c\lambda}\right) dt \leq 1,$$

from which we see that $\|f \mid L_\Phi(\Omega)\| \lesssim K + \alpha$.

If $t_0 = \infty$, then $\alpha = 0$ and from (3.4.14) we have

$$\int_0^\infty \Phi\left(\frac{f^*(t)}{\lambda}\right) dt \leq \int_0^\infty \Phi\left(\gamma \Phi^{-1}\left(\frac{1}{t}\right)\right) dt < \infty,$$

which as before gives $\|f \mid L_\Phi(\Omega)\| \lesssim K = K + \alpha$.

If $0 < t_0 < \infty$ and $\alpha = 0$, then by Lemma 3.4.25, $f^*(t) = 0$ if $t \geq t_0 + L$, so that by (3.4.14),

$$\int_0^\infty \Phi\left(\frac{f^*(t)}{\lambda}\right) dt \lesssim \int_0^\infty \Phi\left(\gamma \Phi^{-1}\left(\frac{1}{t}\right)\right) dt < \infty,$$

which again gives $\|f \mid L_\Phi(\Omega)\| \lesssim K = K + \alpha$.

For the converse, suppose that $f \in L_\Phi(\Omega)$, $f \neq 0$. Then

$$\int_0^\infty \Phi\left(\frac{f^*(t)}{\|f \mid L_\Phi(\Omega)\|}\right) dt \leq 1,$$

and so for each $t \in (0, t_0)$,

$$1 \geq \int_0^t \Phi\left(\frac{f^*(s)}{\|f \mid L_\Phi(\Omega)\|}\right) ds \geq \Phi\left(\frac{f^*(t)}{\|f \mid L_\Phi(\Omega)\|}\right) t;$$

thus

$$f^*(t) \leq \|f \mid L_\Phi(\Omega)\| \, \Phi^{-1}\left(\frac{1}{t}\right).$$

Moreover,

$$\int_{t_0+L}^\infty \Phi\left(\frac{f^*(t)}{\|f \mid L_\Phi(\Omega)\|}\right) dt \leq 1,$$

which gives $\alpha \leq \|f \mid L_\Phi(\Omega)\|$. It is now clear that $K + \alpha \lesssim \|f \mid L_\Phi(\Omega)\|$, and the proof is complete. □

Remark 3.4.27. (i) If $t_0 = \infty$ we see that

$$\|f \mid L_\Phi(\Omega)\| \approx \sup_{0<t<\infty} \frac{f^*(t)}{\Phi^{-1}(1/t)}.$$

(ii) If $0 < t_0 < t_1 < \infty$, then

$$\sup_{0<t<t_0} \frac{f^*(t)}{\Phi^{-1}(1/t)} \approx \sup_{0<t<t_1} \frac{f^*(t)}{\Phi^{-1}(1/t)}.$$

A particularly important case of Theorem 3.4.26 occurs when Ω has finite volume $|\Omega| = \mu_n(\Omega)$.

Corollary 3.4.28. *Suppose that $|\Omega| < \infty$ and that $0 < t_0 \le |\Omega|$. Let $f \in \mathcal{M}_0(\Omega, \mu_n)$ and assume that Φ is a Young function which satisfies the condition*

$$\int_0^{t_0} \Phi\left(\gamma\Phi^{-1}(1/t)\right) dt < \infty \quad \textit{for some } \gamma > 0.$$

Then $f \in L_\Phi(\Omega)$ if, and only if,

$$\sup_{0<t<t_0} \frac{f^*(t)}{\Phi^{-1}(1/t)} < \infty\,. \tag{3.4.15}$$

In addition,

$$\|f \mid L_\Phi(\Omega)\| \approx \sup_{0<t<t_0} \frac{f^*(t)}{\Phi^{-1}(1/t)} \approx \sup_{0<t<|\Omega|} \frac{f^*(t)}{\Phi^{-1}(1/t)}. \tag{3.4.16}$$

Proof. Apply Theorem 3.4.26 with $t_0 = |\Omega|$ and $L = 0$, noting that $f^*(t) = 0$ if $t \ge |\Omega|$. □

Theorem 3.4.26 enables a connection to be made between Orlicz spaces and certain types of GLZ spaces.

Corollary 3.4.29. *Let $|\Omega| < \infty$, $m \in \mathbb{N}$, and let $\alpha = (\alpha_1, ..., \alpha_m) \in \mathbb{R}^m$ be such that $\alpha_k < 0$ for some $k \in \{1, ..., m\}$ and, if $k \ge 2$, $\alpha_i = 0$ for $i = 1, ..., k-1$. If $k = m$, let Ψ_m be a Young function given by*

$$\Psi_m(t) = Exp_m\left(t^{-1/\alpha_m}\right) \textit{for all large enough } t > 0;$$

here $Exp_m(s) = e^{s-1}$ if $m = 1$, and $Exp_m(s) = e^{Exp_{m-1}(s)-1}$ if $m \ge 2$. If $k < m$, let Ψ_k be a Young function given by

$$\Psi_k(t) = Exp_k\left(f_{m-k}(t)\right) \textit{for all large enough } t > 0,$$

where f_{m-k} is the strictly increasing function defined by

$$f_{m-k}(t) = t^{-1/\alpha_k}\vartheta_\beta^{m-k}(t) \quad \textit{for all large enough } t > 0;$$

here $\beta = (\beta_1, ..., \beta_{m-k})$, *where* $\beta_i = -\alpha_{i+k}/\alpha_k$ *for* $i = 1, ..., m-k$ *and* ϑ_β^{m-k} *is defined in (3.4.3). Then*

$$L_{\infty,\infty;\alpha}(\Omega) = L_{\Psi_k}(\Omega)$$

in the sense of equivalent quasi-norms.

Proof. When $k = m$ we use the fact that $\Psi_m^{-1}(1/t) = l_m^{-\alpha_m}(t)$ for all small enough $t > 0$. Hence the condition (3.4.11) is satisfied for some $t_0 \in (0,\infty)$ and some $\gamma \in (0,1)$. When $k < m$, observe that $f_{m-k}^{-1}(t) \approx t^{-\alpha_k}\vartheta_\gamma^{m-k}(t)$ for all large enough $t > 0$, where $\gamma = \alpha_k\beta$, and that consequently

$$\Psi_k^{-1}(1/t) \approx \vartheta_{-\alpha}^m(t) \text{ for all small enough } t > 0.$$

Again this implies that Ψ_k satisfies (3.4.11) for some $t_0 \in (0,\infty)$ and some $\gamma \in (0,1)$. The result now follows from Corollary 3.4.28. □

Remark 3.4.30. In the last Corollary the function Ψ_k may be replaced by a Young function Φ_k given by

$$\Phi_k(t) = \exp_k\left(t^{-1/\alpha_k}\mu_\beta^{m-k}(t)\right) \text{ for all large enough } t > 0,$$

where

$$\mu_\beta^{m-k}(t) = \prod_{i=1}^{m-k} \widetilde{l}_i^{\beta_i}(t),$$

with the agreement that this is 1 if $m = k$; the functions $\widetilde{l}_i$ are given by $\widetilde{l}_0(t) = t$ $(t > 1)$, $\widetilde{l}_i(t) = \log \widetilde{l}_{i-1}(t)$ $(t > \exp_{i-1} 1)$ for $i \in \{1, ..., m\}$, where $\exp_m$ stands for the m-fold composition of the exponential function. To justify this claim we note that Φ_k and Ψ_k are equivalent at infinity and use Theorem 8.12 of [2].

Not only do we have the identity given above between Orlicz spaces of (multiple) exponential type and GLZ spaces, but also very useful equivalent norms can be identified. The next Theorem addresses this point.

Theorem 3.4.31. *Let* Ω *be a domain in* $\mathbb{R}^n$ *with finite volume; let* $p \in (0,\infty]$, $m \in \mathbb{N}$ *and suppose that* $\alpha = (\alpha_1, ..., \alpha_m) \in \mathbb{R}^m$ *is such that either* $\alpha_1 < 0$ *or there exists* $k \in \{2, ..., m\}$ *with* $\alpha_j = 0$ *for* $j = 1, ..., k-1$ *and* $\alpha_k < 0$. *Then for all* $f \in L_{\infty,\infty;\alpha}(\Omega)$,

$$\|f \mid L_{\infty,\infty;\alpha}(\Omega)\| \approx \sup_{q\in[1,\infty)} \|f \mid L_q(\Omega)\| \prod_{i=1}^m l_{i-1}^{\alpha_i}(q),$$

the equivalence constants implicit in the symbol $\approx$ *being independent of* f.

For a proof of this we refer to [183].

Note that these results show that if Ω is a domain in $\mathbb{R}^n$ with finite volume and Φ is a Young function such that

$$\Phi(t) \approx \exp(t^\nu) \text{ for all } t \geq t_0,$$

for some positive numbers ν and t_0, then for all $f \in L_\Phi(\Omega)$, all $k_0 \in \mathbb{N}$ and all $q_0 \in [1, \infty)$,

$$\|f \mid L_\Phi(\Omega)\| \approx \sup_{k \in \mathbf{N}, k \geq k_0} k^{-1/\nu} \|f \mid L_k(\Omega)\| \tag{3.4.17}$$

$$\approx \sup_{q \geq q_0} q^{-1/\nu} \|f \mid L_q(\Omega)\| . \tag{3.4.18}$$

Moreover, if again Ω has finite volume but this time Ψ is a Young function such that

$$\Psi(t) \approx \exp(\exp t^\nu) \text{ for all } t \geq t_0,$$

where ν and t_0 are positive numbers, then for all $f \in L_\Psi(\Omega)$,

$$\|f \mid L_\Psi(\Omega)\| \approx \sup_{q \in (1,\infty)} (e + \log q)^{-1/\nu} \|f \mid L_q(\Omega)\| . \tag{3.4.19}$$

3.4.3 Lorentz-Karamata spaces

The idea here is to introduce a class of spaces which will include even the GLZ spaces as special cases. This is done not in the pursuit of mere generality, but because in this framework it is easier to see what are the central issues and one is less likely to be overwhelmed by technical details. We begin with the notion of a slowly varying function.

Definition 3.4.32. *A Lebesgue-measurable function $b : [1, \infty) \to (0, \infty)$ is said to be* ***slowly varying*** *if given any $\varepsilon > 0$, $t \longmapsto t^\varepsilon b(t)$ is equivalent to a non-decreasing function and $t \longmapsto t^{-\varepsilon} b(t)$ is equivalent to a non-increasing function on $[1, \infty)$.*

Functions of power type obviously do not fit into this class, but it is easy to check that the functions given by the following prescriptions are slowly varying:

(i) $b(t) = \vartheta_\alpha^m(t) = \prod_{i=1}^m l_i^{\alpha_i}(t)$ where $m \in \mathbb{N}$ and $\alpha \in \mathbb{R}^m$. We recall that the functions l_i are given on $[1, \infty)$ by

$$l_0(t) = t, \; l_i(t) = 1 + \log l_{i-1}(t) \;\; \text{for } i = 1, ..., m.$$

(ii) $b(t) = \exp(\log^a t)$, where $0 < a < 1$.
(iii) $b(t) = \exp(l_m^a(t))$, where $0 < a < 1$ and $m \in \mathbb{N}$.
(iv) $b(t) = l_m(t))$, where $m \in \mathbb{N}$.

Given any slowly varying function b, we denote by γ_b the function defined on $(0,\infty)$ by

$$\gamma_b(t) = b\left(\max\{t, 1/t\}\right), \quad t > 0.$$

It is easy to see that if b_1 and b_2 are slowly varying, then so is their product and

$$\gamma_{b_1 b_2}(t) = \gamma_{b_1}(t)\gamma_{b_2}(t), \quad t > 0.$$

Some of the basic properties of slowly varying functions are given in the next Proposition.

Proposition 3.4.33. *Let b be a slowly varying function. Then:*
(i) Given any $r \in \mathbb{R}$, the function b^r is slowly varying and $\gamma_{b^r} = \gamma_b^r$.
(ii) Given any $\varepsilon > 0$, $t \longmapsto t^\varepsilon b(t)$ is equivalent to a non-decreasing function on $(0,\infty)$ and $t \longmapsto t^{-\varepsilon} b(t)$ is equivalent to a non-increasing function on $(0,\infty)$.
(iii) Let $r > 0$. Then $\gamma_b(rt) \approx \gamma_b(t)$ for all $t > 0$.
(iv) If

$$\int_1^\infty s^{-1} b(s)ds < \infty,$$

then the function b_1 defined by

$$b_1(t) = \int_t^\infty s^{-1} b(s)ds, \quad t \geq 1,$$

is slowly varying.
(v) If $a > 0$, then for all $t > 0$,

$$\int_0^t s^{a-1}\gamma_b(s)ds \approx \sup_{0<s<t} s^a \gamma_b(s) \approx t^a \gamma_b(t)$$

and

$$\int_t^\infty s^{-a-1}\gamma_b(s)ds \approx \sup_{t<s<\infty} s^{-a}\gamma_b(s) \approx t^{-a}\gamma_b(t).$$

(vi) If $a < 0$, then for all $t > 0$,

$$\int_0^t s^{a-1}\gamma_b(s)ds \approx \sup_{0<s<t} s^a \gamma_b(s)$$

$$= \int_t^\infty s^{-a-1}\gamma_b(s)ds \approx \sup_{t<s<\infty} s^{-a}\gamma_b(s) = \infty.$$

(vii) Let $\varepsilon > 0$. Then as $t \to \infty$,

$$t^\varepsilon b(t) \to \infty \text{ and } t^{-\varepsilon} b(t) \to 0.$$

(viii) Let $a > 0$ and let b_1 be defined on $[1,\infty)$ by $b_1(t) = b(t^a)$, $t \geq 1$. Then b_1 is slowly varying.

Proof. We omit the simple proofs of (i), (iii), (vi) and (viii).

(ii) Given $\varepsilon > 0$, let f_ε be a non-decreasing function equivalent to $t^\varepsilon b(t)$ on $[1,\infty)$ and let $f_{-\varepsilon}$ be a non-increasing function equivalent to $t^{-\varepsilon} b(t)$ on $[1,\infty)$. Then $t^\varepsilon \gamma_b(t)$ is equivalent to the positive non-decreasing function g_ε on $(0,\infty)$ defined by

$$g_\varepsilon(t) = \frac{f_\varepsilon(1)}{f_{-\varepsilon}(1)} f_{-\varepsilon}\left(\max\{t, 1/t\}\right) \chi_{(0,1)}(t) + f_\varepsilon\left(\max\{t, 1/t\}\right) \chi_{[1,\infty)}(t), \ t > 0,$$

and $t^{-\varepsilon}\gamma_b(t)$ is equivalent to the positive non-increasing function $g_{-\varepsilon}$ on $(0,\infty)$ defined by

$$g_{-\varepsilon}(t) = \frac{f_\varepsilon(1)}{f_{-\varepsilon}(1)} f_{-\varepsilon}\left(\max\{t, 1/t\}\right) \chi_{[1,\infty)}(t) + f_\varepsilon\left(\max\{t, 1/t\}\right) \chi_{(0,1)}(t), \ t > 0.$$

(iv) Given any $\varepsilon > 0$, let f_ε and $f_{-\varepsilon}$ be defined as above. Then $t^\varepsilon b_1(t)$ is equivalent to the non-decreasing function g_ε defined on $[1,\infty)$ by

$$g_\varepsilon(t) = t^\varepsilon \int_t^\infty s^{-1+\varepsilon} f_{-\varepsilon}(s)ds, \ t \geq 1,$$

and $t^{-\varepsilon} b_1(t)$ is equivalent to the non-increasing function $g_{-\varepsilon}$ defined on $[1,\infty)$ by

$$g_{-\varepsilon}(t) = t^{-\varepsilon} \int_t^\infty s^{-1-\varepsilon} f_\varepsilon(s)ds, \ t \geq 1.$$

(v) Put

$$g_1(t) = \int_0^t s^{a-1}\gamma_b(s)ds \text{ and } g_\infty(t) = \sup_{0<s<t} s^a \gamma_b(s), \ t > 0.$$

Let $t > 0$. Then by (ii) ,

$$g_1(t) \gtrsim t^{-1}\gamma_b(t) \int_0^t s^a ds \approx t^a \gamma_b(t);$$

also

$$g_1(t) \lesssim t^{a/2-1}\gamma_b(t) \int_0^t s^{a/2-1} ds \approx t^a \gamma_b(t).$$

As it is clear that $g_\infty(t) \approx t\gamma_b(t)$, $t > 0$, the proof of (v) is complete, taking into account the fact that $\gamma_b(t) = \gamma_b(1/t)$ for all $t > 0$.

(vii) Choose $\varepsilon_1 \in (0,\varepsilon)$ and use the definition of a slowly varying function with ε replaced by $\varepsilon_1 - \varepsilon$. □

In addition to these simple results we shall need certain inequalities of Hardy type, contained in the next two Lemmas. These are given without proof as they are special cases of general inequalities of this nature: see, for example, [191], Theorems 5.9, 5.10, 6.2 and 6.3.

Lemma 3.4.34. *Let $p, q \in [1, \infty]$, $\nu \neq 0$ and let b_1, b_2 be slowly varying functions. Then:*
(i) The inequality

$$\left\| t^{\nu-1/q}\gamma_{b_2}(t) \int_0^t g(s)ds \mid L_q(0,\infty) \right\| \lesssim \left\| t^{\nu+1/p'}\gamma_{b_1}(t)g(t) \mid L_p(0,\infty) \right\| \tag{3.4.20}$$

holds for all $g \in \mathcal{M}^+((0,\infty), m)$ if, and only if, $\nu < 0$ and either

$$1 \leq p \leq q \leq \infty, \quad \sup_{0<t<1} \frac{b_2(1/t)}{b_1(1/t)} < \infty \tag{3.4.21}$$

or

$$1 \leq q < p \leq \infty, \quad \left\| t^{-1/r}\frac{b_2(1/t)}{b_1(1/t)} \mid L_r(0,1) \right\| < \infty, \text{ where } \frac{1}{r} = \frac{1}{q} - \frac{1}{p}. \tag{3.4.22}$$

(ii) The inequality

$$\left\| t^{\nu-1/q}\gamma_{b_2}(t) \int_t^\infty g(s)ds \mid L_q(0,\infty) \right\| \lesssim \left\| t^{\nu+1/p'}\gamma_{b_1}(t)g(t) \mid L_p(0,\infty) \right\| \tag{3.4.23}$$

holds for all $g \in \mathcal{M}^+((0,\infty), m)$ if, and only if, $\nu > 0$ and either (3.4.21) or (3.4.22) holds.

Lemma 3.4.35. *Let $p, q \in [1, \infty]$, $\nu \neq 0$ and let b_1, b_2 be slowly varying functions; for $i = 1, 2$ put $B_i(r,t) = t^{1/r}b_i(1/t)$ when $r \in [1,\infty]$ and $t \in (0,1)$. Then:*
(i) If $1 \leq p \leq q \leq \infty$, the inequality

$$\left\| B_2(-q,t) \int_t^1 g(s)ds \mid L_q(0,1) \right\| \lesssim \| B_1(p',t)g(t) \mid L_p(0,1) \| \tag{3.4.24}$$

holds for all $g \in \mathcal{M}^+((0,\infty), m)$ if, and only if, there is a positive constant c such that for all $x \in (0,1)$,

$$\| B_2(-q,t) \mid L_q(0,x) \| \ \left\| (B_1(p',t))^{-1} \mid L_{p'}(x,1) \right\| \leq c. \tag{3.4.25}$$

(ii) If $1 \leq q < p \leq \infty$, the inequality (3.4.24) holds for all $g \in \mathcal{M}^+((0,\infty), m)$ if, and only if,

$$\int_0^1 \left[\| B_2(-q,t) \mid L_q(0,x) \| \ \left\| (B_1(p',t))^{-1} \mid L_{p'}(x,1) \right\|^{p'/q'} \right]^r (B_1(p',x))^{-p'} dx \tag{3.4.26}$$

is finite. Here $1/r = 1/q - 1/p$.

Some particular cases of these Lemmas will be of interest. To explain these in a concise form it it convenient to introduce some additional notation here. Given α and β in $\mathbb{R}^m$ we write $\beta \prec \alpha$ or $\alpha \succ \beta$ if either

(i) $\beta_1 < \alpha_1$,

or

(*ii*) there exists $k \in \{2, ..., m\}$ such that $\beta_j = \alpha_j$ for $j = 1, ..., k-1$ and $\beta_k < \alpha_k$.

By $\beta \preceq \alpha$ or $\alpha \succeq \beta$ we shall mean that either $\beta \prec \alpha$ or $\beta = \alpha$. If $a \in \mathbb{R}$ we shall write $\alpha + a$ for $(\alpha_1 + a, ..., \alpha_m + a)$. Moreover, given $p \in [1, \infty]$ and $k \in \{1, ..., m\}$, we shall write $\delta_{p;m,k} = (\delta_1, ..., \delta_m) \in \mathbb{R}^m$, where

$$\delta_i = 1/p \text{ for } i = 1, ..., k \text{ and, if } k+1 \leq m, \ \delta_i = 0 \text{ for } i = k+1, ..., m.$$

Corollary 3.4.36. *Let $m \in \mathbb{N}$ and $\alpha, \beta \in \mathbb{R}^m$; suppose that $b_1 = \vartheta_\alpha^m = \prod_{i=1}^m l_i^{\alpha_i}$, $b_2 = \vartheta_\beta^m$.*
(i) If $1 \leq p \leq q \leq \infty$, then (3.4.21) holds if, and only if, $\beta \preceq \alpha$.
(ii) If $1 \leq q < p \leq \infty$, then (3.4.22) holds if, and only if, $\beta + 1/q \prec \alpha + 1/p$.

Corollary 3.4.37. *Let $p, q \in [1, \infty]$, $m \in \mathbb{N}$ and let $\alpha \in \mathbb{R}^m$ be such that for some $k \in \{1, ..., m\}$, $\alpha_k \neq 1/p'$ and, if $k \geq 2$, $\alpha_i = 1/p'$ for $i = 1, ..., k-1$. Let $\beta \in \mathbb{R}^m$ be such that $\beta_k \neq -1/q$ and, if $k \geq 2$, $\beta_i = -1/q$ for $i = 1, ..., k-1$; let $b_1 = \vartheta_\alpha^m$ and $b_2 = \vartheta_\beta^m$.*
(i) If $1 \leq p \leq q \leq \infty$, then (3.4.25) holds if, and only if,

$$\beta_k < -1/q \text{ and } \beta + \delta_{q;m,k} \preceq \alpha - \delta_{p';m,k}. \tag{3.4.27}$$

If the assumption $\beta_k \neq -1/q$ is omitted, (3.4.25) will still hold provided that

$$1 \leq p \leq q = \infty, \ \beta = \mathbf{0} \text{ and } -\alpha_k + 1/p' < 0. \tag{3.4.28}$$

If $k = m$ and we also omit the assumption $\alpha_m \neq 1/p'$, (3.4.25) will still hold provided that

$$p = 1, \ q = \infty \text{ and } \alpha = \beta = \mathbf{0}. \tag{3.4.29}$$

(ii) If $1 \leq q < p \leq \infty$, then (3.4.26) holds if, and only if,

$$\beta_k < -1/q \text{ and } \beta + 1/q \prec \alpha - \delta_{1;m,k} + 1/p. \tag{3.4.30}$$

We can now introduce the promised Lorentz-Karamata spaces.

Definition 3.4.38. *Let $p, q \in (0, \infty]$ and let b be a slowly varying function. The **Lorentz-Karamata** (LK) space $L_{p,q;b}(R, \mu)$ is defined to be the linear space of all functions $f \in \mathcal{M}_0(R, \mu)$ such that*

$$\|f \mid L_{p,q;b}(R, \mu)\| := \left\| t^{1/p-1/q} \gamma_b(t) f^*(t) \mid L_q(0, \infty) \right\| \tag{3.4.31}$$

is finite.

Observe that if $m \in \mathbb{N}$, $\alpha \in \mathbb{R}^m$ and $b = \vartheta_\alpha^m$, then $L_{p,q;b}(R,\mu)$ is just the GLZ space $L_{p,q;\alpha}(R,\mu)$ of Definition 3.4.8, special cases of which are the Lebesgue, Lorentz and Lorentz-Zygmund spaces. Note also that while $L_{p,q;b}(R,\mu)$ is non-trivial when $p < \infty$, it is nontrivial when $p = \infty$ if, and only if, $\left\| t^{-1/q}\gamma_b(t) \mid L_q(0,\infty) \right\| < \infty$, which is so if, and only if, $\left\| t^{-1/q}\gamma_b(t) \mid L_q(0,1) \right\| < \infty$.

Following our procedure in the case of Lorentz spaces we next introduce a functional obtained by replacing f^* in (3.4.31) by f^{**} :

$$\left\| f \mid L_{(p,q;b)}(R,\mu) \right\| := \left\| t^{1/p-1/q}\gamma_b(t) f^{**}(t) \mid L_q(0,\infty) \right\|. \tag{3.4.32}$$

Lemma 3.4.39. *Let $1 < p \le \infty$, $1 \le q \le \infty$ and suppose that b is slowly varying. Then for all $f \in \mathcal{M}_0(R,\mu)$,*

$$\left\| f \mid L_{p,q;b}(R,\mu) \right\| \le \left\| f \mid L_{(p,q;b)}(R,\mu) \right\| \lesssim \left\| f \mid L_{p,q;b}(R,\mu) \right\|,$$

where the constant implicit in $\lesssim$ is independent of f.

Proof. The first inequality is obvious, and the second follows, just as in the case of Lemma 3.4.6, from an appropriate form of Hardy's inequality, this time Lemma 3.4.34 (i). □

This Lemma shows that, under the given restrictions on p and q, we may regard $L_{p,q;b}(R,\mu)$ as the set of all functions f with $\left\| f \mid L_{(p,q;b)}(R,\mu) \right\| < \infty$. Moreover, the subadditivity of $f \mapsto f^{**}$ shows that $\left\| \cdot \mid L_{(p,q;b)}(R,\mu) \right\|$ is a norm if $q \ge 1$.

Our immediate objectives are to show that, under appropriate conditions, $L_{p,q;b}(R,\mu)$ is a rearrangement-invariant Banach function space and to identify its associate space. To do this we shall use the following Lemma.

Lemma 3.4.40. *Let $p,q \in (1,\infty)$, let b be slowly varying, let $g \in \mathcal{M}_0(\mathbb{R}^+,\mu_1)$ and define Φ by*

$$\Phi(t) = t^{q/p-1}\left(\gamma_b(t)\right)^q \left(g^*(t)\right)^{q-1}, \ t > 0.$$

Then

$$\Phi(t) \lesssim \int_{t/2}^{t} \Phi(s)s^{-1}ds, \ s > 0.$$

Proof. Let $\varepsilon > 1$ and note that by Lemma 3.4.33, $t^{q/p-1+\varepsilon}\left(\gamma_b(t)\right)^q$ is equivalent to a positive non-decreasing function on $(0,\infty)$. Thus

$$\begin{aligned} \int_{s/2}^{s} t^{q/p-1+\varepsilon}\left(\gamma_b(t)\right)^q t^{-1-\varepsilon}dt &\gtrsim (t/2)^{q/p-1+\varepsilon}\left(\gamma_b(t/2)\right)^q \varepsilon^{-1}(2^\varepsilon - 1)t^{-\varepsilon} \\ &\approx t^{q/p-1}\left(\gamma_b(t/2)\right)^q \approx t^{q/p-1}\left(\gamma_b(t)\right)^q, \end{aligned}$$

from which the result follows as plainly

$$\int_{t/2}^{t} \Phi(s)s^{-1}ds \gtrsim (g^*(t))^{q-1} \int_{s/2}^{s} t^{q/p-1+\varepsilon} (\gamma_b(t))^q t^{-1-\varepsilon} dt.$$

□

The promised assertions are contained in the next Theorem.

Theorem 3.4.41. *Let* $1 < p < \infty, 1 \leq q \geq \infty$, *let* b *be slowly varying and suppose that* (R, μ) *is a resonant measure space. Then*

$$X := \left(L_{p,q;b}(R), \left\|\cdot \mid L_{(p,q;b)}(R)\right\|\right) \text{ and } Y := \left(L_{p',q';1/b}(R,\mu), \|\cdot \||\right)$$

are rearrangement-invariant Banach function spaces which are mutually associate, up to the equivalence of norms.

Proof. We focus on the matter of associate spaces as everything else is evident. For simplicity we denote the norms on X and Y by $\|\cdot \mid X\|$ and $\|\cdot \mid Y\|$, respectively.

Let $g \in Y$. Then by Theorem 3.2.10, Hölder's inequality, Proposition 3.4.33 (i) and Lemma 3.4.39,

$$\int_R |fg|\, d\mu \leq \int_0^\infty f^*(t)g^*(t)dt \leq \|g \mid Y\| \, \|f \mid X\|, \; f \in X,$$

so that

$$\|g \mid X'\| = \sup_{\|f|X\| \leq 1} \int_R |fg|\, d\mu \leq \|g \mid Y\|.$$

To obtain a reverse inequality, note that by the Luxemburg representation theorem, Theorem 3.3.4, it is enough to deal with the case in which $(R, \mu) = (\mathbb{R}^+, \mu_1)$ and to suppose that $g = g^*$. Let g be a simple function with $g = g^*$; $g \in X'$. First suppose that $1 < q \leq \infty$ and write

$$f(s) = \int_{s/2}^\infty \Phi(t)t^{-1}dt, \text{ where } \Phi(s) = s^{q'/p'-1} \left(\gamma_{1/b}(s)\right)^{q'} g^*s)^{q'-1}, \; s > 0.$$

Since g is simple, it is immediate from Proposition 3.4.33 (v) that

$$\int_0^\infty \Phi(s)g^*(s)ds = \|g \mid Y\|^{q'} < \infty.$$

The function f defined above is non-increasing, so that $f = f^*$. It also belongs to X. To see this, observe that if $1 < q < \infty$, then by Proposition 3.4.33 (i) and (iii) together with Lemma 3.4.39 we have

$$\|f \mid X\| \approx \left\| s^{1/p-1/q}\gamma_b(s) \int_{s/2}^\infty \Phi(t)t^{-1}dt \mid L_q(0,\infty) \right\|$$

$$\approx \left\| s^{1/p-1/q}\gamma_b(s) \int_s^\infty \Phi(t)t^{-1}dt \mid L_q(0,\infty) \right\|$$

$$\approx \|g \mid Y\|^{q'/q} < \infty.$$

If $q = \infty$, then by Proposition 3.4.33 (v),

$$f^*(t) = f(t) = \int_{t/2}^\infty s^{1/p'-1}\gamma_{1/b}(s)s^{-1}ds \approx (t/2)^{1/p'-1}\gamma_{1/b}(t/2) \approx \Phi(t),$$

and so

$$\|f \mid X\| \approx \left\| t^{1/p}\gamma_b(t)t^{1/p'-1}\gamma_{1/b}(t) \mid L_\infty(0,\infty) \right\| = 1.$$

Hence $f \in X$ in all cases.

If $1 < q < \infty$, Lemma 3.4.40 applied to $1/b$ and with p, q replaced by p', q' respectively gives

$$\Phi(s) \lesssim \int_{s/2}^s \Phi(t)t^{-1}dt, \; s > 0,$$

from which we have

$$\begin{aligned} \|g \mid L_{p',q';1/b}(R,\mu)\|^{q'} &\lesssim \int_0^\infty \left(\int_{s/2}^s \Phi(t)t^{-1}dt \right) g^*(s)ds \\ &\leq \int_0^\infty f^*(s)g^*(s)ds \leq \|f \mid X\| \, \|g \mid X'\| \\ &\lesssim \|g \mid Y\|^{q'/q} \, \|g \mid X'\| . \end{aligned}$$

Thus

$$\|g \mid L_{p',q';1/b}(R,\mu)\| \lesssim \|g \mid X'\| .$$

On the other hand, if $q = \infty$, then it is easy to see from the above calculations that

$$\|g \mid L_{p',1;1/b}(R,\mu)\| \lesssim \|f \mid X\| \, \|g \mid X'\| \lesssim \|g \mid X'\| .$$

It remains to deal with the case $q = 1$. However,

$$tg^{**}(t) = \int_0^\infty \chi_{(0,t)}(s)g^*(s)ds \leq \left\|\chi_{(0,t)} \mid X\right\| \|g \mid X'\| , \; t > 0,$$

and

$$\left\|\chi_{(0,t)} \mid X\right\| \approx \left\|\chi_{(0,t)} \mid L_{p,1;b}(R,\mu)\right\| = \int_0^t s^{1/p-1}\gamma_b(s)ds \approx t^{1/p}\gamma_b(t).$$

Thus

$$\left\|g \mid L_{p',\infty;1/b}(R,\mu)\right\| \leq \sup_{t>0} t^{1/p'}\gamma_{1/b}(t)g^{**}(t)$$

$$\lesssim \sup_{t>0} t^{1/p'} \gamma_{1/b}(t) t^{-1/p'} \gamma_b(t) \, \|g \mid X'\|$$
$$= \|g \mid X'\| \, .$$

In summary, we have shown that in all cases, and for all simple functions g with $g = g^*$, $\|g \mid Y\| \lesssim \|g \mid X'\|$ To extend this to all $g \in X'$ we appeal to the Fatou property (P3) of Definition 3.1.1 and rearrangement invariance.

Together with the reverse inequality obtained in the first part of the proof, this inequality shows that Y coincides with X', up to equivalence of norms: X and Y are therefore mutually associate. □

Next we address the question of absolute continuity of the norm in the Lorentz-Karamata spaces. To do this it is convenient to establish the following pointwise estimates from above of f^* and f^{**} for elements f of certain spaces of this type.

Lemma 3.4.42. *Suppose that either $p \in (1, \infty)$ and $q \in [1, \infty]$, or $p = q = \infty$. Let b be slowly varying. Then there is a positive constant $c = c(p, q, b)$ such that for every $f \in L_{p,q;b}(R, \mu)$ and all $t > 0$,*

$$f^*(t) \le f^{**}(t) \le \frac{c}{t^{1/p} \gamma_b(t)} \, \|f \mid L_{p,q;b}(R, \mu)\| \, .$$

Proof. By Lemma 3.4.39,

$$\|f \mid L_{p,q;b}(R, \mu)\| \approx \left\| t^{1/p - 1/q} \gamma_b(t) f^{**}(t) \mid L_q(0, \infty) \right\| ,$$

which gives the result directly if $q = \infty$. If $1 \le q < \infty$, then using Proposition 3.4.33 (v) we see that

$$\|f \mid L_{p,q;b}(R, \mu)\| \gtrsim f^{**}(t) \left\{ \int_0^t \left(s^{1/p - 1/q} \gamma_b(s) \right)^q ds \right\}^{1/q}$$
$$\approx f^{**}(t) t^{1/p} \gamma_b(t), \ t > 0.$$

The rest is clear. □

Lemma 3.4.43. *Let $1 < p < \infty, 1 \le q < \infty$ and let b be slowly varying. Then the space $X := \left(L_{p,q;b}\left(R, \mu\right), \left\| \cdot \mid L_{(p,q;b)}\left(R, \mu\right) \right\| \right)$ has absolutely continuous norm.*

Proof. Let $f \in X$. Then by Lemma 3.4.42,

$$f^*(t) \le f^{**}(t) \le \frac{c}{t^{1/p} \gamma_b(t)} \, \|f \mid L_{p,q;b}(R, \mu)\| < \infty, \ t > 0;$$

together with Proposition 3.4.33 (vii) this shows that $\lim_{t \to \infty} f^*(t) = 0$.

Let $\{E_n\}$ be a sequence of subsets of R such that $\chi_{E_n} \downarrow 0$ μ-*a.e.*, and put $f_n = f \chi_{E_n}$ for all $n \in \mathbb{N}$. Then $\lim_{n \to \infty} f_n = 0$ μ-*a.e.* and

$$|f_n| \leq |f| \text{ for all } n \in \mathbb{N}.$$

By Proposition 3.2.4, $\lim_{n\to\infty} f_n^* = 0^* = 0$. Moreover, $f_n^* \leq f^*$ for all $n \in \mathbb{N}$. Hence by dominated convergence and Lemma 3.4.39,

$$\begin{aligned}\|f\chi_{E_n} \mid X\| &= \|f_n \mid X\| \\ &\approx \left(\int_0^\infty \left\{t^{1/p-1/q}\gamma_b(t).f_n^*(t)\right\}^q dt\right)^{1/q} \\ &\to 0 \text{ as } n \to \infty.\end{aligned}$$

The result now follows from Lemma 3.1.3. □

Together with Theorem 3.4.41 this gives

Corollary 3.4.44. *Let $p, q \in (1, \infty)$ and let b be slowly varying; suppose that (R, μ) is resonant. Then the space*

$$X := \left(L_{p,q;b}(R,\mu), \|\cdot \mid L_{p,q;b}(R,\mu)\|\right)$$

is (equivalent to) a Banach function space with associate space X' (equivalent to)

$$\left(L_{p',q';1/b}(R,\mu), \left\|\cdot \mid L_{p',q';1/b}(R,\mu)\right\|\right),$$

and both X and X' have absolutely continuous norms.

Just as in the case of Lorentz spaces there are embeddings between Lorentz-Karamata spaces.

Theorem 3.4.45. *Let $p, q_1, q_2 \in (0, \infty]$ and let b_1, b_2 be slowly varying. If $p = \infty$, suppose that*

$$\left\|t^{1/p-1/q_1} b_1(1/t) \mid L_{q_1}(0,1)\right\| < \infty. \tag{3.4.33}$$

Then

$$L_{p,q_1;b_1}(R,\mu) \hookrightarrow L_{p,q_2;b_2}(R,\mu) \tag{3.4.34}$$

if either

$$0 < q_1 \leq q_2 \leq \infty, \quad \sup_{0<t<1} \frac{b_2(1/t)}{b_1(1/t)} < \infty \tag{3.4.35}$$

or

$$0 < q_2 < q_1 \leq \infty, \left\|t^{-1/r}\frac{b_2(1/t)}{b_1(1/t)} \mid L_r(0,1)\right\| < \infty, \tag{3.4.36}$$

where $1/r = 1/q_2 - 1/q_1$.

Proof. First we consider the case in which (3.4.35) holds, when plainly $\gamma_{b_2}(t) \lesssim \gamma_{b_1}(t)$, $t > 0$. The result is obvious when $q_1 = \infty$ and so we suppose that $0 < q_1 < \infty$. Let $\varepsilon > 0$ and put $\varepsilon_p = 1/p + \varepsilon$. Then for all $f \in L_{p,q_1;b_1}(R, \mu)$, we see with the aid of Proposition 3.4.33 (v) that

$$t^{1/p}\gamma_{b_2}(t)f^*(t) \approx t^{-\varepsilon} f^*(t) \left(\int_0^t \left(s^{\varepsilon_p - 1/q_1}\gamma_{b_2}(s) \right)^{q_1} ds \right)^{1/q_1}$$

$$\lesssim \left(\int_0^t \left(s^{1/p - 1/q_1}\gamma_{b_1}(s) f^*(s) \right)^{q_1} ds \right)^{1/q_1}$$

$$\lesssim \| f \mid L_{p,q_1;b_1}(R, \mu) \| .$$

This proves (3.4.34) when $q_2 = \infty$.

Next suppose that $0 < q_1 \le q_2 < \infty$ and let $f \in L_{p,q_1;b_1}(R, \mu)$. Then, with $F(s) = s^{1/p}\gamma_{b_2}(s)f^*(s)$,

$$\| f \mid L_{p,q_2;b_2}(R, \mu) \| = \left(\int_0^t (F(s))^{q_2 - q_1} \left(s^{-1/q_1} F(s) \right)^{q_1} ds \right)^{1/q_2}$$

$$\lesssim \| f \mid L_{p,\infty;b_2}(R, \mu) \|^{1 - q_1/q_2} \| f \mid L_{p,q_1;b_1}(R, \mu) \|^{q_1/q_2} ,$$

and the result follows.

Finally we consider the case (3.4.36). We use Hölder's inequality with exponents q_1/q_2 and $q_1/(q_1 - q_2)$ to obtain, when $q_1 < \infty$,

$$\| f \mid L_{p,q_2;b_2}(R, \mu) \| \lesssim \| f \mid L_{p,q_1;b_1}(R, \mu) \| \left\| t^{-1/r} \frac{b_2(t)}{b_1(t)} \mid L_r(0, \infty) \right\| .$$

If $q_1 = \infty$ this is immediate. Since the second factor on the right-hand side is finite if, and only if, $\left\| t^{-1/r} \frac{b_2(1/t)}{b_1(1/t)} \mid L_r(0,1) \right\| < \infty$, the proof is complete. □

To deal with embeddings in which the first index, p, is finite we establish a preliminary result.

Lemma 3.4.46. *Let $0 < p \le \infty$, $0 < q_1 < q_2 < \infty$ and let b_1, b_2 be slowly varying. Then*

$$\| f \mid L_{p,q_2;b_2}(R, \mu) \| \le \| f \mid L_{p,q_1;b_1}(R, \mu) \|^{q_1/q_2} \| f \mid L_{p,\infty;b_3}(R, \mu) \|^{1 - q_1/q_2} \tag{3.4.37}$$

for all $f \in \mathcal{M}_0(R, \mu)$, where b_3 is the slowly varying function defined by

$$b_3(t) = \left\{ \frac{b_2^{q_2}(t)}{b_1^{q_1}(t)} \right\}^{1/(q_2 - q_1)} , \quad t \ge 1.$$

Proof. Let $f \in \mathcal{M}_0(R,\mu)$ and suppose that the right-hand side of (3.4.37) is finite. Then

$$\|f \mid L_{p,q_2;b_2}(R,\mu)\|^{q_2} = \int_0^\infty \left(t^{1/p}\gamma_{b_1}(t)f^*(t)\right)^{q_1} \left(t^{1/p}\gamma_{b_3}(t)f^*(t)\right)^{q_2-q_1} \frac{dt}{t}$$
$$\leq \|f \mid L_{p,q_1;b_1}(R,\mu)\|^{q_1/q_2} \|f \mid L_{p,\infty;b_3}(R,\mu)\|^{1-q_1/q_2}.$$

□

Theorem 3.4.47. *Let $0 < q_1 < \infty$, $0 < q_2 \leq \infty$; suppose that b_1, b_2 are slowly varying and that*

$$\left\|t^{-1/q_1}b_1(1/t) \mid L_{q_1}(0,1)\right\| < \infty.$$

Let I_{b_1,q_1} be the slowly varying function defined by

$$I_{b_1,q_1}(t) = \left(\int_t^\infty \left(s^{-1/q_1}b_1(s)\right)^{q_1} ds\right)^{1/q_1}, \quad t \geq 1.$$

Then

$$L_{\infty,q_1;b_1}(R,\mu) \hookrightarrow L_{\infty,q_2;b_2}(R,\mu) \tag{3.4.38}$$

provided that either

$$0 < q_1 < q_2 = \infty \text{ and } \sup_{0<t<1} \frac{b_2(1/t)}{I_{b_1,q_1}(1/t)} < \infty \tag{3.4.39}$$

or

$$0 < q_1 < q_2 < \infty \text{ and } \sup_{0<t<1} \left[\left\{\frac{b_2^{q_2}(1/t)}{b_1^{q_1}(1/t)}\right\}^{1/(q_2-q_1)} \frac{1}{I_{b_1,q_1}(1/t)}\right] < \infty. \tag{3.4.40}$$

In particular, if $0 < q_1 < \infty$,

$$L_{\infty,q_1;b_1}(R,\mu) \hookrightarrow L_{\infty,\infty;I_{b_1,q_1}}(R,\mu) \hookrightarrow L_{\infty,\infty;b_1}(R,\mu). \tag{3.4.41}$$

Proof. First suppose that (3.4.39) holds. Given $f \in \mathcal{M}_0(R,\mu)$ and $t > 0$,

$$\gamma_{b_2}(t)f^*(t) \lesssim \gamma_{I_{b_1,q_1}}(t)f^*(t) = f^*(t)\left(\int_{\max\{t,1/t\}}^\infty \left(s^{-1/q_1}\gamma_{b_1}(s)\right)^{q_1} ds\right)^{1/q_1}$$

$$= f^*(t)\left(\int_0^{\min\{t,1/t\}} \left(s^{-1/q_1}\gamma_{b_1}(s)\right)^{q_1} ds\right)^{1/q_1}$$

$$\leq \left(\int_0^t \left(f^*(s)s^{-1/q_1}\gamma_{b_1}(s)\right)^{q_1} ds\right)^{1/q_1}$$

$$\leq \|f \mid L_{\infty,q_1;b_1}(R,\mu)\| .$$

This gives the desired embedding (3.4.38) with $q_2 = \infty$.

Now suppose that (3.4.40) holds. By Lemma 3.4.30 with $p = \infty$ we have

$$\|f \mid L_{\infty,q_2;b_2}(R,\mu)\| \leq \|f \mid L_{\infty,q_1;b_1}(R,\mu)\|^{q_1/q_2} \|f \mid L_{\infty,\infty;b_3}(R,\mu)\|^{1-q_1/q_2},$$

where

$$b_3(t) = \left\{\frac{b_2^{q_2}(t)}{b_1^{q_1}(t)}\right\}^{1/(q_2-q_1)}, \quad t \geq 1.$$

Since (3.4.39) holds with b_2 replaced by b_3, the last case shows that for all $f \in L_{\infty,q_1;b_1}(R,\mu)$,

$$\|f \mid L_{\infty,\infty;b_3}(R,\mu)\| \lesssim \|f \mid L_{\infty,q_1;b_1}(R,\mu)\| .$$

This completes the proof of (3.4.38).

As for (3.4.41), the first embedding there follows from (3.4.38) since $0 < q_1 < q_2 = \infty$ and condition (3.4.39) holds with $b_2 = I_{b_1,q_1}$. The second embedding results from Theorem 3.4.45, as

$$\sup_{0<t<1} \frac{b_1(1/t)}{I_{b_1,q_1}(1/t)} < \infty.$$

To see this, let $\varepsilon > 0$ and note that for every $t \geq 1$,

$$I_{b_1,q_1}(t) \gtrsim t^{\varepsilon/q_1} b_1(t) \left(\int_t^\infty s^{-1-\varepsilon} ds\right)^{1/q_1} = \varepsilon^{-1/q_1} b_1(t).$$

The proof is complete. □

To illustrate what happens when the first indices are different we give the following result.

Theorem 3.4.48. *Let (R,μ) be a finite measure space, let $p_1, p_2, q_1, q_2 \in (0,\infty]$ with $p_2 < p_1$, and let b_1, b_2 be slowly varying. Assume also that*

$$\left\|t^{1/p_1 - 1/q_1} b_1(1/t) \mid L_{q_1}(0,1)\right\| < \infty, \quad \text{if } p_1 = \infty.$$

Then

$$L_{p_1,q_1;b_1}(R,\mu) \hookrightarrow L_{p_2,q_2;b_2}(R,\mu).$$

Proof. It is enough to show that $L_{p_1,\infty;b_1}(R,\mu) \hookrightarrow L_{p_2,q_2;b_2}(R,\mu)$ since, by Theorem 3.4.45, $L_{p_1,q_1;b_1}(R,\mu) \hookrightarrow L_{p_1,\infty;b_1}(R,\mu)$. Let $f \in L_{p_1,\infty;b_1}(R,\mu) = X$ and suppose that $q_2 < \infty$. Then by Proposition 3.4.33 (v) and with $\gamma(t) = \gamma_{b_2}(t)/\gamma_{b_1}(t)$,

$$\|f \mid L_{p_2,q_2;b_2}(R,\mu)\| \leq \|f \mid X\| \left(\int_0^{\mu(R)} \left(t^{1/p_2-1/p_1-1/q_2}\gamma(t) \right)^{q_2} dt \right)^{1/q_2}$$

$$\approx \|f \mid X\| \left(\mu(R)\right)^{1/p_2-1/p_1} \gamma\left(\mu(R)\right)$$

$$\approx \|f \mid X\| .$$

The proof when $q_2 = \infty$ is similar. □

For density results we confine ourselves to function spaces over $(\mathbb{R}^n, \mu_n)$.

Lemma 3.4.49. *Let $1 < p < \infty, 1 \leq q < \infty$ and let b be slowly varying. Then $C_0^\infty(\mathbb{R}^n)$ is dense in $L_{p,q;b}(\mathbb{R}^n)$.*

Proof. We know that $L_{p,q;b}(\mathbb{R}^n)$ is a rearrangement-invariant function space with absolutely continuous norm. The result now follows from the usual process of regularisation: see [46], Proposition V.1.8. □

3.4.4 Decompositions

For some of the frequently occurring Orlicz spaces, such as Zygmund or exponential spaces, equivalent norms of an extrapolation character are known. This special technique developed after the quantitative behaviour of norms of various operators between function spaces near L_1 and L_∞ was understood. It goes back to Yano [229]; for abstract extrapolation theory see [130] and [175]. Interpolation constructions for special Orlicz spaces are given in [74]. In all this one usually works with global norms, but in [66] and [39] a decomposition method was given, making it possible to extrapolate towards Orlicz spaces of exponential type by using suitable pieces of the extrapolating Lebesgue norms, based on subsets of the original domain. This has considerable advantages: see, for example, [67] for an application of this technique to the circle of ideas around Moser's lemma.

In this subsection we give a number of results of this type, presenting them in the framework of Lorentz-Karamata spaces for economy of presentation. We begin with a simple lemma.

Lemma 3.4.50. *Let b be slowly varying and let $M > 0$. Then:*

(i) For each $k \in \mathbb{N}$,

$$b(e^k) \approx b(e^{k-1}).$$

(ii) For each $k \in \mathbb{N}$ and each $t \in (Me^{-k}, Me^{-k+1})$,

$$\gamma_b(t) \approx b(e^{k-1}).$$

(iii) For each $k \in \mathbb{N}$ and each $t \in (e^{k-1}, e^k)$,

$$\gamma_b(t) \approx b(e^{k-1}).$$

We omit the easy proof.

Theorem 3.4.51. *Suppose that (R, μ) is a finite measure space, let $p, q \in (0, \infty]$ and let b be slowly varying. Then for every $f \in L_{p,q;b}(R, \mu)$ we have:*
(i) if $0 < q < \infty$,

$$\|f \mid L_{p,q;b}(R,\mu)\| \approx \left\{ \sum_{k=1}^{\infty} \left(e^{-k/p} b(e^{k-1}) f^*(\mu(R)e^{-k}\right)^q \right\}^{1/q} ; \qquad (3.4.42)$$

(ii)

$$\|f \mid L_{p,\infty;b}(R,\mu)\| \approx \sup_{k \geq 1} \left\{ e^{-k/p} b(e^{k-1}) f^*(\mu(R)e^{-k} \right\}. \qquad (3.4.43)$$

Proof. For (i), let $f \in L_{p,q;b}(R, \mu)$ and $q < \infty$. By Lemma 3.4.50,

$$\|f \mid L_{p,q;b}(R,\mu)\|^q \gtrsim \sum_{k=2}^{\infty} \left(e^{-k/p} b(e^{k-1}) f^*(\mu(R)e^{-k+1}\right)^q$$

$$\approx \sum_{k=1}^{\infty} \left(e^{-k/p} b(e^{k-1}) f^*(\mu(R)e^{-k}\right)^q .$$

On the other hand, Lemma 3.4.50 also shows that

$$\|f \mid L_{p,q;b}(R,\mu)\|^q \lesssim \sum_{k=1}^{\infty} \left(e^{-k/p} b(e^{k-1}) f^*(\mu(R)e^{-k}\right)^q ,$$

and so the proof when $q < \infty$ is complete. The case $q = \infty$ can be handled in a similar way. □

As an immediate consequence of this we have

Corollary 3.4.52. *Let Ω be a measurable subset of $\mathbb{R}^n$ with finite volume $|\Omega|_n = \mu_n(\Omega)$ and let $a > 0$. Then for every $f \in L_{\infty,\infty;-1/a}(\Omega)$,*

$$\left\| f \mid L_{\infty,\infty;-1/a}(\Omega) \right\| \approx \sup_{k \geq 1} k^{-1/a} f^*(e^{-k} |\Omega|_n) \approx \sup_{k \geq 2} k^{-1/a} f^*(e^{-k+1} |\Omega|_n);$$

and for every $f \in L_{\infty,\infty;0,-1/a}(\Omega)$,

$$\left\| f \mid L_{\infty,\infty;0,-1/a}(\Omega) \right\| \approx \sup_{k \geq 1} (1 + \log k)^{-1/a} f^*(e^{-k} |\Omega|_n)$$

$$\approx \sup_{k > 2} f^*(e^{-k+1} |\Omega|_n)(\log k)^{-1/a}.$$

Another simple lemma will give us further equivalent quasinorms, this time in terms of norms of f^* on subintervals of $(0, \mu(R))$.

Lemma 3.4.53. . *Let $f \in \mathcal{M}_0(R, \mu)$, suppose that $\mu(R) < \infty$ and put $J_k = (\mu(R)e^{-k}, \mu(R)e^{-k+1})$ for each $k \in \mathbb{N}$. Then*
(i) for each $k \in \mathbb{N}$,

$$c_1 f^*(\mu(R)e^{-k+1}) \leq \|f^* \mid L_k(J_k)\| \leq c_2 f^*(\mu(R)e^{-k}), \tag{3.4.44}$$

where c_1 and c_2 are positive constants independent of f and k;
(ii) for each $k \in \mathbb{N}$, $k \geq 2$,

$$c_1 f^*(\mu(R)e^{-k+2}) \leq \|f^* \mid L_k(J_{k-1})\| \leq c_2 f^*(\mu(R)e^{-k+1}), \tag{3.4.45}$$

where c_1 and c_2 are positive constants independent of f and k.

We omit the elementary proof and go straight on to the matter of equivalent quasinorms.

Theorem 3.4.54. . *Suppose that (R, μ) is a finite measure space, let $p, q \in (0, \infty]$ and let b be slowly varying; for each $k \in \mathbb{N}$ put*
$J_k = (\mu(R)e^{-k}, \mu(R)e^{-k+1})$. Then for every $f \in L_{p,q;b}(R, \mu)$ we have:
(i) if $0 < q < \infty$,

$$\|f \mid L_{p,q;b}(R, \mu)\| \approx \left\{ \sum_{k=1}^{\infty} \left(e^{-k/p} b(e^{k-1}) \, \|f^* \mid L_k(J_k)\| \right)^q \right\}^{1/q} \tag{3.4.46}$$

$$\approx \left\{ \sum_{k=2}^{\infty} \left(e^{-k/p} b(e^{k-1}) \, \|f^* \mid L_k(J_{k-1})\| \right)^q \right\}^{1/q} ; \tag{3.4.47}$$

(ii) for the case $q = \infty$,

$$\|f \mid L_{p,\infty;b}(R, \mu)\| \approx \sup_{k \geq 1} \left\{ e^{-k/p} b(e^{k-1}) \, \|f^* \mid L_k(J_k)\| \right\} \tag{3.4.48}$$

$$\approx \sup_{k \geq 2} \left\{ e^{-k/p} b(e^{k-1}) \, \|f^* \mid L_k(J_{k-1})\| \right\}. \tag{3.4.49}$$

Proof. First suppose that $0 < q < \infty$ and let $f \in L_{p,q;b}(R, \mu)$. Then by Theorem 3.4.51 and Lemma 3.4.53,

$$\|f \mid L_{p,q;b}(R, \mu)\|^q \gtrsim \sum_{k=1}^{\infty} \left(e^{-k/p} b(e^{k-1}) \, \|f^* \mid L_k(J_k)\| \right)^q .$$

Moreover, the same results plus Lemma 3.4.50 show that

$$\|f \mid L_{p,q;b}(R,\mu)\|^q \approx \sum_{k=1}^{\infty} \left(e^{-k/p} b(e^k) f^*(\mu(R)e^{-k})\right)^q$$

$$\gtrsim \sum_{k=2}^{\infty} \left(e^{-k/p} b(e^{k-1}) \|f^* \mid L_k(J_k)\|\right)^q.$$

In the opposite direction we have, in the same way,

$$\|f \mid L_{p,q;b}(R,\mu)\|^q \lesssim \sum_{k=2}^{\infty} \left(e^{-k/p} b(e^{k-1}) f^*(\mu(R)e^{-k+1})\right)^q$$

$$\lesssim \sum_{k=1}^{\infty} \left(e^{-k/p} b(e^{k-1}) \|f^* \mid L_k(J_k)\|\right)^q.$$

Proceeding similarly we then obtain

$$\|f \mid L_{p,q;b}(R,\mu)\|^q \lesssim \sum_{k=3}^{\infty} \left(e^{-k/p} b(e^{k-2}) f^*(\mu(R)e^{-k+2})\right)^q$$

$$\lesssim \sum_{k=2}^{\infty} \left(e^{-k/p} b(e^{k-1}) \|f^* \mid L_k(J_{k-1})\|\right)^q,$$

and the proof is complete, when $q < \infty$.

When $q = \infty$ the proof is similar. □

As an immediate consequence of this Theorem we have

Corollary 3.4.55. *Let Ω be a measurable subset of $\mathbb{R}^n$ with finite volume $|\Omega|_n$ and let $a > 0$; write $J_k = (|\Omega|_n e^{-k}, |\Omega|_n e^{-k+1})$ for each $k \in \mathbb{N}$. Then for every $f \in L_{\infty,\infty;-1/a}(\Omega)$,*

$$\left\|f \mid L_{\infty,\infty;-1/a}(\Omega)\right\| \approx \sup_{k\geq 1} k^{-1/a} \|f^* \mid L_k(J_k)\| \approx \sup_{k\geq 2} k^{-1/a} \|f^* \mid L_k(J_{k-1})\|;$$

and for every $f \in L_{\infty,\infty;0,-1/a}(\Omega)$,

$$\left\|f \mid L_{\infty,\infty;0,-1/a}(\Omega)\right\| \approx \sup_{k\geq 1} (1+\log k)^{-1/a} \|f^* \mid L_k(J_k)\|$$

$$\approx \sup_{k\geq 2} (\log k)^{-1/a} \|f^* \mid L_k(J_{k-1})\|.$$

We remind the reader that $L_{\infty,\infty;-1/a}(\Omega)$ and $L_{\infty,\infty;0,-1/a}(\Omega)$ are Orlicz spaces with Young functions which behave at infinity like $\exp(t^a)$ and $\exp(\exp t^a)$ respectively; see Corollary 3.4.29.

For results of this kind see also [66] and [39]; when the underlying measure space has infinite measure see [183].

3.5 Operators of joint weak type

3.5.1 Definitions

Definition 3.5.1. *Let (R_1, ν_1) and (R_2, ν_2) be σ-finite measure spaces, let $D(T)$ be a linear subspace of $\mathcal{M}_0(R_1, \nu_1)$ and let $T : D(T) \to \mathcal{M}(R_2, \nu_2)$. The map T is said to be* ***quasilinear*** *if there is a constant $k \geq 1$ such that for all $f, g \in D(T)$ and all scalars λ,*

$$|T(f+g)| \leq k(|Tf| + |Tg|) \text{ and } |T(\lambda f)| = |\lambda| \, |Tf|$$

ν_2-a.e. on R_2.

Remark 3.5.2. (i) If $T : D(T) \to \mathcal{M}(R_2, \nu_2)$ is linear, it is obviously quasilinear, with $k = 1$.
(ii) Let Ω be a measurable subset of $\mathbb{R}^n$, let $D(T) = L_1^{loc}(\Omega)$ and let $\mathbb{M} : D(T) \to \mathcal{M}(\Omega, \mu_n)$ be the Hardy-Littlewood maximal operator defined by

$$(\mathbb{M}f)(x) = \sup (\mu_n(Q))^{-1} \int_{Q \cap \Omega} |f(y)| \, dy, \ x \in \Omega,$$

where the supremum is taken over all cubes Q in $\mathbb{R}^n$, with sides parallel to the coordinate axes, and to which x belongs. Then $\mathbb{M}$ is quasilinear with $k = 1$. For convenience, we shall sometimes replace the cubes Q by balls $B(x, r)$ and take the supremum over $r > 0$. This has no effect on the essential properties of $\mathbb{M}$.

Definition 3.5.3. *Let $1 \leq p_1 < p_2 \leq \infty$ and $q_1, q_2 \in [1, \infty]$, $q_1 \neq q_2$. Define the* ***interpolation segment*** *σ to be the line segment in $\mathbb{R}^2$ with end-points $(1/p_j, 1/q_j)$ $(j = 1, 2)$, and write*

$$\sigma = [(1/p_1, 1/q_1), (1/p_2, 1/q_2)]. \tag{3.5.1}$$

Put

$$l = \left(\frac{1}{q_1} - \frac{1}{q_2} \right) / \left(\frac{1}{p_1} - \frac{1}{p_2} \right). \tag{3.5.2}$$

For each $g \in \mathcal{M}^+(0, \infty)$ and each $t \in (0, \infty)$, the ***Calderón operator*** *S_σ associated with the interpolation segment σ is given by*

$$(S_\sigma g)(t) = t^{-1/q_1} \int_0^{t^l} s^{(1/p_1)-1} g(s) ds + t^{-1/q_2} \int_{t^l}^{\infty} s^{(1/p_2)-1} g(s) ds. \tag{3.5.3}$$

Proposition 3.5.4. *Let S_σ be defined by (3.5.3), let g be a non-negative measurable function on $(0, \infty)$ and let $t, u > 0$. Then*

$$(S_\sigma g)(t) \leq \max \left\{ (u/t)^{1/q_1}, (u/t)^{1/q_2} \right\} (S_\sigma g)(u). \tag{3.5.4}$$

In particular, $S_\sigma g$ is non-increasing and

$$(S_\sigma g)(t) = (S_\sigma g)^*(t) \leq (S_\sigma g^*)(t) \tag{3.5.5}$$

for each $t > 0$.

Proof. First note that

$$(S_\sigma g)(t) = \int_0^\infty g(s) \min\left(\frac{s^{1/p_1}}{t^{1/q_1}}, \frac{s^{1/p_2}}{t^{1/q_2}}\right) \frac{ds}{s}. \tag{3.5.6}$$

Together with the inequality

$$\min\left(\frac{s^{1/p_1}}{t^{1/q_1}}, \frac{s^{1/p_2}}{t^{1/q_2}}\right) \le \min\left(\frac{s^{1/p_1}}{u^{1/q_1}}, \frac{s^{1/p_2}}{u^{1/q_2}}\right) \max\left(\left(\frac{u}{t}\right)^{1/q_1}, \left(\frac{u}{t}\right)^{1/q_2}\right),$$

this gives (3.5.4). That $S_\sigma g$ is non-increasing is now immediate, and so the inequality in (3.5.5) follows. To prove the rest of (3.5.5), observe that the kernel k_t $(t > 0)$ defined by

$$k_t(s) = s^{-1} \min\left(\frac{s^{1/p_1}}{t^{1/q_1}}, \frac{s^{1/p_2}}{t^{1/q_2}}\right)$$

is a non-increasing function of s. Thus by the Hardy-Littlewood inequality given in Theorem 3.2.10 we have

$$(S_\sigma g)(t) = \int_0^\infty g(s) k_t(s) ds \le \int_0^\infty g^*(s) k_t^*(s) ds = (S_\sigma g^*)(t).$$

□

Definition 3.5.5. *Let $1 \le p_1 < p_2 \le \infty$ and $q_1, q_2 \in [1, \infty]$, with $q_1 \ne q_2$. Let T be a quasilinear map (with respect to σ-finite measure spaces (R_1, ν_1) and (R_2, ν_2)) with domain containing all those $f \in \mathcal{M}(R_1, \nu_1)$ such that*

$$(S_\sigma f^*)(1) = \int_0^1 s^{1/p_1} f^*(s) \frac{ds}{s} + \int_1^\infty s^{1/p_2} f^*(s) \frac{ds}{s} < \infty. \tag{3.5.7}$$

We say that T is of ***joint weak type*** $(p_1, q_1; p_2, q_2)$ *if there is a constant c such that for all f satisfying (3.5.7),*

$$(Tf)^*(t) \le c (S_\sigma f^*)(t), \quad 0 < t < \infty; \tag{3.5.8}$$

the least such c is called the ***weak-type*** $(p_1, q_1; p_2, q_2)$ ***norm*** *of T.*

To provide examples of operators T of this nature it is convenient to have available the following result due to O'Neil.

Lemma 3.5.6. *Let f, g and h be measurable functions on $\mathbb{R}^n$ such that $h = f * g$. Then for all $t > 0$,*

$$h^{**}(t) \le t f^{**}(t) g^{**}(t) + \int_t^\infty f^*(s) g^*(s) ds.$$

Proof. See [231], Lemma 1.8.8 and also [142]. □

Remark 3.5.7. (i) Let $1 < p < \infty$, let $g \in L_{p,\infty}(\mathbb{R}^n)$ and let T be the convolution operator given by

$$(Tf)(x) = (f * g)(x), \ x \in \mathbb{R}^n.$$

Then T is of joint weak type $(1, p; p', \infty)$. To prove this, note that $g^*(t) \leq t^{-1/p} \|g \mid L_{p,\infty}\|$. Hence by O'Neil's result,

$$(Tf)^*(t) \leq p' \|g \mid L_{p,\infty}\| \int_0^t f^*(s)ds + \int_t^\infty f^*(s)s^{-1/p} \|g \mid L_{p,\infty}\| \, ds,$$

and the claim follows.

As a special case of this there is the Riesz potential map I_η, defined by

$$(I_\eta f)(x) = \int_\Omega \frac{f(y)}{|x-y|^{n-\eta}} dy, \ 0 < \eta < n;$$

here Ω is a measurable subset of $\mathbb{R}^n$. We may suppose that f is extended by 0 outside Ω, so that the integral can be written as a convolution. Since the kernel g, given by $g(y) = |y|^{\eta - n}$, belongs to $L_{n/(n-\eta),\infty}(\mathbb{R}^n)$, we see that I_η is of joint weak type $(1, \frac{n}{n-\eta}; \frac{n}{\eta}, \infty)$.

(ii) Let Ω be a measurable subset of $\mathbb{R}^n$, let $0 \leq \eta < n$ and let $\mathbb{M}_{\Omega,\eta}$ be the fractional Hardy-Littlewood maximal operator, defined for all $f \in L_{1,loc}(\Omega)$ by

$$(\mathbb{M}_{\Omega,\eta} f)(x) = \sup \frac{1}{(\mu_n(Q))^{1-\eta/n}} \int_{\Omega \cap Q} |f(y)| \, dy, \ x \in \Omega,$$

where the supremum is taken over all cubes Q in $\mathbb{R}^n$ containing x, with sides parallel to the coordinate axes. When $\eta = 0$, this is the classical Hardy-Littlewood maximal operator $\mathbb{M}$ (see Remark 3.5.2(ii)). It is easy to see that $\mathbb{M}_{\Omega,\eta}$ is of joint weak type $\left(1, \frac{n}{n-\eta}; \frac{n}{\eta}, \infty\right)$; we interpret this when $\eta = 0$ as meaning that $\mathbb{M}_\Omega$ is of joint weak type $(1, 1; \infty, \infty)$. To prove this, let $f \in L_{1,loc}(\Omega)$ be extended by 0 to all of $\mathbb{R}^n$ and observe that for some constant $c = c(\eta, n)$,

$$(\mathbb{M}_{\Omega,\eta} f)(x) \leq c \int_{\mathbb{R}^n} \frac{f(y)}{|x-y|^{n-\eta}} dy, \ x \in \Omega.$$

Our claim now follows from Example (i).

(iii) The Hilbert transform H is defined on $L_{1,loc}(\mathbb{R})$ by the principal-value integral

$$\begin{aligned}(Hf)(x) &= p.v. \frac{1}{\pi} \int_{\mathbb{R}} \frac{f(t)}{x-t} dt \\ &= \lim_{\varepsilon \to 0} \frac{1}{\pi} \int_{|x-t| \geq \varepsilon} \frac{f(t)}{x-t} dt, \ x \in \mathbb{R}.\end{aligned}$$

It turns out that H is of joint weak type $(1, 1; \infty, \infty)$: for a proof of this see [15], Theorem 3.5.6.

3.5.2 Operators of strong and weak type

In connection with the ideas just introduced, certain versions of boundedness are useful.

Definition 3.5.8. *Let (R, μ) and (S, ν) be $\sigma-$finite measure spaces and let $p, q \in [1, \infty]$. Let T be a linear operator defined on all $\mu-$simple functions on R and with values in the $\nu-$measurable functions on S. The map T is said to be of **strong type** (p, q) if there is a constant M such that for all $\mu-$simple functions f,*

$$\|Tf \mid L_q(\nu)\| \leq M \|f \mid L_p(\mu)\| ;$$

the least such M is called the strong-type (p, q) norm of T.

An important result concerning such operators T is the celebrated Riesz-Thorin theorem.

Theorem 3.5.9. ***(Riesz-Thorin).*** *Let (R, μ) and (S, ν) be $\sigma-$finite measure spaces, let $p_0, p_1, q_0, q_1 \in [1, \infty]$, let $\theta \in [0, 1]$ and put*

$$\frac{1}{p} = \frac{1-\theta}{p_0} + \frac{\theta}{p_1}, \frac{1}{q} = \frac{1-\theta}{q_0} + \frac{\theta}{q_1}.$$

Let T be a linear operator of strong types (p_0, q_0) and (p_1, q_1), with corresponding strong-type norms M_0 and M_1. Then if the underlying spaces are complex, T is of strong type (p, q) and its strong-type (p, q) norm M_θ satisfies

$$M_\theta \leq M_0^{1-\theta} M_1^\theta.$$

If the underlying spaces are real, the same result holds if $p_j \leq q_j (j = 0, 1)$; in the absence of this last condition it still holds but with the weakened inequality

$$M_\theta \leq 2M_0^{1-\theta} M_1^\theta.$$

For a proof of this result see, for example, [15], Theorem 4.2.2

Note that if $p_0, p_1 < \infty$, the operator T in Theorem 3.5.9 may be extended to all of L_{p_0} and L_{p_1}, so that Theorem 3.5.9 may be expressed in terms of a map from L_{p_j} to L_{q_j} $(j = 0, 1)$. In terms of interpolation theory (see [15] or [218]), this implies that (L_p, L_q) is an interpolation pair for the couples (L_{p_0}, L_{q_0}) and (L_{p_1}, L_{q_1}), and it shows easily that L_p is an intermediate space for the couple (L_{p_0}, L_{p_1}).

Use of Lorentz spaces rather than Lebesgue spaces enables another handy version of boundedness to be given.

Definition 3.5.10. *Let (R, μ) and (S, ν) be $\sigma-$finite measure spaces and let $p, q \in [1, \infty]$. A map $T : L_{p,1}(R, \mu) \to \mathcal{M}_0(S, \nu)$ is said to be of **weak type** (p, q) if it is bounded from $L_{p,1}(R, \mu)$ to $L_{q,\infty}(S, \nu)$; that is, if there is a constant M such that for all $f \in L_{p,1}(R, \mu)$,*

$$\|Tf \mid L_{q,\infty}(S, \nu)\| \leq M \|f \mid L_{p,1}(R, \mu)\| . \tag{3.5.9}$$

*The least such constant M is called the **weak-type** (p, q) **norm** of T.*

Remark 3.5.11. (i) Note that (3.5.9) means that for all $t > 0$,

$$(Tf)^*(t) \leq Mt^{-1/q} \|f \mid L_{p,1}\|$$

and that for all $\lambda > 0$,

$$\nu(\{x \in S : |(Tf)(x)| > \lambda\}) \leq \left(M\lambda^{-1} \|f \mid L_{p,1}\|\right)^q .$$

(ii) Let $p < \infty$ and let T be a linear map of strong type (p,q) from the μ–simple functions on R with values in the ν–measurable functions on S. Then T has a unique linear extension to a bounded map from $L_p(R)$ to $L_q(S)$. By Proposition 3.4.3,

$$L_{p,1}(R) \hookrightarrow L_p(R), \; L_q(S) \hookrightarrow L_{q,\infty}(S);$$

hence T is of weak type (p,q).

Lemma 3.5.12. *Let* $1 \leq p_1 < p_2 < \infty$ *and* $q_1, q_2 \in [1,\infty]$; *let* S_σ *be the Calderón operator associated with the interpolation segment*

$$\sigma = [(1/p_1, 1/q_1), (1/p_2, 1/q_2)]$$

(see Definition 3.5.3). Then for all $f \in L_{p_i,1}(\mathbb{R}^+)$,

$$t^{1/q_i} (S_\sigma f^*)(t) \leq \int_0^\infty s^{1/p_i} f^*(s) \frac{ds}{s}, \quad t > 0, i = 1,2. \tag{3.5.10}$$

In particular, S_σ *is of weak type* (p_i, q_i) *for* $i = 1,2$.

Proof. The definition of S_σ shows that

$$t^{1/q_1} (S_\sigma f^*)(t) \leq \int_0^{t^l} s^{1/p_1} f^*(s) \frac{ds}{s} + t^{1/q_1 - 1/q_2} \int_{t^l}^\infty s^{1/p_2} f^*(s) \frac{ds}{s},$$

where l is defined by (3.5.2). Since $p_1 < p_2$,

$$t^{1/q_1 - 1/q_2} = \left(t^l\right)^{1/p_1 - 1/p_2} \leq s^{1/p_1 - 1/p_2}$$

if $t^l \leq s$. The result (3.5.10) for $i = 1$ now follows immediately; that when $i = 2$ is proved in a similar manner. The remainder of the Lemma follows directly from (3.5.10) and Proposition 3.5.4. □

We can now give a useful characterisation of operators of joint weak type.

Theorem 3.5.13. *Let* $1 \leq p_1 < p_2 < \infty$ *and* $q_1, q_2 \in [1,\infty], q_1 \neq q_2$. *A quasilinear operator* T *is of joint weak type* $(p_1, q_1; p_2, q_2)$ *if, and only if, it is of weak types* (p_1, q_1) *and* (p_2, q_2).

Proof. If T is of joint weak type $(p_1, q_1; p_2, q_2)$, then by Lemma 3.5.12 it is of weak types (p_1, q_1) and (p_2, q_2). Conversely, suppose that T is of weak types (p_1, q_1) and (p_2, q_2). Let $f \in L_{p_1,1}(R,\mu) + L_{p_2,1}(R,\mu)$, let $t > 0$ and define l by (3.5.2). For each $x \in R$ put

$$f_2(x) = \min\left\{|f(x)|, f^*(t^l)\right\} \text{ sgn } f(x)$$

and

$$f_1(x) = f(x) - f_2(x) = \left\{|f(x)| - f^*(t^l)\right\}^+ \text{sgn } f(x).$$

Since

$$f_2^*(s) = \min\left\{f^*(s), f^*(t^l)\right\}, \ s > 0,$$

it follows that

$$\|f_2 \mid L_{p_2,1}\| = p_2 t^{l/p_2} f^*(t^l) + \int_{t^l}^{\infty} s^{1/p_2} f^*(s) \frac{ds}{s}. \tag{3.5.11}$$

In the same way it can be shown that

$$\|f_1 \mid L_{p_1,1}\| = \int_0^{t^l} s^{1/p_1} f^*(s) \frac{ds}{s} - p_1 t^{l/p_1} f^*(t^l). \tag{3.5.12}$$

Since T is quasilinear, let k be the constant which occurs in Definition 3.5.1. As $f = f_1 + f_2$, by Proposition 3.2.4(vi),

$$\begin{aligned}(Tf)^*(t) &\le \{k(|Tf_1| + |Tf_2|)\}^*(t) \\ &\le k\left\{(Tf_1)^*(t/2) + (Tf_2)^*(t/2)\right\}.\end{aligned}$$

Moreover, because T is of weak type (p_i, q_i) $(i = 1, 2)$,

$$(Tf_i)^*(t/2) \le (t/2)^{-1/q_i} M_i \|f_i \mid L_{p_i,1}\|, \ i = 1, 2. \tag{3.5.13}$$

Hence

$$(Tf)^*(t) \le c\left\{t^{-1/q_1} \|f_1 \mid L_{p_1,1}\| / p_1 + t^{-1/q_2} \|f_2 \mid L_{p_2,1}\| / p_2\right\}, \tag{3.5.14}$$

where $c = k \max\left\{p_i M_i 2^{1/q_i} : i = 1, 2\right\}$. Together with (3.5.11) and (3.5.12) this shows that

$$(Tf)^*(t) \le c(S_\sigma f^*)(t),$$

and the proof is complete. □

Remark 3.5.14. The same argument as above shows that if T is of joint weak type $(p_1, q_1; \infty, q_2)$ with $p_1 < \infty$, then it is of weak type (p_1, q_1). In the opposite direction, it can be shown that if T is of weak type (p_1, q_1) with $p_1 < \infty$, and of strong type (∞, q_2), then T is of joint weak type $(p_1, q_1; \infty, q_2)$. The modifications of the proof needed to establish this are to use

$$\|f_2 \mid L_\infty\| = f^*\left(t^l\right)$$

instead of (3.5.11), and to replace (3.5.13) with $i = 2$ by

$$(Tf_2)^*(t/2) \leq (t/2)^{-1/q_2} M_2 \|f_2 \mid L_\infty\|,$$

which arises from the strong-type hypothesis.

We can now give a variant of the Marcinkiewicz interpolation theorem (see [15], Theorem 4.4.13 and Corollary 4.4.14) adapted to Lorentz-Karamata spaces defined on $\mathbb{R}^n$.

Theorem 3.5.15. *Suppose that $1 \leq p_1 < p_2 \leq \infty$ and $1 \leq q_1, q_2 \leq \infty$, with $q_1 \neq q_2$; let $1 \leq r \leq s \leq \infty$ and let b be slowly varying. Suppose that $0 < \theta < 1$ and define p, q by*

$$\frac{1}{p} = \frac{1-\theta}{p_1} + \frac{\theta}{p_2}, \frac{1}{q} = \frac{1-\theta}{q_1} + \frac{\theta}{q_2}.$$

Let T be a quasilinear operator of joint weak type $(p_1, q_1; p_2, q_2)$, with values in $\mathcal{M}(\mathbb{R}^n, \mu_n)$ and defined for all those $f \in \mathcal{M}(\mathbb{R}^n, \mu_n)$ for which $(S_\sigma f^*)(1) < \infty$ (see (3.5.7)). Suppose that

$$\sup_{0<t<1} \frac{b_l(1/t)}{b(1/t))} < \infty,$$

where l is given by (3.5.2) and b_l is the slowly varying function defined by $b_l(t) = b(t^{1/|l|})$, $t \geq 1$. Then T maps $L_{p,r;b}(\mathbb{R}^n)$ boundedly into $L_{q,s;b}(\mathbb{R}^n)$.

Proof. Let $f \in L_{p,r;b}(\mathbb{R}^n)$. From (3.5.7) and (3.5.8),

$$\|Tf \mid L_{q,s;b}(\mathbb{R}^n)\| \lesssim N_1 + N_2,$$

where

$$N_1 = \left\| t^{1/q-1/q_1-1/s} \gamma_b(t) \int_0^{t^l} u^{1/p_1-1} f^*(u)\, du \mid L_s(0,\infty) \right\|$$

and

$$N_2 = \left\| t^{1/q-1/q_2-1/s} \gamma_b(t) \int_{t^l}^{\infty} u^{1/p_2-1} f^*(u)\, du \mid L_s(0,\infty) \right\|.$$

Recall that l is given by (3.5.2), so that

$$1/q - 1/q_j = l\,(1/p - 1/p_j) \text{ for } j = 1, 2.$$

Moreover, $\gamma_b(t^{1/l}) = \gamma_b(t^{1/|l|}) = \gamma_{b_l}(t)$, $t > 0$. The change of variables $\tau = t^l$ shows that

$$N_1 \lesssim \left\| \tau^{1/p-1/p_1-1/s}\gamma_{b_l}(\tau) \int_0^{\tau} u^{1/p_1-1} f^*(u)\,du \mid L_s(0,\infty) \right\|$$

and

$$N_2 \lesssim \left\| \tau^{1/p-1/p_2-1/s}\gamma_{b_l}(\tau) \int_{\tau}^{\infty} u^{1/p_2-1} f^*(u)\,du \mid L_s(0,\infty) \right\|.$$

By Hardy's inequality in the form given in Lemma 3.4.34(i) we have

$$N_1 \lesssim \left\| \tau^{1/p-1/r} f^*(\tau)\gamma_b(\tau) \mid L_r(0,\infty) \right\|;$$

part (ii) of the same Lemma gives the same upper estimate for N_2. This completes the proof of the Theorem. □

Corollary 3.5.16. *Let T be a quasilinear operator such that for all $q \in (1,\infty)$,*

$$T : L_q(\mathbb{R}^n) \to L_q(\mathbb{R}^n)$$

is bounded. Let $1 < p < \infty, 1 \leq r \leq \infty$ and let b be slowly varying. Then

$$T : L_{p,r;b}(\mathbb{R}^n) \to L_{p,r;b}(\mathbb{R}^n)$$

is bounded.

Proof. Choose p_1, p_2 so that $1 < p_1 < p < p_2 < \infty$ and define $\theta \in (0,1)$ by $1/p = (1-\theta)/p_1 + \theta/p_2$. By assumption, T is of strong types (p_1,p_1) and (p_2,p_2). Hence by Remark 3.5.11 (ii), T is of weak types (p_1,p_1) and (p_2,p_2), so that it is of joint weak type $(p_1,p_1;p_2,p_2)$, by Theorem 3.5.13. The result now follows from Theorem 3.5.15. □

Remark 3.5.17. If $1 < p \leq \infty$, the Hardy-Littlewood maximal operator $\mathbb{M}$ defined in Remark 3.5.2 maps $L_p(\mathbb{R}^n)$ boundedly into itself: there is a constant c, depending only on n and p, such that for all $f \in L_p(\mathbb{R}^n)$,

$$\|\mathbb{M}f \mid L_p(\mathbb{R}^n)\| \leq c \|f \mid L_p(\mathbb{R}^n)\|.$$

For this famous result see, for example, [211], Theorem I.1. Its proof relies on (part of) the estimate

$$(\mathbb{M}f)^*(t) \approx f^{**}(t),\ t > 0,\ f \in L_{1,loc}(\mathbb{R}^n)$$

(see [15], 3.3). It follows that $\mathbb{M}$ maps $L_{p,r;b}(\mathbb{R}^n)$ boundedly into itself whenever $1 < p < \infty$, $1 \leq r \leq \infty$ and b is slowly varying. For a comprehensive account of the mapping properties of the maximal operator, its fractional variant and even more general operators between Lorentz spaces of classical and weak type, see [70].

3.6 Bessel-Lorentz-Karamata-potential spaces

In this section we introduce spaces of Bessel-potential type modelled on Lorentz-Karamata spaces and give an account of the principal embedding results involving them.We begin with a discussion of abstract Sobolev spaces.

3.6.1 Abstract Sobolev spaces

Let Ω be a domain in $\mathbb{R}^n$ and let $X = X(\Omega)$ be a Banach function space on (Ω, μ_n) with norm $\|\cdot \mid X\|$. By Definition 3.1.1, condition (P5), we see that $X(\Omega) \subset L_1^{loc}(\Omega)$: hence any $f \in X(\Omega)$ has distributional derivatives on Ω of all orders. Note also that in view of (P2) and (P4), $C_0^\infty(\Omega) \subset X(\Omega)$. Following [73] (see also [72]) we introduce abstract Sobolev spaces as follows.

Definition 3.6.1. *Let $X = X(\Omega)$ and $Y = Y(\Omega)$ be Banach function spaces on (Ω, μ_n) and let $k \in \mathbb{N}$. The* ***abstract Sobolev space*** *$W^k(X,Y) = W^k(X(\Omega), Y(\Omega))$ (also written as $W^k(X,Y)(\Omega)$) is defined by*

$$W^k(X,Y) = \{f \in X(\Omega) : D^\alpha f \in Y(\Omega) \text{ for all } \alpha \in \mathbb{N}_0^n \text{ with } |\alpha| \le k\},$$

and is equipped with the norm

$$\|f \mid W^k(X,Y)\| := \sum_{0<|\alpha|\le k} \|D^\alpha f \mid Y(\Omega)\| + \|f \mid X(\Omega)\|.$$

When $k = 1$ we may, on occasion, omit the superscript 1. We write $W^k(X)$ or even $W^k X$ for $W^k(X,X)$.

When $X(\Omega) = Y(\Omega) = L_p(\Omega)$ $(1 \le p \le \infty)$, $W^k(X,Y)$ becomes the familiar Sobolev space $W_p^k(\Omega)$.

While $W^k(X,Y)$ is not a Banach function space, we nevertheless have

Lemma 3.6.2. *The abstract Sobolev space $W^k(X,Y)$ is a Banach space.*

Proof. Let (u_j) be a Cauchy sequence in $W^k(X,Y)$ and let $\alpha \in \mathbb{N}_0^n$, $|\alpha| \le k$. Then there are functions u and g_α $(|\alpha| \le k)$ such that

$$u_j \to u \text{ in } X \text{ and } D^\alpha u_j \to g_\alpha \text{ in } Y \quad (|\alpha| \le k).$$

Let $\phi \in C_0^\infty(\Omega)$, put $S = \operatorname{supp} \phi$ and define

$$L_0(g) = \int_\Omega g\phi \, dx \ (g \in Y), \ L_\alpha(f) = \int_\Omega f D^\alpha \phi \, dx \ (f \in X).$$

By (P5),

$$|L_0(g)| \le \left(\max_\Omega |\phi|\right) \int_S |g| \, dx \le C(S) \left(\max_\Omega |\phi|\right) \|g \mid Y\|$$

for all $g \in Y(\Omega)$. In the same way,

$$|L_\alpha(f)| \le C(S)\left(\max_\Omega |\phi|\right) \|f \mid X\|$$

for all $f \in X$. Thus $L_0 \in Y^*$ and $L_\alpha \in X^*$, and it follows that

$$L_0\left(D^\alpha u_j\right) \to L_0\left(g_\alpha\right), \; L_\alpha\left(u_j\right) \to L_\alpha\left(u\right)$$

as $j \to \infty$. Hence

$$\begin{aligned}
(-1)^{|\alpha|} \int_\Omega u D^\alpha \phi dx &= (-1)^{|\alpha|} L_\alpha(u) = (-1)^{|\alpha|} \lim_{j\to\infty} L_\alpha(u_j) \\
&= (-1)^{|\alpha|} \lim_{j\to\infty} \int_\Omega u_j D^\alpha \phi dx \\
&= \lim_{j\to\infty} \int_\Omega \phi D^\alpha u_j dx \\
&= \lim_{j\to\infty} L_0\left(D^\alpha u_j\right) = L_0\left(g_\alpha\right) \\
&= \int_\Omega g_\alpha \phi dx.
\end{aligned}$$

Hence $g_\alpha = D^\alpha u$ and so $u_j \to u$ in $W^k(X, Y)$. The proof is complete. □

Remark 3.6.3. If $\phi \in C_0^\infty(\Omega)$, then obviously $D^\alpha \phi \in C_0^\infty(\Omega)$ for all $\alpha \in \mathbb{N}_0^n$. Thus $C_0^\infty(\Omega) \subset W^k(X(\Omega), Y(\Omega))$. We therefore define the subspace $\mathring{W}^k(X(\Omega), Y(\Omega))$ of $W^k(X(\Omega), Y(\Omega))$ to be the closure of $C_0^\infty(\Omega)$ in $W^k(X(\Omega), Y(\Omega))$. For shortness we shall write $\mathring{W}^k(X)$ or $\mathring{W}^k X$ instead of $\mathring{W}^k(X, X)$. If $X(\Omega) = L_p(\Omega)$, it is denoted by $\mathring{W}_p^k(\Omega)$.

3.6.2 Bessel-Lorentz-Karamata-potential spaces

Given any $\sigma \in (0, \infty)$, the **Bessel kernel** g_σ is defined to be that function whose Fourier transform $\hat{g}_\sigma$ is given by

$$\hat{g}_\sigma(x) = (2\pi)^{-n/2}\left(1 + |x|^2\right)^{-\sigma/2}, \; x \in \mathbb{R}^n; \tag{3.6.1}$$

we recall that the Fourier transform $\hat{f}$ (or $F(f)$) of a function f is defined by

$$\hat{f}(x) = (2\pi)^{-n/2} \int_{\mathbb{R}^n} e^{-ix.y} f(y) dy,$$

where $x.y = \sum_{k=1}^n x_k y_k$. We summarise the basic properties of g_σ in the following

Proposition 3.6.4. *Let $\sigma > 0$ and let g_σ be the Bessel kernel. Then:*
(i) g_σ is a positive function in $L_1(\mathbb{R}^n)$ which is analytic except at 0 and is given by

$$g_\sigma(\xi) = \frac{1}{(4\pi)^{\sigma/2}\Gamma(\sigma/2)} \int_0^\infty e^{-\pi|\xi|^2/x} e^{-x/(4\pi)} x^{(\sigma-n)/2} \frac{dx}{x}, \quad \xi \neq 0;$$

(ii) $g_\sigma * g_\tau = g_{\sigma+\tau}$ if $\tau > 0$;
(iii) as $|x| \to 0$,

$$g_\sigma(x) \approx \begin{cases} |x|^{\sigma-n} & \text{if } 0 < \sigma < n, \\ \log(1/|x|) & \text{if } \sigma = n, \\ 1 & \text{if } \sigma > n; \end{cases} \tag{3.6.2}$$

(iv) as $|x| \to \infty$,

$$g_\sigma(x) \approx |x|^{(\sigma-n-1)/2} e^{-|x|}; \tag{3.6.3}$$

(v) there exists $c > 0$ such that for all $x \in \mathbb{R}^n$ and all $\sigma \in (0, n)$,

$$g_\sigma(x) \approx |x|^{\sigma-n} e^{-c|x|}. \tag{3.6.4}$$

These results are well known: see [211], [231] and (for $\sigma \geq n$) [8]. What we now need is an estimate for the non-increasing rearrangement of the Bessel kernel. This is provided by the following lemma (see [50] and [52]).

Lemma 3.6.5. *(i) Suppose that $0 < \sigma < n$. Then there are positive constants A and B such that for all $t > 0$,*

$$g_\sigma^*(t) \leq A t^{(\sigma/n)-1} \exp\left(-B t^{1/n}\right) \tag{3.6.5}$$

and

$$g_\sigma^{**}(t) \leq \frac{n}{\sigma} A t^{(\sigma/n)-1}. \tag{3.6.6}$$

(ii) Suppose that $n \leq \sigma < \infty$. Then there are positive constants C and D such that

$$g_\sigma^*(t) = \begin{cases} C \exp\left(-D t^{1/n}\right) \log(e/t) & \text{if } \sigma = n \text{ and } 0 < t < 1, \\ C \exp\left(-D t^{1/n}\right) & \text{if } \sigma > n \text{ and } 0 < t < 1, \\ C \exp\left(-D t^{1/n}\right) & \text{if } \sigma \geq n \text{ and } 1 \leq t < \infty. \end{cases} \tag{3.6.7}$$

Proof. (i) By Proposition 3.6.4, (i) and (iv),

$$0 \leq g_\sigma(x) \leq H(x), \quad x \in \mathbb{R}^n,$$

where $H(x) = h(|x|)$ and $h(t) = c_1 t^{\sigma-n} \exp(-ct)$ $(t \geq 0)$ for some $c_1 > 0$. It is therefore enough to show that (3.6.5) and (3.6.6) hold with g_σ replaced by H.

Since h is decreasing, its distribution function μ_h satisfies $\mu_h(\lambda) = h^{-1}(\lambda)$ for all $\lambda > 0$. Hence

$$\mu_H(\lambda) = \mu_n\left(B_n\left(0, h^{-1}(\lambda)\right)\right) = \omega_n\left(h^{-1}(\lambda)\right)^n, \ \lambda > 0;$$

and

$$\begin{aligned} H^*(t) &= \inf\left\{\lambda > 0 : \omega_n\left(h^{-1}(\lambda)\right)^n \leq t\right\} \\ &= h\left((t/\omega_n)^{1/n}\right) = At^{(\sigma/n)-1}\exp\left(-Bt^{1/n}\right), \ t > 0, \end{aligned}$$

where

$$A = c_1\omega_n^{1-(\sigma/n)}, B = c\omega_n^{-1/n}.$$

This gives (3.6.5); (3.6.6) follows directly since

$$H^{**}(t) = t^{-1}\int_0^t As^{(\sigma/n)-1}\exp\left(-Bs^{1/n}\right)ds \leq (n/\sigma)\,At^{(\sigma/n)-1}.$$

(ii) Define a function h_σ on $(0,\infty)$ by

$$h_\sigma(t) = \begin{cases} \log(e/t) & \text{if } \sigma = n \text{ and } 0 < t < 1, \\ 1 & \text{if } \sigma > n \text{ and } 0 < t < 1, \\ 1 & \text{if } \sigma \geq n \text{ and } 1 \leq t < \infty. \end{cases}$$

Then by (3.6.2) and (3.6.3), there exists $c > 0$ such that

$$0 \leq g_\sigma(x) \lesssim h_\sigma(|x|)\exp(-c\,|x|) := H_\sigma(|x|)$$

for all $x \in \mathbb{R}^n \setminus \{0\}$ and all $\sigma \geq n$. Hence

$$g_\sigma^*(t) \lesssim H_\sigma^*(t) \text{ if } t > 0, \sigma \geq n.$$

Define a function f by $f(t) = h_\sigma(t)\exp(-ct)$, $t > 0$. Since f is decreasing, its distribution function satisfies

$$\mu_f(\lambda) = f^{-1}(\lambda) \text{ for } \lambda > 0.$$

Thus

$$\mu_{H_\sigma}(\lambda) = \omega_n\left(f^{-1}(\lambda)\right)^n, \ \lambda > 0,$$

and so

$$H_\sigma^*(t) = \inf\left\{\lambda > 0 : \omega_n\left(f^{-1}(\lambda)\right)^n \leq t\right\} = f\left((t/\omega_n)^{1/n}\right), \ t > 0.$$

The estimate (3.6.7) now follows immediately. □

We note that Gurka and Opic [112] have obtained the following sharpening of the estimate (3.6.6):

$$g_\sigma^{**}(t) \lesssim t^{-1} \text{ for } t > 1. \tag{3.6.8}$$

Definition 3.6.6. *Let $\sigma > 0, 1 \leq p < \infty$, $1 \leq q \leq \infty$ and let b be slowly varying. The* ***Bessel-Lorentz-Karamata-potential space*** *$H^{\sigma}L_{p,q;b}(\mathbb{R}^n)$, is defined to be*

$$\{u : u = g_{\sigma} * f \text{ for some } f \in L_{p,q;b}(\mathbb{R}^n)\}.$$

It is equipped with the (quasi-)norm

$$\|u \mid H^{\sigma}L_{p,q;b}(\mathbb{R}^n)\| := \|f \mid L_{p,q;b}(\mathbb{R}^n)\|$$

where $u = g * f$. By $H^0L_{p,q;b}(\mathbb{R}^n)$ we shall mean $L_{p,q;b}(\mathbb{R}^n)$.

Taking $b = \vartheta_{\alpha}^{m}$, where $m \in \mathbb{N}$ and $\alpha \in \mathbb{R}^m$ the space defined above is the logarithmic Bessel-potential space $H^{\sigma}L_{p,q;\alpha}(\mathbb{R}^n)$ considered in [52] and which reduces to the (fractional) Sobolev space $H_p^{\sigma}(\mathbb{R}^n)$ of order σ when $\alpha = \mathbf{0}$.

We now turn our attention to the connection between the spaces just introduced and the spaces

$$W^kL_{p,q;b}(\mathbb{R}^n) := \{u : D^{\alpha}u \in L_{p,q;b}(\mathbb{R}^n) \text{ if } |\alpha| \leq k\},$$

with norm

$$\|u \mid W^kL_{p,q;b}(\mathbb{R}^n)\| = \sum_{|\alpha|\leq k} \|D^{\alpha}u \mid L_{p,q;b}(\mathbb{R}^n)\|,$$

($k \in \mathbb{N}$) of classical Sobolev type based on $L_{p,q;b}(\mathbb{R}^n)$: see Definition 3.6.1. It is well known that if $1 < p < \infty$, the Bessel potential space $H_p^k(\mathbb{R}^n)$ coincides with the Sobolev space $W_p^k(\mathbb{R}^n)$ when $k \in \mathbb{N}$. To establish an analogous result in general circumstances we begin with some density results.

Lemma 3.6.7. *Let $1 < p < \infty$, $1 \leq q < \infty$ and suppose that b is slowly varying. Then*
(i) $C_0^{\infty}(\mathbb{R}^n)$ is dense in $W^1L_{p,q;b}(\mathbb{R}^n)$;
(ii) for all $\sigma \geq 0$, the Schwartz space $\mathcal{S}$ is dense in $H^{\sigma}L_{p,q;b}(\mathbb{R}^n)$.

Proof. We know that $L_{p,q;b}(\mathbb{R}^n)$ is a rearrangement-invariant Banach function space with absolutely continuous norm. Thus (i) follows from the standard regularisation process, as pointed out in [52], Remark 3.13.

As for (ii), first we show that $\mathcal{S} \subset L_{p,q;b}(\mathbb{R}^n)$. Given $f \in \mathcal{S}$, for every $m \in \mathbb{N}$ there is a positive constant c_m such that for all $x \in \mathbb{R}^n$,

$$|f(x)| \leq c_m(1 + |x|^2)^{-m/2}. \tag{3.6.9}$$

Application of this with $m = 0$ shows that $f^*(t) \leq c_0$ for all $t \in (0, \infty)$, and so

$$\int_0^1 \left(t^{1/p-1/q}\gamma_b(t)f^*(t)\right)^q dt \leq c_0^q \int_0^1 t^{q/p-1}\gamma_b^q(t)dt < \infty.$$

Another application of (3.6.9), with an m to be chosen later, shows that

$$f^*(t) \leq c_m \omega_n^{m/n} (\omega_n^{2/n} + t^{2/n})^{-m/2},$$

where ω_n is the volume of the unit ball in $\mathbb{R}^n$. Hence

$$\int_1^\infty \left(t^{1/p-1/q} \gamma_b(t) f^*(t) \right)^q dt \lesssim \int_1^\infty t^{-qm/n+q/p-1} \gamma_b^q(t) dt < \infty$$

if we choose $m > n/p$. Hence $\| f \mid L_{p,q;b}(\mathbb{R}^n) \| < \infty$, as required.

Now let $\sigma > 0$, so that by [211], p. 135, $\{g_\sigma * h : h \in \mathcal{S}\} = \mathcal{S}$. Given $f \in \mathcal{S}$, $f = g_\sigma * h$ for some $h \in \mathcal{S} \subset L_{p,q;b}(\mathbb{R}^n)$: hence $\mathcal{S} \subset H^\sigma L_{p,q;b}(\mathbb{R}^n)$. The density now follows from (i). □

Lemma 3.6.8. *Let $\sigma \in [1, \infty)$, suppose that $p \in (1, \infty)$, $q \in [1, \infty)$ and let b be slowly varying. Then $f \in H^\sigma L_{p,q;b}(\mathbb{R}^n)$ if, and only if, $f \in H^{\sigma-1} L_{p,q;b}(\mathbb{R}^n)$ and the distributional derivatives $\frac{\partial f}{\partial x_j}$ belong to $H^{\sigma-1} L_{p,q;b}(\mathbb{R}^n)$ for $j = 1, ..., n$. The (quasi-)norms*

$$\| f \mid H^\sigma L_{p,q;b}(\mathbb{R}^n) \|$$

and

$$\| f \mid H^{\sigma-1} L_{p,q;b}(\mathbb{R}^n) \| + \sum_{j=1}^n \left\| \frac{\partial f}{\partial x_j} \mid H^{\sigma-1} L_{p,q;b}(\mathbb{R}^n) \right\|$$

are equivalent on $H^\sigma L_{p,q;b}(\mathbb{R}^n)$.

Proof. In view of Lemma 3.6.7 and Corollary 3.5.16 an obvious adaptation of the proof for the L_p case (see [211], Chapter V, Lemma 3) establishes the result. □

The connection between Bessel-Lorentz-Karamata- potential spaces and Sobolev spaces modelled on $L_{p,q;b}(\mathbb{R}^n)$ now follows.

Theorem 3.6.9. *Let $k \in \mathbb{N}$, suppose that $p \in (1, \infty)$, $q \in [1, \infty)$ and let b be slowly varying. Then*

$$H^k L_{p,q;b}(\mathbb{R}^n) = W^k L_{p,q;b}(\mathbb{R}^n),$$

and the corresponding (quasi-)norms are equivalent.

Proof. This is similar to that of [211], Chapter V, Theorem 3. □

We now embark on a detailed study of embeddings of Bessel-Lorentz-Karamata-potential spaces.

3.6.3 Sub-limiting embeddings

The first result is a natural extension of the standard Sobolev embedding theorem..

Theorem 3.6.10. *Let $\sigma \in (0, n)$, $1 < p < n/\sigma$, $1/r = 1/p - \sigma/n$, $q \in [1, \infty]$, $m \in \mathbb{N}$ and let b be slowly varying . Then*

$$H^{\sigma} L_{p,q;b}(\mathbb{R}^n) \hookrightarrow L_{r,q;b}(\mathbb{R}^n).$$

Proof. Let $X = H^{\sigma} L_{p,q;b}(\mathbb{R}^n)$ and let $u \in X$, so that $u = g_{\sigma} * f$ for some $f \in L_{p,q;b}(\mathbb{R}^n)$ and $\|f \mid L_{p,q;b}(\mathbb{R}^n)\| = \|u \mid X\|$. By O'Neil's inequality (Lemma 3.5.6),

$$u^*(t) \le u^{**}(t) \le t g_{\sigma}^{**}(t) f^{**}(t) + \int_t^{\infty} g_{\sigma}^*(s) f^*(s) ds, \ t > 0. \tag{3.6.10}$$

From this and (3.6.6) we see that for every $t > 0$,

$$u^*(t) \lesssim t^{\sigma/n-1} \int_0^t f^*(s) ds + \int_t^{\infty} s^{\sigma/n-1} f^*(s) ds,$$

and so

$$\|u \mid L_{r,q;b}(\mathbb{R}^n)\| \lesssim N_1 + N_2,$$

where

$$N_1 = \left\| t^{1/p-1-1/q} \gamma_b(t) \int_0^t f^*(s) ds \mid L_q(0, \infty) \right\|$$

and

$$N_2 = \left\| t^{1/p-\sigma/n-1/q} \gamma_b(t) \int_t^{\infty} s^{\sigma/n-1} f^*(s) ds \mid L_q(0, \infty) \right\|.$$

By Lemma 3.4.34 (i) ,

$$N_1 \lesssim \left\| t^{1/p-1/q} \gamma_b(t) f^*(t) \mid L_q(0, \infty) \right\| = \|f \mid L_{p,q;b}(\mathbb{R}^n)\|$$

and by Lemma 3.4.34 (ii) the same upper estimate holds also for N_2. The result follows. □

We note that the result may be obtained from the observation that the map T defined by $Tf = u = g_{\sigma} * f$ is of joint weak type $(p_1, q_1; p_2, q_2)$, with $1/q_1 = (n - \sigma)/n$, $p_1 = 1$, $q_2 = \infty$ and $1/p_2 = \sigma/n$. Since $1/p \in (\sigma/n, 1) = (1/p_2, 1/p_1)$, we have $1/p = (1 - \theta)/p_1 + \theta/p_2$ with $\theta = (1 - 1/p)/(1 - \sigma/n) \in (0, 1)$. Moreover, the number r given by $1/r = (1 - \theta)/q_1 + \theta/q_2$ satisfies $1/r = 1/p - \sigma/n$. The result now follows from Theorem 3.5.15.

When $b = 1$ and $p = q$, Theorem 3.6.10 tells us that $H^{\sigma} L_p(\mathbb{R}^n) \hookrightarrow L_{r,p}(\mathbb{R}^n)$, and since $L_{r,p}(\mathbb{R}^n) \hookrightarrow L_r(\mathbb{R}^n)$ as $r > p$, we recover the classical Sobolev embedding theorem.

Note also that in the theorem, $\sigma p < n$: this is what we mean by the term 'sublimiting'. In this case the nature of the embedding is unaffected by the particular b and q : the target spaces are of the same GLZ type. If $\sigma p \ge n$ the situation is quite different, as we shall see in the next sections.

3.6.4 Limiting embeddings

Here we examine the spaces $H^\sigma L_{p,q;b}(\mathbb{R}^n)$ with $\sigma p = n$, so that the power exponent p has a limiting value.

Lemma 3.6.11. *Let $\sigma \in (0,n)$, suppose that $p,q \in [1,\infty]$ and let b_1, b_2 be slowly varying, with*

$$\left\| t^{-1/q} b_2(1/t) \mid L_q(0,1) \right\| < \infty. \tag{3.6.11}$$

Assume also that either conditions (3.4.21) and (3.4.25), or (3.4.22) and (3.4.26) hold. Then

$$\| u^* \mid L_{\infty,q;b_2}(0,1) \| \lesssim \left\| u \mid H^\sigma L_{n/\sigma,q;b_2}(0,1) \right\| \tag{3.6.12}$$

for all $u \in H^\sigma L_{n/\sigma,p;b_1}(\mathbb{R}^n)$.

Proof. For shortness denote $L_{n/\sigma,p;b_1}(\mathbb{R}^n)$ by X, Let $u \in H^\sigma L_{n/\sigma,p;b_1}(\mathbb{R}^n)$, so that $u = g_\sigma * f$ for some $f \in X$ with $\|f \mid X\| = \left\| u \mid H^\sigma L_{n/\sigma,p;b_1}(\mathbb{R}^n) \right\|$. O'Neil's inequality (3.6.10) and (3.6.6) show that for all $t \in (0,1)$,

$$u^*(t) \le n\sigma^{-1} A t^{\sigma/n-1} \int_0^t f^*(s)ds + \int_t^1 g_\sigma^*(s) f^*(s)ds + \int_1^\infty g_\sigma^*(s) f^*(s)ds.$$

Since, by Lemma 3.4.42, there is a positive constant c such that for all $t > 0$,

$$f^*(t) \le f^{**}(t) \le \frac{c}{t^{\sigma/n}\gamma_{b_1}(t)} \|f \mid X\|, \; t > 0,$$

we see with the help of (3.6.5) that

$$\begin{aligned} \int_1^\infty g_\sigma^*(s) f^*(s)ds &\le C \|f \mid X\| \int_1^\infty s^{\sigma/n-1} \exp(-Bs^{1/n}) \frac{s^{-\sigma/n}}{b_1(s)} ds \\ &= C_1 \|f \mid X\|. \end{aligned}$$

It thus follows that

$$\|u^* \mid L_{\infty,q;b_2}(0,1)\| \lesssim N_1 + N_2 + N_3 \|f \mid X\|,$$

where

$$N_1 = \left\| t^{\sigma/n-1-1/q} b_2(1/t) \int_0^t f^*(s)ds \mid L_q(0,1) \right\|,$$

$$N_2 = \left\| t^{-1/q} b_2(1/t) \int_t^1 f^*(s)ds \mid L_q(0,1) \right\|$$

and

$$N_3 = \left\| t^{-1/q} b_2(1/t) \mid L_q(0,1) \right\|.$$

By (3.6.11), $N_3 < \infty$. Use of Lemma 3.4.34 gives

$$N_1 \lesssim \left\| t^{\sigma/n-1+1/p'} \gamma_{b_1}(t) f^*(t) \mid L_p(0,\infty) \right\| = \|f \mid X\|,$$

and Lemma 3.4.35 plus (3.6.5) produce

$$N_2 \lesssim \left\| t^{1/p'} \gamma_{b_1}(t) t^{\sigma/n-1} f^*(t) \mid L_p(0,1) \right\| \lesssim \|f \mid X\|.$$

The proof is complete. □

By taking b_1 and b_2 to be of the form ϑ_α^m and ϑ_β^m we immediately obtain from this lemma:

Corollary 3.6.12. *Let $\sigma \in (0,n)$, $m \in \mathbb{N}$, $\alpha \in \mathbb{R}^m$ and $p,q \in [1,\infty]$; let $k \in \{1,...,m\}$ be such that $\alpha_k \neq 1/p'$ and, if $k \geq 2$, $\alpha_i = 1/p'$ for $i = 1,...,k-1$. Let $\beta \in \mathbb{R}^m$ satisfy $\beta_k \neq -1/q$ and, if $k \geq 2$, $\beta_i = 1/q$ for $i = 1,...,k-1$. Then for all $u \in H^\sigma L_{n/\sigma,p;\alpha}(\mathbb{R}^n)$,*

$$\|u^* \mid L_{\infty,q;\beta}(0,1)\| \lesssim \left\|u \mid H^\sigma L_{n/\sigma,p;\alpha}(\mathbb{R}^n)\right\| \tag{3.6.13}$$

if either

$$1 \leq p \leq q \leq \infty,\ \beta_k < -1/q,\ \beta + \delta_{q;m,k} \preceq \alpha - \delta_{p';m,k} \tag{3.6.14}$$

or

$$1 \leq q < p \leq \infty,\ \beta_k < -1/q,\ \beta + 1/q \prec \alpha - \delta_{1;m,k} + 1/p. \tag{3.6.15}$$

If the assumption $\beta \neq -1/q$ is omitted the result still holds if

$$1 \leq p \leq q = \infty,\ \beta = \mathbf{0} \text{ and } -\alpha_k + 1/p' < 0. \tag{3.6.16}$$

When $k = m$, if we also omit the assumption $\alpha_m \neq 1/p'$, the result remains true if

$$p = 1, q = \infty,\ \alpha = \beta = \mathbf{0}. \tag{3.6.17}$$

As a special case of this we see that when $\Omega \subset \mathbf{R}^n$ has finite volume, $p \in (1,\infty)$ and $k \in \mathbf{N}$, $kp = n$,

$$W_p^k(\mathbf{R}^n) \hookrightarrow L_{\infty,p}(\log L)_{-1}(\Omega).$$

Results of this type were established by Hansson [117] and Brézis and Wainger [22].

Another result of limiting form is provided by

Theorem 3.6.13. *Let Ω be a measurable subset of $\mathbb{R}^n$ with finite volume $|\Omega|$, let $\sigma \in (0,n)$, suppose that $p,q \in [1,\infty]$ and let b_1, b_2 be slowly varying. Suppose that*

$$\left\| t^{-1/q} b_2(1/t) \mid L_q(0,1) \right\| < \infty$$

and that either conditions (3.4.21) and (3.4.25) or (3.4.22) and (3.4.26) hold. Define a slowly varying function $I_{b_2,q}$ by

$$I_{b_2,q}(t) = \left(\int_t^\infty \left(s^{-1/q} b_2(s) \right)^q ds \right)^{1/q}, \quad t \geq 1,$$

if $q < \infty$, and by $I_{b_2,q}(t) = b_2(t)$, $t \geq 1$, if $q = \infty$. Then

$$H^\sigma L_{n/\sigma,p;b_1}\left(\mathbb{R}^n\right) \hookrightarrow L_{\infty,q;b_2}\left(\Omega\right) \hookrightarrow L_{\infty,\infty;I_{b_2,q}}\left(\Omega\right).$$

Proof. We deal with the case in which $|\Omega| = 1$. By Lemma 3.6.11, with u_Ω as the restriction of u to Ω,

$$\begin{aligned} \|u \mid L_{\infty,q;b_2}(\Omega)\| &= \left\| t^{-1/q} \gamma_{b_2}(t) u_\Omega^*(t) \mid L_q(0,1) \right\| \\ &\leq \left\| t^{-1/q} \gamma_{b_2}(t) u^*(t) \mid L_q(0,1) \right\| \\ &\lesssim \left\| u \mid H^\sigma L_{n/\sigma,p;b_1}\left(\mathbb{R}^n\right) \right\| \end{aligned}$$

for all $u \in H^\sigma L_{n/\sigma,p;b_1}\left(\mathbb{R}^n\right)$. Since by Theorem 3.4.47 we have $L_{\infty,q;b_2}(\Omega) \hookrightarrow L_{\infty,\infty;I_{b_2,q}}(\Omega)$ the result follows. □

From this result with $b_1 = \vartheta_\alpha^m$ and $b_2 = \vartheta_\beta^m$ we easily obtain

Corollary 3.6.14. *Let Ω be a measurable subset of $\mathbb{R}^n$ with finite volume $|\Omega|$, let $\sigma \in (0,n)$, suppose that $p,q \in [1,\infty]$ and let $\alpha, \beta \in \mathbb{R}^m$ (for some $m \in \mathbb{N}$) and $k \in \{1,...,m\}$ be as in Corollary 3.6.12. Let $\nu \in \mathbb{R}^m$ be such that $\nu_k = \beta_k + 1/q$ and, if $k \geq 2$, $\nu_j = 0$ for $j = 1,...,k-1$ and, if $k \leq m-1$, $\nu_j = \beta_j$ for $j = k+1,...,m$. Then*

$$H^\sigma L_{n/\sigma,p;\alpha}\left(\mathbb{R}^n\right) \hookrightarrow L_{\infty,q;\beta}\left(\Omega\right) \hookrightarrow L_{\infty,\infty;\nu}\left(\Omega\right)$$

if one of the conditions (3.6.14)-(3.6.17) is satisfied.

As an additional Corollary we have

Corollary 3.6.15. *Let Ω be a measurable subset of $\mathbb{R}^n$ with finite volume $|\Omega|$, let $\sigma \in (0,n)$, suppose that $p,q \in [1,\infty]$ and let $\alpha, \beta \in \mathbb{R}^m$ (for some $m \in \mathbb{N}$) and $k \in \{1,...,m\}$ be such that $\alpha_k < 1/p', \beta_k = \alpha_k - 1/p'$ and, if $k \geq 2$, $\alpha_j = 1/p'$ and $\beta_j = 0$ for $j = 1,...,k-1$ while, if $k \leq m-1$, $\beta_j = \alpha_j$ for $j = k+1,...,m$. Then*

$$H^\sigma L_{n/\sigma,p;\alpha}\left(\mathbb{R}^n\right) \hookrightarrow L_{\Phi_k}\left(\Omega\right),$$

where Φ_k is a Young function given by

$$\Phi_k(t) = \exp_k \left(t^{-1/\beta_k} \prod_{i=1}^{m-k} \widetilde{l}_i^{\gamma_i}(t) \right) \text{ for all large enough } t > 0.$$

Here $\widetilde{l}_0(t) = t$ *and* $\widetilde{l}_i(t) = \log \widetilde{l}_{i-1}(t)$ *if* $i > 1$; *if* $k < m$, $\gamma \in \mathbb{R}^{m-k}$ *and* $\gamma_i = -\beta_{i+k}/\beta_k$ *for* $i = 1, ..., m-k$; *if* $k = m$, $\Phi_m(t) = \exp_m(t^{-1/\beta_m})$.

Proof. From Corollary 3.6.14 with $q = \infty$ we have $H^\sigma L_{n/\sigma,p;\alpha}(\mathbb{R}^n) \hookrightarrow L_{\infty,\infty;\beta}(\Omega)$. Now use Remark 3.4.30 to identify $L_{\infty,\infty;\beta}(\Omega)$ with $L_{\Phi_k}(\Omega)$. □

Yet another result in the same general direction is provided by

Theorem 3.6.16. *Let Ω be a measurable subset of $\mathbb{R}^n$ with finite volume $|\Omega|$, let $\sigma \in (0,n)$, suppose that $p \in [1,\infty]$ and let b be slowly varying; let Φ be a Young function such that the restriction of Φ^{-1} to $[1,\infty)$ is slowly varying. Suppose also that*

$$\int_0^1 \Phi\left(\gamma \Phi^{-1}(1/t)\right) dt < \infty \text{ for some } \gamma > 0; \tag{3.6.18}$$

$$\sup_{0<t<1} \frac{1}{\Phi^{-1}(1/t) b(1/t)} < \infty; \tag{3.6.19}$$

and

$$\sup_{0<t<1} \frac{1}{\Phi^{-1}(1/t)} \left\| \left(t^{1/p'} b(1/t) \right)^{-1} \mid L_{p'}(t,1) \right\| < \infty. \tag{3.6.20}$$

Then

$$H^\sigma L_{n/\sigma,p;b}(\mathbb{R}^n) \hookrightarrow L_\Phi(\Omega).$$

Proof. There will be no loss of generality if we assume that $|\Omega| = 1$. Let $b_1 = b$ and $b_2(t) = 1/\Phi^{-1}(1/t)$, $t \geq 1$. Then b_2 is slowly varying. Since Φ^{-1} is increasing and positive,

$$\sup_{0<t<1} b_2(1/t) = 1/\Phi^{-1}(1) < \infty,$$

so that (3.6.11), with $q = \infty$, holds. Condition (3.6.19) is just (3.4.21), and as

$$\sup_{0<t<s} b_2(1/t) = 1/\Phi^{-1}(s),$$

condition (3.4.25), with $q = \infty$, coincides with (3.6.20). Hence by Theorem 3.6.13, $H^\sigma L_{n/\sigma,p;b}(\mathbb{R}^n) \hookrightarrow L_{\infty,\infty;b_2}(\Omega)$. The proof is now completed by appealing to Corollary 3.4.28. □

In these results we have focused on the case in which $|\Omega| < \infty$. If Ω has infinite volume, corresponding results are available, for the details of which we refer to [112], [193], [194], [52] and [182].

3.6.5 Super-limiting embeddings

Here we discuss embeddings of $H^{\sigma}L_{p,q;b}(\mathbb{R}^n)$ when either $p > \max\{1, n/\sigma\}$, with $0 < \sigma < \infty$, or $p = \sigma/n$ and b satisfies a certain condition. The following Lemma, from [52], will be useful. In it, and subsequently, we shall understand by $C(\mathbb{R}^n)$ the space of all scalar-valued, bounded, continuous functions on $\mathbb{R}^n$, endowed with the $L_\infty(\mathbb{R}^n)$ norm. The space $H^{\sigma}Y$ is defined in the obvious way by analogy with Definition 3.6.6.

Lemma 3.6.17. *Let $Y = Y(\mathbb{R}^n)$ be a Banach function space, with absolutely continuous norm, which is rearrangement-invariant (with respect to Lebesgue measure μ_n on $\mathbb{R}^n$) Suppose that $\sigma > 0$ and $\|g_\sigma \mid Y'\| < \infty$, where g_σ is the Bessel kernel. Then*

$$H^{\sigma}Y \hookrightarrow C(\mathbb{R}^n). \tag{3.6.21}$$

Proof. Let $u \in H^{\sigma}Y$, so that $u = g_\sigma * f$ for some $f \in Y$. Then

$$|u(x)| \le \int_{\mathbb{R}^n} |g_\sigma(y)|\,|f(x-y)|\,dy \le C\,\|f \mid Y\|,$$

where $C = \|g_\sigma \mid Y'\|$. Hence

$$H^{\sigma}Y \hookrightarrow L_\infty(\mathbb{R}^n).$$

Now let $x, h \in \mathbb{R}^n$. Then

$$\begin{aligned}|u(x+h) - u(x)| &\le \|g_\sigma \mid Y'\|\,\|f(x+h-\cdot) - f(x-\cdot) \mid Y\| \\ &= \|g_\sigma \mid Y'\|\,\left\|\widetilde{f}(\cdot - h) - \widetilde{f}(\cdot) \mid Y\right\|,\end{aligned}$$

where $\widetilde{f}(y) = f(x-y)$, $y \in \mathbb{R}^n$. Since $f \in Y$ if, and only if, $\widetilde{f} \in Y$, and $\|f \mid Y\| = \left\|\widetilde{f} \mid Y\right\|$, it follows from the properties of Y that given $\varepsilon > 0$, there exists $\delta > 0$ such that

$$|u(x+h) - u(x)| < \varepsilon \text{ if } |h| < \delta.$$

Hence $u \in C(\mathbb{R}^n)$ and the proof is complete. □

The next result extends [52], Corollary 4.6 and Remark 4.7.

Theorem 3.6.18. *Let $q \in [1, \infty)$, let b be slowly varying and suppose that either*

$$\sigma \in (0, n),\ p = n/\sigma \ \text{ and } \left\|t^{-1/q'}\gamma_{1/b}(t) \mid L_{q'}(0,1)\right\| < \infty, \tag{3.6.22}$$

or

$$0 < \sigma < \infty,\ \max\{1, n/\sigma\} < p < \infty. \tag{3.6.23}$$

Then

$$H^{\sigma}L_{p,q;b}(\mathbb{R}^n) \hookrightarrow C(\mathbb{R}^n).$$

Proof. In view of Lemma 3.6.17 all we have to do is to take $Y = L_{p,q;b}(\mathbb{R}^n)$ and prove that $\|g_\sigma \mid Y'\| \approx \|g_\sigma \mid L_{p',q';1/g}(\mathbb{R}^n)\| < \infty$. Suppose that (3.6.22) holds. Then

$$Y' = L_{n/(n-\sigma),q';1/b}(\mathbb{R}^n),$$

and so with the help of (3.6.5) we have

$$\begin{aligned}\|g_\sigma \mid Y'\| &\approx \left\|t^{(n-\sigma)/n-1/q'} g_\sigma^*(t)\gamma_{1/b}(t) \mid L_{q'}(0,\infty)\right\| \\ &\lesssim I_1 + I_2,\end{aligned}$$

where

$$I_1 = \left\|t^{-1/q'}\gamma_{1/b}(t)\exp\left(-Bt^{1/n}\right) \mid L_{q'}(0,1)\right\|$$

and

$$I_2 = \left\|t^{-1/q'}\left\|s^{-1/q'}\gamma_{1/b}(s)\exp\left(-Bs^{1/n}\right) \mid L_{q'}(0,1)\right\|\exp\left(-Bt^{1/n}\right) \mid L_{q'}(1,\infty)\right\|.$$

Because of the exponential factor it is clear that $I_2 < \infty$; we also have $I_1 < \infty$ since $I_1 \le \left\|t^{-1/q'}\gamma_{1/b}(t) \mid L_{q'}(0,1)\right\| < \infty$. Hence $\|g_\sigma \mid Y'\| < \infty$.

On the other hand, if (3.6.23) holds, then

$$Y' = L_{p',q';1/b}(\mathbb{R}^n).$$

If $0 < \sigma < n$, use of (3.6.5) again shows that

$$\|g_\sigma \mid Y'\| \lesssim J_1 + J_2,$$

where

$$J_1 = \left\|t^{1/p'-1/q'+\sigma/n-1}\gamma_{1/b}(t)\exp\left(-Bt^{1/n}\right) \mid L_{q'}(0,1)\right\|$$

and J_2 is the $L_{q'}(1,\infty)$–norm of the same expression. Plainly $J_2 < \infty$; since our assumptions imply that $\sigma/n + 1/p' - 1 > 0$, we also have

$$J_1 \le \left\|t^{1/p'-1/q'+\sigma/n-1}\gamma_{1/b}(t) \mid L_{q'}(0,1)\right\| < \infty.$$

Thus $\|g_\sigma \mid Y'\| < \infty$.

Finally, suppose that $n \le \sigma < \infty$. Use of Lemma 3.6.5(ii) now enables us to prove that $\|g_\sigma \mid Y'\| \lesssim K_1 + K_2$, where

$$K_1 = \begin{cases} \left\|t^{1/p'-1/q'}\gamma_{1/b}(t)\widetilde{l}_1(t)\exp\left(-Bt^{1/n}\right) \mid L_{q'}(0,1)\right\| & \text{if } \sigma = n, \\ \left\|t^{1/p'-1/q'}\gamma_{1/b}(t)\exp\left(-Bt^{1/n}\right) \mid L_{q'}(0,1)\right\| & \text{if } \sigma > n, \end{cases}$$

and

$$K_2 = \left\|t^{1/p'-1/q'}\gamma_{1/b}(t)\exp\left(-Bt^{1/n}\right) \mid L_{q'}(1,\infty)\right\|.$$

Evidently $K_2 < \infty$. To deal with K_2, put $b_1(t) = \widetilde{l}_1(t)/b(t)$, note that b_1 is slowly varying and $\gamma_{b_1}(t) = \gamma_{1/b}(t)\widetilde{l}_1(t)$ and use

$$K_1 = \begin{cases} \left\| t^{1/p'-1/q'}\gamma_{1/b_1}(t) \mid L_{q'}(0,1) \right\| & \text{if } \sigma = n, \\ \left\| t^{1/p'-1/q'}\gamma_{1/b}(t) \mid L_{q'}(0,1) \right\| & \text{if } \sigma > n. \end{cases}$$

Lemma 3.4.33 (v) now shows that $K_1 < \infty$ and the proof is complete. □

Remark 3.6.19. Let $q \in [1,\infty)$, $m \in \mathbb{N}$, $\alpha \in \mathbb{R}^m$ and $b = \vartheta_\alpha^m$. Then

$$\left\| t^{-1/q'}\gamma_{1/b}(t) \mid L_{q'}(0,1) \right\| < \infty$$

if, and only if, either $1 < q < \infty$ and $\alpha \succ 1/q' + \mathbf{0}$ or $q = 1$ and $\alpha \succeq \mathbf{0}$.

In the classical situation, it is well-known that if $p > n$ then $W_p^1(\mathbb{R}^n)$ is embedded in spaces of Hölder-continuous functions. We next show that there are analogous results for the spaces $H^\sigma L_{p,q;b}(\mathbb{R}^n)$.

Theorem 3.6.20. *Let $1 \le q < \infty, 1 \le \sigma < n+1, \max\{1, n/\sigma\} < p < n/(\sigma-1)$ and let b be slowly varying. Then there is a constant $C > 0$ such that for all $u \in H^\sigma L_{p,q;b}(\mathbb{R}^n) := X_\sigma$ and all $x, y \in \mathbb{R}^n$,*

$$|u(x) - u(y)| \le C \|u \mid X_\sigma\| \, |x-y|^{\sigma - n/p} \gamma_{1/b}(|x-y|^n), \tag{3.6.24}$$

where $\lambda = \sigma - n/p \in (0,1)$.

Proof. Since the Schwartz space $\mathcal{S}$ is dense in X_σ (Lemma 3.6.7), it follows from Theorem 3.6.18 that it is enough to prove (3.6.24) when $u \in \mathcal{S}$.

Thus let $u \in \mathcal{S}$ and let $x, y \in \mathbb{R}^n$, with $0 < |x-y| < \rho$. Then there is a cube $Q_\rho = Q_\rho(x,y)$ with side length ρ and with $x, y \in \overline{Q}_\rho$. Given $z \in Q_\rho$,

$$\begin{aligned} |u(x) - u(z)| &= \left| \int_0^1 \sum_{j=1}^n \frac{\partial u}{\partial x_j}(x + t(z-x))(z_j - x_j)\, dt \right| \\ &\le \rho\sqrt{n} \sum_{j=1}^n \int_0^1 \left| \frac{\partial u}{\partial x_j}(x + t(z-x)) \right| dt. \end{aligned}$$

Hence

$$\begin{aligned} \left| u(x) - \rho^{-n} \int_{Q_\rho} u(z)dz \right| &\le \rho^{1-n}\sqrt{n} \sum_{j=1}^n \int_0^1 \int_{Q_\rho} \left| \frac{\partial u}{\partial x_j}(x + t(z-x)) \right| dzdt \\ &= \sqrt{n} \int_0^1 \rho^{1-n} t^{-n} \left(\sum_{j=1}^n \int_{Q_{t\rho}^x} \left| \frac{\partial u}{\partial x_j}(s) \right| ds \right) dt, \end{aligned} \tag{3.6.25}$$

where $Q_{t\rho}^x$ denotes the subcube of Q_ρ with faces parallel to those of Q_ρ and given by

$$Q_{t\rho}^x = \{s \in Q_\rho : s = x + t(z - x), z \in Q_\rho\}.$$

Since $u \in \mathcal{S} \subset X_\sigma$, Lemma 3.6.8 shows that

$$\frac{\partial u}{\partial x_j} \in X_{\sigma-1} \ (j = 1, ..., n). \tag{3.6.26}$$

If $\sigma > 1$, then by Theorem 3.6.10, with $\sigma - 1$ instead of σ,

$$X_{\sigma-1} \hookrightarrow Y := L_{r,q;b}(\mathbb{R}^n),$$

where $1/r = 1/p - (\sigma - 1)/n$. If $\sigma = 1$, then $r = p$ and $X_{\sigma-1} = Y$. Hence in both cases,

$$\sum_{j=1}^n \int_{Q_{t\rho}^x} \left| \frac{\partial u}{\partial x_j}(s) \right| ds \leq \sum_{j=1}^n \left\| \frac{\partial u}{\partial x_j} \mid Y \right\| \left\| \chi_{Q_{t\rho}^x} \mid Y' \right\|. \tag{3.6.27}$$

Moreover,

$$\sum_{j=1}^n \left\| \frac{\partial u}{\partial x_j} \mid Y \right\| \lesssim \sum_{j=1}^n \left\| \frac{\partial u}{\partial x_j} \mid X_{\sigma-1} \right\| \lesssim \|u \mid X_\sigma\| \tag{3.6.28}$$

and by Lemma 3.4.33 (v),

$$\begin{aligned} \left\| \chi_{Q_{t\rho}^x} \mid Y' \right\| &= \left\| \tau^{1/r' - 1/q'} \gamma_{1/b}(\tau) \chi_{(0,(t\rho)^n)}(\tau) \mid L_{q'}(0, \infty) \right\| \\ &\approx (t\rho)^{n/r'} \gamma_{1/b}((t\rho)^n), \ t > 0; \end{aligned}$$

in the final step it is important that $r' > 0$. Hence

$$\int_0^1 \rho^{1-n} t^{-n} \left(\sum_{j=1}^n \int_{Q_{t\rho}^x} \left| \frac{\partial u}{\partial x_j}(s) \right| ds \right) dt \lesssim \|u \mid X_\sigma\| \, I(\rho),$$

where

$$\begin{aligned} I(\rho) &= \int_0^1 \rho(\rho t)^{-n+n/r'} \gamma_{1/b}((t\rho)^n) dt = n^{-1} \int_0^{\rho^n} \tau^{\sigma/n - 1/p - 1} \gamma_{1/b}(\tau) d\tau \\ &\approx \rho^{\sigma - n/p} \gamma_{1/b}(\rho^n), \end{aligned}$$

the final step following from Lemma 3.4.33 (v) again. Putting these estimates together we finally obtain

$$\left| u(x) - \rho^{-n} \int_{Q_\rho} u(z) dz \right| \leq C \|u \mid X_\sigma\| \rho^{\sigma - n/p} \gamma_{1/b}(\rho^n)$$

for some constant C independent of u and x. The desired inequality (3.6.24) follows immediately. □

This result extends Theorem 4.9 of [52] and refines Theorem 5.7.8 (i) of [148].

Conditions under which Lipschitz continuity is assured are provided in the next Theorem.

Theorem 3.6.21. *Let $1 \leq q < \infty$ and let b be slowly varying. Suppose that either*

$$\sigma \in (1, n+1), \ p = n/(\sigma - 1) \ \textit{and} \ \left\| t^{-1/q'} \gamma_{1/b}(t) \mid L_{q'}(0,1) \right\| < \infty \quad (3.6.29)$$

or

$$\sigma \in (1, \infty) \ \textit{and} \ \max\{1, n/(\sigma - 1)\} < p < \infty. \quad (3.6.30)$$

Then there is a constant C such that for all $u \in H^\sigma L_{p,q;b}(\mathbb{R}^n)$ and all $x, y \in \mathbb{R}^n$,

$$|u(x) - u(y)| \leq C |x - y| \, \|u \mid H^\sigma L_{p,q;b}(\mathbb{R}^n)\| .$$

Proof. As in the proof of the last Theorem, it is enough to prove the desired inequality when $u \in \mathcal{S}$. Let $x, y \in \mathbb{R}^n$, $x \neq y$, and let $\rho = |x - y| > 0$. Then (3.6.25) holds. By (3.6.26) and Theorem 3.6.18, applied with $\sigma - 1$ instead of σ, we see that $\frac{\partial u}{\partial x_i} \in L_\infty(\mathbb{R}^n)$. With the same notation as in the proof of Theorem 3.6.20, application of Hölder's inequality gives

$$\sum_{j=1}^n \int_{Q_{t\rho}^x} \left| \frac{\partial u}{\partial x_j}(s) \right| ds \leq \sum_{j=1}^n \left\| \frac{\partial u}{\partial x_j} \mid L_\infty(\mathbb{R}^n) \right\| \left\| \chi_{Q_{t\rho}^x} \mid L_1(\mathbb{R}^n) \right\| .$$

Moreover, with $X_\sigma = H^\sigma L_{p,q;b}(\mathbb{R}^n)$,

$$\sum_{j=1}^n \left\| \frac{\partial u}{\partial x_j} \mid L_\infty(\mathbb{R}^n) \right\| \lesssim \sum_{j=1}^n \left\| \frac{\partial u}{\partial x_j} \mid X_{\sigma-1} \right\| \lesssim \|u \mid X_\sigma\|$$

and

$$\left\| \chi_{Q_{t\rho}^x} \mid L_1(\mathbb{R}^n) \right\| = (t\rho)^n .$$

It follows that

$$\left| u(x) - \rho^{-n} \int_{Q_\rho} u(z) dz \right| \lesssim \|u \mid X_\sigma\| \int_0^1 \rho^{1-n} t^{-n} (t\rho)^n dt$$

$$= \rho \, \|u \mid X_\sigma\| .$$

This complete the proof. □

Before giving further embedding results it is convenient to introduce some notation concerning the modulus of smoothness of a function. We begin with differences: given any scalar-valued function f on $\mathbb{R}^n$, any $x, h \in \mathbb{R}^n$ and any $m \in \mathbb{N}$, we define difference maps Δ_h^m by $(\Delta_h^1 f)(x) = f(x+h) - f(x)$,

$(\Delta_h^{m+1} f)(x) = \Delta_h^1(\Delta_h^m f)(x)$. Given any $r \in \mathbf{N}$, the rth **modulus of smoothness** of a function $f \in L_p(\mathbb{R}^n)$, when $1 \le p < \infty$, is defined by

$$\omega_r(f,t)_p = \sup_{|h| \le t} \|\Delta_h^r f \mid L_p(\mathbb{R}^n)\|, \; t \ge 0,$$

and the rth **modulus of smoothness** of a function $f \in C(\mathbb{R}^n)$ is defined by

$$\omega_r(f,t)_\infty = \sup_{|h| \le t} \|\Delta_h^r f \mid L_\infty(\mathbb{R}^n)\|, \; t \ge 0.$$

When $r = 1$ the corresponding subscript may be omitted.

For each $f \in L_p(\mathbb{R}^n)$, $1 \le p \le \infty$, $\omega_r(f,\cdot)_p$ is a non-negative, non-decreasing function on $[0,\infty)$; for each fixed $t > 0$, $\omega_r(\cdot,t)_p$ is a semi-norm on $L_p(\mathbb{R}^n)$ $(C(\mathbb{R}^n)$, if $p = \infty)$. Finally, we define

$$\widetilde{\omega}(f,t)_p = t^{-1}\omega(f,t)_p, \; t > 0;$$

$\widetilde{\omega}(f,\cdot)_p$ is equivalent to a non-increasing function on $(0,\infty)$.

We shall also need the subspace $C(\overline{\mathbb{R}^n})$ of $C(\mathbb{R}^n)$ consisting of all those uniformly continuous functions in $C(\mathbb{R}^n)$, and the space $C^1(\overline{\mathbb{R}^n})$ of all those $f \in C(\overline{\mathbb{R}^n})$ which have all their first-order partial derivatives $\partial f/\partial x_i$ also in $C(\overline{\mathbb{R}^n})$; this latter space is endowed with the norm

$$\|f \mid L_\infty(\mathbb{R}^n)\| + \sum_{i=1}^n \|\partial f/\partial x_i \mid L_\infty(\mathbb{R}^n)\|.$$

In addition, we shall use the Besov-Lipschitz-Karamata spaces $\Lambda_{p,q}^{\lambda,b}(\mathbb{R}^n)$ introduced by Neves [186]. To define these, let $p \in [1,\infty]$, $q \in (0,\infty]$, $\lambda \in (0,1)$ and let b be a slowly varying function which, if $\lambda = 1$, satisfies the condition

$$\left\| t^{-1/q}/b(1/t) \mid L_q(0,1) \right\| < \infty.$$

Then $\Lambda_{p,q}^{\lambda,b}(\mathbb{R}^n)$ is the space of all functions $f \in L_p(\mathbb{R}^n)$ $(C(\mathbb{R}^n)$, if $p = \infty)$ such that $\left\|f \mid \Lambda_{p,q}^{\lambda,b}(\mathbb{R}^n)\right\| < \infty$, where

$$\left\|f \mid \Lambda_{p,q}^{\lambda,b}(\mathbb{R}^n)\right\| = \|f \mid L_p(\mathbb{R}^n)\| + \left(\int_0^1 \left(\frac{\omega(f,t)_p}{t^\lambda b(1/t)}\right)^q \frac{dt}{t}\right)^{1/q}$$

if $q < \infty$, and

$$\left\|f \mid \Lambda_{p,\infty}^{\lambda,b}(\mathbb{R}^n)\right\| = \|f \mid L_p(\mathbb{R}^n)\| + \sup_{0<t<1} \frac{\omega(f,t)_p}{t^\lambda b(1/t)}$$

if $q = \infty$. Note that if $\lambda = 1$, $p = q = \infty$ and b =1, then this space reduces to the space of Lipschitz-continuous functions on $\mathbb{R}^n$. When $b(t) = |\log\, t|^\alpha$, with $\alpha > 1/q$ ($\alpha \ge 0$ if $q = \infty$), the space $\Lambda_{p,q}^{\lambda,b}(\mathbb{R}^n)$ corresponds to the space $Lip_{p,q}^{(1,-\alpha)}(\mathbb{R}^n)$ considered in [56], [57] and [120].

It can be shown that $\Lambda^{\lambda,b}_{p,q}(\mathbb{R}^n)$ is a quasi-Banach space when endowed with the quasi-norm $\left\|\cdot \mid \Lambda^{\lambda,b}_{p,q}(\mathbb{R}^n)\right\|$, and that $\Lambda^{\lambda,b}_{\infty,\infty}(\mathbb{R}^n) \hookrightarrow C(\overline{\mathbb{R}^n})$. Moreover, for all $f \in \Lambda^{\lambda,b}_{\infty,\infty}(\mathbb{R}^n)$,

$$\left\|f \mid \Lambda^{\lambda,b}_{\infty,\infty}(\mathbb{R}^n)\right\| \approx \|f \mid L_\infty(\mathbb{R}^n)\| + \sup_{x,y\in\mathbb{R}^n,\ 0<|x-y|<1} \frac{|f(x)-f(y)|}{|x-y|^\lambda \gamma_b(|x-y|)}.$$

We remark that $\Lambda^{\lambda,b}_{\infty,\infty}(\mathbb{R}^n)$ is just the space $C^{0,\nu(t)}(\mathbb{R}^n)$, in the notation of [148], 7.2.12, where $\nu(t) = t^\lambda \gamma_b(t)$. For economy of presentation we shall sometimes prefer this notation.

Various embeddings hold between spaces of this type. For example, we have that $\Lambda^{\lambda,b_1}_{p,q_1}(\mathbb{R}^n) \hookrightarrow \Lambda^{\lambda,b_2}_{p,q_2}(\mathbb{R}^n)$ if either

$$0 < q_1 \le q_2 \le \infty \text{ and } \sup_{0<t<1} b_1(1/t)/b_2(1/t) < \infty$$

or

$$0 < q_2 < q_1 \le \infty \text{ and } \left\|\frac{t^{-1/r} b_1(1/t)}{b_2(1/t)} \mid L_r(0,1)\right\| < \infty,$$

where $1/r = 1/q_2 - 1/q_1$. Moreover, if $0 < \lambda_2 < \lambda_1 \le 1$, then $\Lambda^{\lambda_1,b_1}_{p,q_1}(\mathbb{R}^n) \hookrightarrow \Lambda^{\lambda_2,b_2}_{p,q_2}(\mathbb{R}^n)$ for all $q_1, q_2 \in (0,\infty]$ and all $p \in [1,\infty]$; while if $1 \le p_1 \le p_2 < \infty$, $1 \le q_1, q_2 \le \infty$ and $\lambda_1, \lambda_2 \in (0,1)$ are such that $\lambda_1 - n/p_1 = \lambda_2 - n/p_2$, we have $\Lambda^{\lambda_1,b_1}_{p_1,q_1}(\mathbb{R}^n) \hookrightarrow \Lambda^{\lambda_2,b_2}_{p_2,q_2}(\mathbb{R}^n)$. For the proofs of these and other related results we refer to [182] and [186].

After this brief diversion we present without proof two propositions, the first due to Triebel [221], the second resulting from it and due to Neves [186] following a special case given in [221].

Proposition 3.6.22. *Let $\varepsilon \in (0,1)$. Then there exists $c > 0$ such that for all $t \in (0,\varepsilon)$ and all $f \in C^1(\overline{\mathbb{R}^n})$,*

$$\widetilde{\omega}(f,t)_\infty \le c\,|\nabla f|^{**}(t^{2n-1}) + 3 \sup_{0<\tau\le t^2} \tau^{-1/2}\omega(f,t)_\infty.$$

Here $|\nabla f| = \left(\sum_{i=1}^n |\partial f/\partial x_i|^2\right)^{1/2}$.

Proposition 3.6.23. *Let $q \in [1,\infty]$ and let b be a slowly varying function such that*

$$\left\|t^{-1/q}/b(1/t) \mid L_q(0,1)\right\| < \infty.$$

Let b_1 be the slowly varying function defined by $b_1(t) = b(t^{2n-1})$, $t \ge 1$. Then for all $f \in C^1(\overline{\mathbb{R}^n})$,

$$\begin{aligned}\left\|t^{-1/q}\gamma_{b_1}(t)\widetilde{\omega}(f,t)_\infty \mid L_q(0,1)\right\| &\lesssim \left\|t^{-1/q}\gamma_b(t)\,|\nabla f|^{**}(t) \mid L_q(0,1)\right\| \\ &\approx \left\|t^{-1/q}\gamma_b(t)\,|\nabla f|^{*}(t) \mid L_q(0,1)\right\|.\end{aligned}$$

The next theorem extends the celebrated result of Brézis and Wainger [22] about 'almost' Lipschitz continuity.

Theorem 3.6.24. *Let $p \in [1,\infty), q \in [1,\infty]$ and $\sigma \in (1, n+1)$. Let b_1, b_2 be slowly varying functions such that*

$$\|t^{-1/p'}\gamma_{1/b_1}(t) \mid L_{p'}(0,1)\| = \infty, \qquad \left\|t^{-1/q}b_2(1/t) \mid L_q(0,1)\right\| < \infty, \quad (3.6.31)$$

and suppose that either conditions (3.4.21) and (3.4.25) or (3.4.22) and (3.4.26) hold. Let b_3 be the slowly varying function defined by $b_3(t) = 1/b_2(t^{2n-1})$. Then

$$H^\sigma L_{n/(\sigma-1),p;b_1}(\mathbb{R}^n) \hookrightarrow \Lambda^{1,b_3}_{\infty,q}(\mathbb{R}^n).$$

Proof. Let $u \in \mathcal{S} \subset H^\sigma L_{n/(\sigma-1),p;b_1}(\mathbb{R}^n) = X$. Then

$$\partial u/\partial x_i \in H^{\sigma-1}L_{n/(\sigma-1),p;b_1}(\mathbb{R}^n) = Y$$

and, by Lemma 3.6.11 with $\sigma - 1$ instead of σ and by noting that $v^* = |v|^*$ for all $v \in \mathcal{M}_0(\mathbb{R}^n, \mu_n)$,

$$\left\|\left|\frac{\partial u}{\partial x_i}\right|^* \mid L_{\infty,q;b_2}(0,1)\right\| \lesssim \left\|\frac{\partial u}{\partial x_i} \mid Y\right\|.$$

Thus by Lemma 3.6.8,

$$\sum_{i=1}^n \left\|\left|\frac{\partial u}{\partial x_i}\right|^* \mid L_{\infty,q;b_2}(0,1)\right\| \lesssim \sum_{i=1}^n \left\|\frac{\partial u}{\partial x_i} \mid Y\right\| \lesssim \|u \mid X\|.$$

Together with Proposition 3.6.23 and inequality (3.4.21) of Hardy type, with $\nu = -1$, this gives

$$\begin{aligned}
\left\|t^{-1/q}\gamma_{b_3}(t)\widetilde{\omega}(u,t)_\infty \mid L_q(0,1)\right\| &\lesssim \left\|t^{-1/q}\gamma_{b_2}(t)\,|\nabla u|^{**}(t) \mid L_q(0,1)\right\| \\
&\approx \left\|t^{-1/q}\gamma_{b_2}(t)\left(\sum_{i=1}^n\left|\frac{\partial u}{\partial x_i}\right|\right)^{**}(t) \mid L_q(0,1)\right\| \\
&\le \sum_{i=1}^n \left\|t^{-1/q}\gamma_{b_2}(t)\left|\frac{\partial u}{\partial x_i}\right|^{**}(t) \mid L_q(0,1)\right\| \\
&\approx \sum_{i=1}^n \left\|t^{-1/q}\gamma_{b_2}(t)\left|\frac{\partial u}{\partial x_i}\right|^{*}(t) \mid L_q(0,1)\right\| \\
&\le \|u \mid X\|.
\end{aligned}$$

By Theorem 3.6.18, $\|u \mid L_\infty(\mathbb{R}^n)\| \lesssim \|u \mid X\|$. Hence

$$\left\|u \mid \Lambda^{1,b_3}_{\infty,q}(\mathbb{R}^n)\right\| \le \|u \mid X\| \quad (3.6.32)$$

for all $u \in \mathcal{S}$. This may be extended to the whole of X by a standard approximation argument.The proof is complete. □

Remark 3.6.25. If $\left\| t^{-1/p'}\gamma_{1/b_1}(t) \mid L_{p'}(0,1) \right\| < \infty$, it follows from Theorem 3.6.21 that

$$H^{\sigma} L_{n/(\sigma-1),p;b_1}(\mathbb{R}^n) \hookrightarrow \Lambda^{1,1}_{\infty,\infty}(\mathbb{R}^n) = Lip(\mathbb{R}^n),$$

which is a better result than that given by Theorem 3.6.24 since it can be shown that (see [182]) $\Lambda^{1,1}_{\infty,\infty}(\mathbb{R}^n) \hookrightarrow \Lambda^{1,b_3}_{\infty,q}(\mathbb{R}^n)$.

Corollary 3.6.26. *Let* $p \in [1,\infty), q \in [1,\infty]$ *and* $\sigma \in (1, n+1)$; *let* $m \in \mathbf{N}$, $\alpha \in \mathbb{R}^m$ *and let* $k \in \{1,...,m\}$ *be such that* $\alpha_k < 1/p'$ *and, if* $k \geq 2$, $\alpha_i = 1/p'$, $i = 1,...,k-1$. *Let* $\beta \in \mathbb{R}^m$ *with* $\beta_k \neq -1/q$ *and, if* $k \geq 2, \beta_i = -1/q$, $i = 1,...,k-1$. *Then*

$$H^{\sigma} L_{n/(\sigma-1),p;\alpha}(\mathbb{R}^n) \hookrightarrow \Lambda^{(1,-\beta)}_{\infty,q}(\mathbb{R}^n) = Lip^{(1,\beta)}_{\infty,q}(\mathbb{R}^n), \tag{3.6.33}$$

provided that either

$$1 \leq p \leq q \leq \infty,\ \beta_k < -1/q, \beta + \delta_{q;m,k} \preceq \alpha - \delta_{p';m,k}; \tag{3.6.34}$$

or

$$1 \leq q < p < \infty,\ \beta_k < -1/q, \beta + 1/q \prec \alpha - \delta_{1;m,k} + 1/p. \tag{3.6.35}$$

Proof. Just use Theorem 3.6.24 with $b_1 = \vartheta^m_\alpha$ and $b_2 = \vartheta^m_\beta$, noting that $b_3(t) \approx \vartheta^m_{-\beta}(t)$. □

Note that the situation when $\alpha_1 = ... = \alpha_m = 1/p'$ is also covered by this Corollary: simply take $\overline{\alpha} = (\widetilde{\alpha}_1, ..., \widetilde{\alpha}_{m+1}) \in \mathbb{R}^{m+1}$ in place of $\alpha \in \mathbb{R}^m$, where $\widetilde{\alpha}_j = 1/p'$ for $j = 1,...,m$ and $\widetilde{\alpha}_{m+1} = 0$, with $k = m+1$.

Remark 3.6.27. When $p = q = n/(\sigma-1)$, $k = m = 1$, $\alpha = 0$ and $\beta = -1/p'$ this gives the result of Brézis and Wainger about the 'almost Lipschitz continuity' of elements of $H^{1+n/p}_p$.

3.7 Examples

Following [52] we now give some illustrations of these results. The first set of examples is concerned with first-order spaces of Sobolev type, and it is supposed that $n > 1$.

Examples 3.7.1

(i) Suppose that $1 < p < n$, $1/r = 1/p - 1/n$ and $\beta \in \mathbb{R}$. Then

$$W^1 L_p(\log L)_\beta(\mathbb{R}^n) \hookrightarrow L_{r,p;\beta}(\mathbb{R}^n).$$

This is a consequence of Theorem 3.6.10. When $\beta = 0$ this gives

$$W^1_p(\mathbb{R}^n) \hookrightarrow L_{r,p}(\mathbb{R}^n),$$

which is the familiar refinement (found in [195], for instance) already mentioned of the classical Sobolev embedding theorem, since $L_{r,p}(\mathbb{R}^n) \hookrightarrow L_r(\mathbb{R}^n)$. Note that in this example the power exponent p is sub-critical and that change of the logarithmic exponent β does not alter the essential nature of the embedding. This is not so when p has the limiting value n : the next three examples illustrate this.
(ii) Let $\Omega \subset \mathbb{R}^n$ have finite n-dimensional volume. Suppose that $\beta < 1/n'$, and let Φ be a Young function such that for large t,

$$\Phi(t) = \exp\left(t^{n'/(1-\beta n')}\right).$$

Then

$$W^1 L_n(\log L)_\beta(\mathbb{R}^n) \hookrightarrow L_\Phi(\Omega).$$

This is a special case of Corollary 3.6.15. Note that when $\beta = 0$ we have

$$W_n^1(\mathbb{R}^n) \hookrightarrow L_\Phi(\Omega),$$

which was proved in [195]; see also [200], [213], [222] and [230]. The case $\beta < 0$ corresponds to that dealt with by Fusco, Lions and Sbordone [96].
(iii) Here we allow the logarithmic exponent β to take the limiting value $1/n'$. Let $\Omega \subset \mathbb{R}^n$ have finite n-dimensional volume and let Ψ be a Young function such that for large t,

$$\Psi(t) = \exp\left(\exp t^{n'}\right).$$

Then

$$W^1 L_n(\log L)_{1/n'}(\mathbb{R}^n) \hookrightarrow L_\Psi(\Omega).$$

This is the special case $m = k = 2, \sigma = 1, p = n, \alpha_2 = 0$ of Corollary 3.6.15 and corresponds to a result given in [52]. We see that the increase of β to $1/n'$ results in a target space of double, rather than single, exponential type.
(iv) Let $\beta > 1/n'$. Then

$$W^1 L_n(\log L)_\beta(\mathbb{R}^n) \hookrightarrow C(\mathbb{R}^n).$$

This is a special case of Theorem 3.6.18: see Remark 3.6.19. The importance of the logarithmic term here is underlined by the fact that, as is well known, $W_n^1(\mathbb{R}^n)$ is not embedded in $L_\infty(\mathbb{R}^n)$.

If the power exponent is increased beyond the critical value n we have an embedding in a space of Hölder type.
(v) Let $n < p < \infty, \beta \in \mathbb{R}$ and $\lambda(t) = t^{1-n/p} l_1^{-\beta}(t), t > 0$. Then in the notation of [148],

$$W^1 L_p(\log L)_\beta(\mathbb{R}^n) \hookrightarrow C^{0,\lambda(t)}(\mathbb{R}^n).$$

This results from Theorem 3.6.20, and extends the classical result (corresponding to $\beta = 0$) that

$$W_p^1(\mathbb{R}^n) \hookrightarrow C^{1-n/p}(\mathbb{R}^n).$$

The next set of examples deals with second-order spaces.

Examples 3.7.2

(i) Suppose that $1 < p < n/2, 1/r = 1/p - 2/n$ and $\beta \in \mathbf{R}$. Then by Theorem 3.6.10,

$$W^2 L_p(\log L)_\beta (\mathbb{R}^n) \hookrightarrow L_{r,p;\beta} (\mathbb{R}^n) .$$

Here the power exponent p is sub-critical, and we have an extension of the classical Sobolev embedding theorem.

The following three examples consider the situation when the power exponent p has the critical value $n/2$, and illustrate how the size of the logarithmic exponent determines the nature of the embedding.

(ii) Let $\Omega \subset \mathbb{R}^n$ have finite n-dimensional volume, suppose that $\beta < (n-2)/2$ and let Φ be a Young function such that for large $t, \Phi(t) = \exp\left(t^{n/(n-2-n\beta)}\right)$. Then by Corollary 3.6.15,

$$W^2 L_{n/2}(\log L)_\beta (\mathbb{R}^n) \hookrightarrow L_\Phi (\Omega) .$$

(iii) Let $n > 2$, let $\Omega \subset \mathbb{R}^n$ have finite n-dimensional volume and let Ψ be a Young function such that for large $t, \Psi(t) = \exp\left(\exp t^{n/(n-2)}\right)$. Then by Corollary 3.6.15, with $m = 2$ and $\alpha_2 = 0$,

$$W^2 L_{n/2}(\log L)_{(n-2)/2} (\mathbb{R}^n) \hookrightarrow L_\Psi (\Omega) .$$

We observe that the critical value $(n-2)/2$ of the logarithmic exponent results in an embedding in a double-exponential space.

(iv) Further increase in the logarithmic exponent produces an embedding into the space of bounded continuous functions. Thus if $n > 2$ and $\beta > (n-2)/2$, we see from Theorem 3.6.18 (see also Remark 3.6.19) that

$$W^2 L_{n/2}(\log L)_\beta (\mathbb{R}^n) \hookrightarrow C (\mathbb{R}^n) .$$

(v) When the power exponent p exceeds the critical value $n/2$, embeddings into spaces of Hölder type result. In fact, Theorem 3.6.20 shows that if $n > 1, n/2 < p < n$ and $\beta \in \mathbb{R}$, then

$$W^2 L_p(\log L)_\beta (\mathbb{R}^n) \hookrightarrow C^{0,\lambda(t)} (\mathbb{R}^n) ,$$

where

$$\lambda(t) = t^{2-n/p} l_1^{-\beta}(t), t > 0.$$

When the power exponent p reaches the next critical value n, the size of the logarithmic exponent affects the embedding. The next three examples illustrate this.

(vi) Suppose that $n > 1, \beta < 1/n'$ and $\lambda(t) = t l_1^{(1-n'\beta)/n'}(t), t > 0$. Then Corollary 3.6.26 with $m = 1$ gives

$$W^2L_n(\log L)_\beta(\mathbb{R}^n) \hookrightarrow C^{0,\lambda(t)}(\mathbb{R}^n).$$

Note that when $\beta = 0$ this shows that

$$W_n^2(\mathbb{R}^n) \hookrightarrow C^{0,\lambda(t)}(\mathbb{R}^n),$$

which corresponds to the result of Brézis and Wainger [22].

(vii) When the power exponent p is at the critical value n and the logarithmic exponent β has the critical value $1/n'$, the target space involves a double logarithm. For by Corollary 3.6.26 with $m = 2$,

$$W^2L_n(\log L)_{1/n'}(\mathbb{R}^n) \hookrightarrow C^{0,\lambda(t)}(\mathbb{R}^n),$$

where

$$\lambda(t) = tl_2^{1/n'}(t), t > 0.$$

(viii) Further increase of the logarithmic exponent gives embeddings in spaces of Lipschitz-continuous functions. In fact, by Corollary 3.6.26, if $\beta > 1/n'$ and $n > 1$, then

$$W^2L_n(\log L)_\beta(\mathbb{R}^n) \hookrightarrow C^{0,1}(\mathbb{R}^n) := C^{0,id(t)}(\mathbb{R}^n),$$

where id is the function $t \longmapsto t$.

(ix) When the power exponent $p > n$, the significance of the logarithmic exponent is lost and we always have an embedding in a space of Lipschitz-continuous functions, no matter what the value of β is. Thus if $1 < n < p < \infty$ and $\beta \in \mathbb{R}$, then by Corollary 3.6.26,

$$W^2L_p(\log L)_\beta(\mathbb{R}^n) \hookrightarrow C^{0,1}(\mathbb{R}^n).$$

3.8 Other spaces

An interesting development stemming from the Lorentzian scale of spaces is given by the replacement of f^{**} (or f^*) by $f^{**} - f^*$ in the definition of the norm of $L_{\infty,q}(R,\mu)$. This idea comes from Alvino, Trombetti and Lions [5] and Bennett, De Vore and Sharpley [14]; it has been followed up by Bastero, Milman and Ruiz Blaser [11], and in an equivalent form by Malý and Pick [164]. We give below a consequence of this idea for the Sobolev embedding theorem in a limiting situation, following the treatment of [11].

Definition 3.8.1 *Let* (R,μ) *be a* σ*−finite measure space and let* $q \in (0,\infty]$. *Then by* $W_q = W_q(R)$ *we shall mean the set*

$$\{f \in \mathcal{M}_0(R,\mu) : \|f \mid W_q\| < \infty\},$$

where

$$\|f \mid W_q\| := \left\{ \int_0^{\mu(R)} t^{-1}(f^{**}(t) - f^*(t))^q dt \right\}^{1/q}$$

if $q < \infty$, *and*

$$\|f \mid W_\infty\| := \sup_{0<t<\mu(R)} \{f^{**}(t) - f^*(t)\}.$$

The weak-L_∞ space of [14] is just W_∞. Note that in general the W_q are not linear spaces and $\|\cdot \mid W_q\|$ is not a norm: see [164] and [11] . Observe also that if E is any subset of R with $0 < \mu(E) < \infty$, and $f = \chi_E$, then

$$f^{**}(t) - f^*(t) = \mu(E)t^{-1}\chi_{(\mu(E),\infty)}(t),$$

so that

$$\|\chi_E \mid W_q\| = \begin{cases} q^{-1/q} & \text{if } 0 < q < \infty, \\ 1 & \text{if } q = \infty. \end{cases}$$

Hence W_q is non-trivial. Moreover, if $\mu(E) < \infty$ and $1 < p < q < \infty$, then (see [11])

$$W_1 \subset W_p \subset W_q \subset W_\infty.$$

That these inclusions are strict, in general, can be seen by consideration of the functions f_α, where

$$f_\alpha(t) = (t^\alpha \log^\alpha(1/t))', \; t \in (0, e^{-1}).$$

Note the identity

$$\frac{d}{dt} f^{**}(t) = -t^{-1}\left(f^{**}(t) - f^*(t)\right).$$

After these general remarks we turn to the consideration of the Sobolev space $\overset{\circ}{W}{}^1_n(\Omega)$, where Ω is an open subset of $\mathbb{R}^n$ and begin with a useful inequality.

Lemma 3.8.2 *Let* $f \in C_0^\infty(\mathbb{R}^n)$ *and suppose that* $f = f^\star$, *the symmetric rearrangement of* f *(see Definition 3.2.18). Then*

$$f^{**}(t) - f^*(t) \le \left(n\omega_n^{1/n}\right)^{-1} t^{1/n} \left(|\nabla f|\right)^{**}(t), \; 0 < t < |\text{supp } f|.$$

Proof. Since $f(x) = f^\star(x) = f^*(\omega_n |x|^n) = g(|x|)$, say, it follows that

$$f^*(t) = g\left((t/\omega_n)^{1/n}\right) = \int_{(t/\omega_n)^{1/n}}^\infty |\nabla g(s)|\, ds.$$

Hence

$$f^{**}(t) = t^{-1} \int_0^t \int_{(r/\omega_n)^{1/n}}^\infty |\nabla g(s)|\, ds dr,$$

and so, changing the order of integration, we have

$$f^{**}(t) - f^*(t) = \omega_n t^{-1} \int_0^{(t/\omega_n)^{1/n}} s^n \left|\nabla g(s)\right| ds$$

$$= (nt)^{-1} \int_{|y| \leq (t/\omega_n)^{1/n}} |y| \left|\nabla f(y)\right| dy$$

$$\leq n^{-1} (t/\omega_n)^{1/n} t^{-1} \int_{|y| \leq (t/\omega_n)^{1/n}} \left|\nabla f(y)\right| dy$$

$$\leq n^{-1} (t/\omega_n)^{1/n} t^{-1} \int_0^t \left(|\nabla f|\right)^* (s) ds.$$

This completes the proof. □

We remark that the inequality of Lemma 3.8.2 holds (for almost all t) under weaker conditions. It is plainly enough, for example, that f should belong to $\overset{\circ}{W}{}_n^1(B)$, where B is any ball centred at the origin and f is assumed to be extended by 0 to the whole of $\mathbb{R}^n$.

The main result can now be given.

Theorem 3.8.3 *Let Ω be an open subset of $\mathbb{R}^n$. Then*

$$\overset{\circ}{W}{}_n^1(\Omega) \subset W_n(\Omega).$$

Proof. First suppose that Ω is a ball centred at 0, with measure $|\Omega|$ ($< \infty$), and let $f \in \overset{\circ}{W}{}_n^1(\Omega)$ be such that $f = f^\star$. By Lemma 3.8.2 and the remark following it,

$$t^{-1}\{f^{**}(t) - f^*(t)\}^n \leq n^{-n} \omega_n^{-1} \{(|\nabla f|)^{**}(t)\}^n,$$

and so

$$\|f \mid W_n(\Omega)\| \leq (n\omega_n^{1/n})^{-1} \|\nabla f \mid L_n(\Omega)\|. \tag{3.8.1}$$

Now let Ω be any open subset of $\mathbb{R}^n$ with $|\Omega| < \infty$, and let $f \in \overset{\circ}{W}{}_n^1(\Omega)$. We may suppose that $f \geq 0$ as $\|\nabla(|f|) \mid L_n(\Omega)\| \leq \|\nabla f \mid L_n(\Omega)\|$. By the Polyá-Szegö principle (see Remark 3.2.23), $f^\star \in \overset{\circ}{W}{}_n^1(\Omega^\star)$, where $\Omega^\star$ is the ball centred at the origin and with the same volume as Ω; moreover,

$$\left\|\nabla f^\star \mid L_n(\Omega^\star)\right\| \leq \|\nabla f \mid L_n(\Omega)\|. \tag{3.8.2}$$

Since $\left(f^\star\right)^* = f^*$, use of (3.8.1) and (3.8.2) gives

$$\|f \mid W_n(\Omega)\| = \left\|f^\star \mid W_n(\Omega^\star)\right\| \leq (n\omega_n^{1/n})^{-1} \left\|\nabla f^\star \mid L_n(\Omega^\star)\right\|$$

$$\leq (n\omega_n^{1/n})^{-1} \|\nabla f \mid L_n(\Omega)\|. \tag{3.8.3}$$

If $|\Omega| = \infty$, the procedure is similar, with $\Omega^\star = \mathbb{R}^n$. □

Remarks about the relationships between the embedding of $\overset{\circ}{W}{}^1_n(\Omega)$ given by this theorem and other embeddings of this Sobolev spaces are provided by the Notes in the next section.

3.9 Notes

3.1. The material here is now quite standard. A fuller account is given in the book by Bennett and Sharpely [15]. In particular, this book gives useful information about the subspace X_a of a Banach function space X consisting of all those functions in X with absolutely continuous norm. It is shown that X is separable if, and only if, $X_a = X$ and the underlying measure μ is separable (Corollary 1.5.6); moreover, X_a is a closed subspace of X contained in the closure in X of the set of all bounded functions supported on sets of finite measure (Theorem 1.3.11). For the interesting notion of a continuous norm and more results in this direction, see [149] and [152].

3.2. The theory of rearrangements goes back to Hardy and Littlewood in the 1920s; the famous book [119] did much to make this technique widely available. Many accounts now exist, including those in [10], [15], [136], [142] and [231]. Numerous isoperimetric theorems and sharp Sobolev inequalities have been proved by rearrangement methods, great stimulus being provided by the classic book [201]; see, for example, [9] and [214]. The reason for this success is that n-dimensional questions are reduced to corresponding 1-dimensional problems, which are easier to handle. A crucial result in this area is Theorem 3.2.21, of Pólya-Szegö type.The proof given is that of Talenti [216]. Lemma 3.2.20 and Theorem 3.2.22 are due to Talenti (see [215] and [216]); in this connection see also [23]. They have proved to be very useful in various contexts, such as the study of optimal Sobolev embeddings. For this study see [32], [35], [40] and [64]. The relationship between the behaviour of a function and that of its rearrangement is best understood when the function involved belongs to a space of Sobolev type or has a certain kind of modulus of continuity. Less is known for functions of bounded variation, but for this see [34]. For a version of the Pólya-Szegö principle for second-order derivatives see [33].

3.4. Lorentz-Zygmund spaces were introduced in 1980 by Bennett and Rudnick [13]; GLZ spaces and spaces of Sobolev type modelled on them were the subject of the series of papers [49] to [54]. For a comprehensive account of GLZ spaces, and their relationship with Orlicz spaces, we refer to [192], while for interpolation properties of GLZ spaces see [88] and [89]. Theorem 3.4.26 is due to Neves [182] and arose from the corresponding result, with $t = \infty$, in [192]. Particular cases of it, and of Corollary 3.4.28, may be found in [13] and in [49]-[53]. Theorem 3.4.31, and even more general results, are found in [184]. These include as special cases the equivalent norms on single and double exponential Orlicz spaces which are contained in (3.4.17)-(3.4.19) and which appear in [219] and [53], respectively.

For the definition and basic properties of slowly varying functions we refer to [232], Chapter 2, p. 184 and to [18]. Our treatment of Lorentz-Karamata spaces is based on that of Neves [182], [185]; for discussions concerning Lorentz-Karamata quasinorms see [64]. The identification of the associate space of a Lorentz-Karamata space is due to Neves [182]; the proof follows the pattern of Theorem IV.4.7 in [15] and Lemma 3.4 in [52]. The proof of absolute continuity of the norm is similar to that given in [52] for GLZ spaces. Special cases of some of the embedding results can be found in [13] and [51].

The decomposition results in 3.4.43 are given in [182], where a discussion of the case of infinite measure spaces is also provided; for earlier work in this direction see [66] and [39], where application of these results to the Brézis-Wainger [22] theorem concerning the limiting case of the Sobolev embedding theorem may be found.

3.5. For variants of the Marcinkiewicz interpolation theorem in GLZ spaces see [52] and [88]; the Lorentz-Karamata version given in Theorem 3.5.15 is in [182].

3.6. Bessel-Lorentz-Karamata-potential spaces were introduced by Neves (see [182]) and the form of the results given in this section is largely due to him. The sublimiting embeddings of 3.6.3 have their roots in [52], where GLZ spaces were considered, and also in [148], Theorem 5.7.7 (i). For an earlier version of Corollary 3.6.12 see [52]; see also [193] and [194]. The prototype of the limiting embeddings discussed in 3.6.4, in which the target spaces are Orlicz spaces of exponential type, is the embedding of the classical Sobolev space $W_p^{n/p}(\Omega)$ $(1 < p < \infty,\ n/p \in \mathbb{N})$ in the Orlicz space Y with Young function $\exp(t^{p'}) - 1$. The history of this special result, and of the similar result when n/p is not required to be an integer, is quite rich: among those who made notable contributions are Peetre [195], Pohozaev [200], Strichartz [213], Trudinger [222] and Yudovich [230]. While this target space is optimal in the class of Orlicz spaces, we know from the work of Hansson [117] and Brézis and Wainger [22] (see also Corollary 3.6.12) that $W_p^{n/p}(\Omega)$ can be embedded in a rearrangement-invariant space which is smaller than Y. In concrete terms this means that

$$\int_0^1 \left(\frac{f^*(t)}{1+|\log t|}\right)^p \frac{dt}{t} \le c \left\| f \mid W_p^{n/p}(\Omega) \right\|^p, \quad f \in W_p^{n/p}(\Omega).$$

There is a connection between inequalities of this type and capacity estimates in function spaces: see [64] for some discussion of this with regard to Maz'ya's results in this connection. Hansson's optimality results were extended by Cwikel and Pustylnik [40] and in [64]; this latter paper (see also

[196] and [197]) additionally considers the question of determining pairs of spaces which are optimal in the sense that the domain spaces cannot be enlarged nor the target made smaller. A substantial contribution to the matter of the optimality of the target space was made by Netrusov [179] in the setting of Lizorkin-Triebel spaces: he used a reduction of the problem to one involving Hardy operators which is similar to that used later (and independently) in [64]. In this paper Netrusov also extended earlier results in the Russian literature by Brudni, Kalyabin and Goldman, for a description of which we refer to Lizorkin [158], D.1.8 and D.1.9, pp. 398-404. For further work on optimality, together with many references, see [221]; additional interesting ideas and results are contained in [101], [102], [103] and [194]. We also note that the papers [76] and [220], dealing with optimality questions in limiting and sub-limiting situations, established some of their results by means of atomic decomposition techniques; for an alternative and more elementary approach see [113].

Turning next to the super-limiting embeddings of 3.6.5, we note that there is now an extensive literature concerning results of Brézis-Wainger type in the sense of Remark 3.6.26: for this and related material see [56], [57], [120], [121], [146] and [221], the last reference giving a comprehensive account of the matter with many references.

3.8. For considerations involving $f^{**} - f^*$ we refer to Bennett, De Vore and Sharpley [14]. For Lemma 3.8.2, see Alvino, Trombetti and Lions [5]. and Kolyada [141]. More recent work includes that of Bastero, Milman and Ruiz Blaser [11] and Malý and Pick [164]: in these papers it is shown that if the requirement that the target space should be a linear space is abandoned, and the limiting Sobolev embedding are expressed in terms of rearrangement-invariant inequalities, then Theorem 3.8.3 gives a sharper result than the embedding due to Hansson [117] and Brézis and Wainger [22].

Of course, there are many interesting aspects of the theory of function spaces which are not covered by this chapter: we mention in particular the excellent book [3] by Adams and Hedberg which gives an authoritative exposition of topics of major importance. We also draw special attention to that development involving spaces with generalised smoothness. These spaces were introduced in the 1970s and are now enjoying a period of considerable study. The work of Goldman ([104], [105], [106], [108]) and Kalyabin ([134], [135]) was influential in this area and indeed also in some of the topics treated in our chapter. Function spaces of generalised smoothness defined on certain ideal spaces instead of on L_p were considered in [107], [109], [180] and [181]. For more recent work we refer to [75], [91],[92] [113], [154], [176], [193] and [194].

4

Poincaré and Hardy inequalities

4.1 Introduction

Spectral matters form a powerful motivation for the study of these inequalities. Thus let Ω be an open subset of $\mathbb{R}^n$ with volume $|\Omega|$ and let $\Delta_{D,\Omega}$, $\Delta_{N,\Omega}$ be respectively the Dirichlet Laplacian and the Neumann Laplacian on Ω. We recall that $-\Delta_{D,\Omega}$ is the Friedrichs extension of -Δ on $C_0^\infty(\Omega)$ and that its domain $\mathcal{D}(-\Delta_{D,\Omega})$ is contained in the Sobolev space $\overset{\circ}{W}{}_2^1(\Omega)$: v is the Dirichlet Laplacian of $u \in \mathcal{D}(-\Delta_{D,\Omega})$, $v = \Delta u$, if it belongs to $L_{1,loc}(\Omega)$ and for all $\phi \in C_0^\infty(\Omega)$,

$$\int_\Omega \nabla u.\nabla\phi \, dx = -\int_\Omega v\phi \, dx \; .$$

Analogously, $u \in \mathcal{D}(-\Delta_{N,\Omega})$ $(\subset W_2^1(\Omega))$, the domain of the Neumann Laplacian, if there exists $v \in L_{1,loc}(\overline{\Omega})$ such that for all $\phi \in C_0^\infty(\mathbb{R}^n)$,

$$\int_\Omega \nabla u.\nabla\phi \, dx = -\int_\Omega v\phi \, dx,$$

and in this event, v is called the Neumann Laplacian of u.

When $|\Omega| < \infty$, the Poincaré inequality asserts that

$$\|f - f_\Omega \mid L_2(\Omega)\| \le K(\Omega, n) \|\nabla f \mid L_2(\Omega)\|, \quad f \in W_2^1(\Omega). \tag{4.1.1}$$

Here $f_\Omega = |\Omega|^{-1} \int_\Omega f dx$ and the constant $K(\Omega, n)$ depends only on Ω and n. It holds, for example, if Ω is a bounded convex domain and in particular if Ω is a cube Q, when

$$\|f - f_Q \mid L_2(Q)\| \le K(n) |Q|^{1/n} \|\nabla f \mid L_2(Q)\|, \quad f \in W_2^1(Q)$$

and $K(n)$ depends only on n. Corresponding to the Poincaré inequality we have, for elements f of $\overset{\circ}{W}{}_2^1(\Omega)$ whenever Ω is bounded, the Friedrichs inequality

$$\|f \mid L_2(\Omega)\| \leq c(\Omega, n) \|\nabla f \mid L_2(\Omega)\|,$$

where $c(\Omega, n)$ depends only on Ω and n. The significance of the constant $c(\Omega, n)$ becomes clearer in connection with the Rayleigh quotient

$$R(f) = \frac{\|\nabla f \mid L_2(\Omega)\|}{\|f \mid L_2(\Omega)\|}.$$

For it is well known that the spectrum of the Dirichlet Laplacian on any bounded Ω is purely discrete and consists of a countable number of positive eigenvalues λ_k, arranged in increasing order, each repeated according to its (finite) multiplicity. The smallest of these, λ_1, is given by the variational characterisation $\lambda_1 = \inf R(f)$, where the infimum is taken over all non-zero f in $\overset{\circ}{W}{}_2^1(\Omega)$. Thus knowledge of the best constant in the Friedrichs inequality will give us the least eigenvalue of the Dirichlet Laplacian.

A similar situation holds for the Poincaré inequality. Here, however, it is convenient to introduce the space

$$W_{2,M}^1(\Omega) := \{f \in W_2^1(\Omega) : f_\Omega = 0\}$$

and to endow it with the norm $\left\|\cdot \mid W_{2,M}^1(\Omega)\right\|$, where

$$\left\|f \mid W_{2,M}^1(\Omega)\right\| = \|\nabla f \mid L_2(\Omega)\|.$$

We assume that the Poincaré inequality holds on Ω. Then $\left\|\cdot \mid W_{2,M}^1(\Omega)\right\|$ is equivalent to $\left\|\cdot \mid W_2^1(\Omega)\right\|$ on $W_{2,M}^1(\Omega)$. Moreover, any $f \in W_2^1(\Omega)$ can be written as $f = f_\Omega + (f - f_\Omega)$, and $f - f_\Omega \in W_{2,M}^1(\Omega)$. Since $\mathcal{C} \cap W_{2,M}^1(\Omega) = \{0\}$, where $\mathcal{C}$ is the set of all constant functions on Ω, and $\|\cdot \mid L_2(\Omega)\| + \left\|\cdot \mid W_{2,M}^1(\Omega)\right\|$ is a norm on $\mathcal{C} \oplus W_{2,M}^1(\Omega)$ which is equivalent to $\left\|\cdot \mid W_2^1(\Omega)\right\|$, we have the topological isomorphism

$$W_2^1(\Omega) \simeq \mathcal{C} \oplus W_{2,M}^1(\Omega).$$

The variational characterisation of the eigenvalues of compact operators in a Hilbert space thus shows that the Neumann Laplacian, in a bounded open set Ω for which the Poincaré inequality holds and for which the natural embedding

$$E : W_2^1(\Omega) \to L_2(\Omega)$$

is compact, has spectrum consisting of a countable family of eigenvalues λ_k, arranged in increasing order, each repeated according to its (finite) multiplicity. The smallest of these is $\lambda_1 = 0$, corresponding to a constant eigenfunction; and λ_2 is given by

$$\lambda_2 = \inf \frac{\|\nabla f \mid L_2(\Omega)\|}{\|f \mid L_2(\Omega)\|},$$

where the infimum is taken over all $f \in W_{2,M}^1(\Omega) \setminus \{0\}$. We see that knowledge of the best constant in the Poincaré inequality will give us λ_2.

Another important fact is that spectral properties of the Neumann Laplacian on an open set Ω are determined by the embedding E just mentioned even when E is not compact. Of especial importance in this context is the measure of non-compactness of E, which is given by

$$\alpha(E) = \inf \ \|E - F\|,$$

where the infimum is taken over all those bounded linear maps from $W_2^1(\Omega)$ to $L_2(\Omega)$ which have finite rank. Plainly $0 \leq \alpha(E) \leq 1$; and $\alpha(E) = 0$ if, and only if, E is compact, in which case the spectrum of $\Delta_{N,\Omega}$ is discrete. If Ω has infinite volume, $\alpha(E) = 1$ (see [46], Proposition V.5.10), but the converse is false (see [46], Theorem V.4.21). It was shown by Amick (see [6]) that $0 \leq \alpha(E) < 1$ if, and only if, the Poincaré inequality (4.1.1) is satisfied; and in this case 0 is an isolated point (an eigenvalue) of the spectrum of $\Delta_{N,\Omega}$. When $0 < \alpha(E) < 1$ the spectrum of $-\Delta_{N,\Omega}$ is not discrete and the least point of the essential spectrum is $1/K_0^2$, where K_0 is the infimum of the constants K in (4.1.1) Moreover, 0 belongs to the essential spectrum of $-\Delta_{N,\Omega}$ if, and only if, $\alpha(E) = 1$.

As for the Hardy inequality, the classical form of this asserts that with $d(x) = \text{dist}\ (x, \partial\Omega)$, there exists $C > 0$ such that for all $f \in \mathring{W}_2^1(\Omega)$,

$$\int_\Omega \left| \frac{f(x)}{d(x)} \right|^2 dx \leq C \int_\Omega |\nabla f(x)|^2 \, dx.$$

This is known to hold if Ω is bounded and has Lipschitz boundary, for example. If this inequality holds, then the important subspace $\mathring{W}_2^1(\Omega)$ of $W_2^1(\Omega)$ may be characterised as the space of those $f \in W_2^1(\Omega)$ such that $f/d \in L_2(\Omega)$ (see [46], Theorem V.3.4). Applications of the Hardy inequality to gain information about the spectrum of the Dirichlet Laplacian are given in [46], X.6.

We shall now develop the theory of Poincaré and Hardy inequalities in fairly general settings; applications will be given in the next Chapter. In most of this development the nature of the domain Ω is important and so we give an account of some of the many conditions on Ω which are currently in use in this connection. This leads naturally to many results of interest, such as the Sobolev-Poincaré inequality which is established when Ω is merely required to be a John domain, which is a weak restriction. The Hardy inequality

$$\|u/d \ |L_p(\Omega)\| \leq C \|\nabla u \ |L_p(\Omega)\|, \quad u \in \mathring{W}_p^1(\Omega),$$

where $d(x) = \text{dist}(x, \partial\Omega)$, is shown to hold for all proper open subsets Ω of $\mathbb{R}^n$ if $n < p < \infty$, and for all Ω satisfying a mild condition if $1 < p \leq n$. We also present the result of Kinnunen and Martio that if Ω is any proper open subset of $\mathbb{R}^n$ and $u \in W_p^1(\Omega)$ is such that $u/d \in$ weak-$L_p(\Omega)$, then $u \in \mathring{W}_p^1(\Omega)$. The chapter concludes with a discussion of the best constant in the Hardy inequality.

4.2 Poincaré inequalities in BFSs

4.2.1 Poincaré and Friedrichs inequalities

Our main object here is to show that under appropriate conditions, inequalities of this type hold if, and only if, the measure of non-compactness of an appropriate embedding is less than 1.The setting is reasonably general, enabling specific concrete cases to be obtained by specialisation.

Let Ω be a domain in $\mathbb{R}^n$ and let $\{\Omega_k\}$ be a fixed sequence of bounded domains such that

$$\Omega_k \subset \overline{\Omega}_k \subset \Omega_{k+1} \subset \Omega \text{ for all } k \in \mathbb{N} \tag{4.2.1}$$

and

$$\Omega = \cup_{k=1}^{\infty} \Omega_k. \tag{4.2.2}$$

$\Omega = \cup_{k=1}^{\infty} \Omega_k$. For each $k \in \mathbb{N}$ we put $\Omega^k = \Omega \setminus \Omega_k$. Let $X = X(\Omega) = X(\Omega, \|\cdot \mid X(\Omega)\|)$ be a Banach function space. Note that as a consequence of the definition, $C_0^{\infty}(\Omega) \subset X(\Omega) \subset L_{1,loc}(\Omega)$. Given any $f \in \mathcal{M}(\Omega_k)$, the set of all Lebesgue-measurable functions on Ω_k, we define a function $E_k f \in M(\Omega)$ (the **extension** of f to Ω) by

$$(E_k f)(x) = \begin{cases} f(x) \; if \; x \in \Omega_k, \\ 0 \quad if \; x \in \Omega^k. \end{cases}$$

Observe that for each $f \in \mathcal{M}(\Omega_k)$,

$$|E_k f| = E_k |f| \tag{4.2.3}$$

on Ω. The **restriction** $R_k f \in \mathcal{M}(\Omega_k)$ of a function $f \in \mathcal{M}(\Omega)$ to Ω_k is defined to be

$$R_k f = f \mid_{\Omega_k} .$$

Now we define

$$X(\Omega_k) = \{f \in \mathcal{M}(\Omega_k) : \|E_k f \mid X(\Omega)\| < \infty\}$$

and endow this space with the norm

$$\|f \mid X(\Omega_k)\| := \|E_k f \mid X(\Omega)\| .$$

It is easy to verify that $(X(\Omega_k), \|\cdot \mid X(\Omega_k)\|)$ is a Banach function space. Note that while, in general, $X(\Omega)$ does not contain $C^{\infty}(\overline{\Omega})$, we do have $C^{\infty}(\overline{\Omega}_k) \subset X(\Omega_k)$ since $\chi_{\Omega_k} \in X(\Omega)$.

As in Definition 3.6.1, given Banach function spaces $X(\Omega)$ and $Y(\Omega)$ we introduce the abstract Sobolev space $W(X, Y)$ by

$$W(X, Y) = \{f \in X(\Omega) : D_i f \in Y(\Omega) \text{ for } i = 1, ..., n\} ,$$

and give this space the norm

$$\|f \mid W(X,Y)\| = \|f \mid X\| + \|\nabla f \mid Y\|,$$

where by $\|\nabla f \mid Y\|$ we mean $\sum_{i=1}^{n} \|D_i f \mid Y\|$. We know from Lemma 3.6.2 that $W(X(\Omega), Y(\Omega))$ is a Banach space, as is $W(X(\Omega_k), Y(\Omega_k))$, for each $k \in \mathbb{N}$. Observe that the restriction operator R_k has a useful property in this connection, for given any $u \in W(X(\Omega), Y(\Omega))$,

$$\|R_k u \mid X(\Omega_k)\| = \|u\chi_{\Omega_k} \mid X(\Omega)\| \le \|u \mid X(\Omega)\|$$

and in fact $R_k u \in W(X(\Omega_k), Y(\Omega_k))$, with

$$\|R_k u \mid W(X(\Omega_k), Y(\Omega_k))\| \le \|u \mid W(X(\Omega), Y(\Omega))\|.$$

As in Remark 3.6.3 we define the subspace $\overset{\circ}{W}(X(\Omega), Y(\Omega))$ of $W(X(\Omega), Y(\Omega))$ to be the closure of $C_0^\infty(\Omega)$ in $W(X(\Omega), Y(\Omega))$.

To introduce the Poincaré inequality we let $X = X(\Omega)$ and $Y = Y(\Omega)$ be Banach function spaces and put

$$\mathcal{W}(\Omega) = \{w \in \mathcal{M}(\Omega) : 0 < w < \infty \text{ a.e. on } \Omega\};$$

the elements of $\mathcal{W}(\Omega)$ are called **weight functions** or **weights**. Let $w \in \mathcal{W}(\Omega)$ be such that

$(P1)$ $\quad F_w \in W(X,Y)^*$,

where

$$F_w(u) = \int_\Omega uw\,dx \quad (u \in W(X,Y)).$$

We say that (w, X, Y) **supports the (weighted) Poincaré inequality** if there is a positive constant K_1 such that for all $u \in W(X,Y)$,

$$\|u \mid X(\Omega)\| \le K_1 \left\{ \left| \int_\Omega uw\,dx \right| + \|\nabla u \mid Y(\Omega)\| \right\}. \tag{4.2.4}$$

Define

$$A_k = \sup_{\|u|W(X,Y)\| \le 1} \|u\chi_{\Omega^k} \mid X(\Omega)\| \quad (k \in \mathbb{N}). \tag{4.2.5}$$

Since for all $k \in \mathbb{N}$, $0 \le A_{k+1} \le A_k \le 1$, it follows that the limit

$$A = \lim_{k\to\infty} A_k \tag{4.2.6}$$

exists and lies in $[0,1]$. We shall see eventually that under certain conditions A is related to the measure of non-compactness of the natural embedding of $W(X,Y)$ in X.

Lemma 4.2.1. *Suppose that (P1) holds and that*
(P2) $W(X(\Omega_k), Y(\Omega_k))$ *is compactly embedded in* $X(\Omega_k)$, *for all* $k \in \mathbb{N}$.
If $A < 1$, *then* (w, X, Y) *supports the Poincaré inequality.*

This follows from the more general result which we now give.

Lemma 4.2.2. *Let (P2) hold and let* F *be a functional on* $W(X(\Omega), Y(\Omega))$ *such that*
(i) F *is continuous on* $W(X(\Omega), Y(\Omega))$,
(ii) $F(\lambda u) = \lambda F(u)$ *for all* $\lambda > 0$ *and all* $u \in W(X(\Omega), Y(\Omega))$, *and*
(iii) if $u \in W(X(\Omega), Y(\Omega))$ *is constant on* Ω *and* $F(u) = 0$, *then* $u = 0$.
If $A < 1$, *then there is a constant* K *such that for all* $u \in W(X(\Omega), Y(\Omega))$,

$$\|u \mid X(\Omega)\| \leq K \left\{ |F(u)| + \|\nabla u \mid Y(\Omega)\| \right\}. \tag{4.2.7}$$

Proof. Let $\alpha \in (A, 1)$. Then for all large enough k, we have for all $u \in W(X(\Omega), Y(\Omega))$,

$$\|u\chi_{\Omega^k} \mid X(\Omega)\| \leq \alpha \left\{ \|u\chi_{\Omega^k} \mid X(\Omega)\| + \|u\chi_{\Omega_k} \mid X(\Omega)\| + \|\nabla u \mid Y(\Omega)\| \right\},$$

and hence

$$\|u\chi_{\Omega^k} \mid X(\Omega)\| \leq \frac{\alpha}{1-\alpha} \left\{ \|R_k u \mid X(\Omega_k)\| + \|\nabla u \mid Y(\Omega)\| \right\}. \tag{4.2.8}$$

Suppose that the result is false. Then there is a sequence $\{u_j\}$ in $W(X(\Omega), Y(\Omega))$ such that

$$\|u_j \mid X(\Omega)\| = 1 \text{ for all } j \in \mathbb{N}, \tag{4.2.9}$$

$$\lim_{j \to \infty} F(u_j) = 0 \tag{4.2.10}$$

and

$$\lim_{j \to \infty} \|\nabla u_j \mid Y(\Omega)\| = 0. \tag{4.2.11}$$

Hence for fixed and sufficiently large $k \in \mathbb{N}$, the sequence $\{R_k u_j\}$ is bounded in $W(X(\Omega_k), Y(\Omega_k))$, and so by (P2) there is a subsequence, again denoted by $\{R_k u_j\}$, which is a Cauchy sequence in $X(\Omega_k)$. Using the estimate

$$\|u \mid X(\Omega)\| \leq \|u\chi_{\Omega^k} \mid X(\Omega)\| + \|R_k u \mid X(\Omega_k\|,$$

we therefore see that $\{u_j\}$ must be a Cauchy sequence in $X(\Omega)$. Thus there exists $u \in X(\Omega)$ such that $u_j \to u$ in $X(\Omega)$ as $j \to \infty$. It follows that $\nabla u = 0$ and so u is constant in $\Omega : u = c$, say. Plainly $\|c \mid X(\Omega)\| = 1$: hence $c \neq 0$ and $\chi_\Omega \in X(\Omega)$. This shows that $u_j \to c$ in $W(X(\Omega), Y(\Omega))$ and by (i) we have $F(u_j) \to F(c)$. Accordingly $F(c) = 0$, which by (iii) implies that $c = 0$, and we have a contradiction. □

Remark 4.2.3. (a) Suppose that (P2) holds and that

$$\chi_\Omega \notin X(\Omega). \tag{4.2.12}$$

If $A < 1$ then there is a constant K such that for all $u \in W(X(\Omega), Y(\Omega))$,

$$\|u \mid X(\Omega)\| \leq K \|\nabla u \mid Y(\Omega)\|. \tag{4.2.13}$$

To see this, just proceed as in the proof of the last lemma and note that (iii) is automatically satisfied: hence if (4.2.13) were false, then $\chi_\Omega \in X(\Omega)$ and we would have a contradiction. It follows that the norms $\|u \mid W(X(\Omega), Y(\Omega))\|$ and $\|\nabla u \mid Y(\Omega)\|$ are equivalent on $W(X(\Omega), Y(\Omega))$.
(b) If we replace condition (i) in the last lemma by the stronger condition
(i*) for all $u_1, u_2 \in W(X(\Omega), Y(\Omega))$,

$$|F(u_1) - F(u_2)| \leq C \leq \|u_1 - u_2 \mid W(X(\Omega), Y(\Omega))\|,$$

and suppose in the lemma additionally that
(iv) $|F(u_1 + u_2)| \leq |F(u_1) + F(u_2)|$ for all $u_1, u_2 \in W(X(\Omega), Y(\Omega))$,
then the last lemma implies that

$$|F(u)| + \|\nabla u \mid Y(\Omega)\|$$

is a norm on $W(X(\Omega), Y(\Omega))$ equivalent to $\|u \mid W(X(\Omega), Y(\Omega))\|$.
(c) Let F be a functional on $W(X(\Omega), Y(\Omega))$ which satisfies conditions (i)-(iii) of the last lemma, and suppose that assumption (P2) is strengthened to the requirement that

$$W(X(\Omega), Y(\Omega)) \text{ is compactly embedded in } X(\Omega).$$

Then an easy adaptation of the proof of the lemma shows that its conclusion still holds.

We note that part (a) of the Remark above can be strengthened as follows:

Lemma 4.2.4. *Suppose that (P2) holds and that $A < 1$. Then (4.2.13) holds for all $u \in W(X(\Omega), Y(\Omega))$ if, and only if, $\chi_\Omega \notin X(\Omega)$..*

Proof. Assume that (4.2.13) holds and that $\chi_\Omega \in X(\Omega)$. Then constant functions belong to $W(X(\Omega), Y(\Omega))$, and this contradicts (4.2.13). The remainder of the proof follows from (a) in the remark above. □

Lemma 4.2.5. *Suppose that the following condition holds:*
(P3) If $\chi_\Omega \in X(\Omega)$, then there exists $k \in \mathbb{N}$ such that

$$\|\chi_{\Omega^k} \mid X(\Omega)\| < \|\chi_\Omega \mid X(\Omega)\|.$$

Let $F \in W(X(\Omega), Y(\Omega))^*$ and suppose that (4.2.7) holds for all $u \in W(X(\Omega), Y(\Omega))$. Then $A < 1$.

Proof. Suppose that $1 = A = \lim_{k\to\infty} A_k$. Then $A_k = 1$ for all $k \in \mathbb{N}$ and so there is a sequence $\{u_k\}$, with $\|u_k \mid W(X,Y)\| \le 1$, such that

$$\lim_{k\to\infty} \|u_k\chi_{\Omega^k} \mid X(\Omega)\| = 1. \tag{4.2.14}$$

Since

$$\|u_k\chi_{\Omega^k} \mid X(\Omega)\| \le \|u_k \mid X(\Omega)\| \le \|u_k \mid W(X,Y)\| \le 1, \tag{4.2.15}$$

we see that as $k \to \infty$,

$$\|u_k \mid X(\Omega)\| \to 1 \tag{4.2.16}$$

and

$$\|u_k \mid W(X,Y)\| \to 1, \tag{4.2.17}$$

so that

$$\|\nabla u_k \mid Y(\Omega)\| \to 0. \tag{4.2.18}$$

Since $\{u_k\}$ is bounded in $W(X,Y)$, $\{F(u_k)\}$ is bounded in $\mathbb{C}$ and thus contains a Cauchy subsequence, again denoted by $\{F(u_k)\}$. Now (4.2.7) shows that

$$\|u_k - u_l \mid X(\Omega)\| \le K\left\{|F(u_k - u_l)| + \|\nabla u_k \mid Y(\Omega)\| + \|\nabla u_l \mid Y(\Omega)\|\right\};$$

hence there exists $u \in X(\Omega)$ such that

$$\lim_{k\to\infty} u_k = u \text{ in } X(\Omega). \tag{4.2.19}$$

From (4.2.18) and (4.2.19) it now follows that u is constant on Ω, say $u = c$.

If $\chi_\Omega \notin X(\Omega)$, then $c = 0$, which contradicts (4.2.16) and (4.2.18). However, if $\chi_\Omega \in X(\Omega)$, then (4.2.14) and (4.2.19), together with

$$\left|\|(u_k - u)\chi_{\Omega^k} \mid X(\Omega)\| - \|u_k\chi_{\Omega^k} \mid X(\Omega)\|\right| \le \|u\chi_{\Omega^k} \mid X(\Omega)\|$$

show that

$$\lim \inf_{k\to\infty} \|u\chi_{\Omega^k} \mid X(\Omega)\| \ge 1,$$

which gives $\lim_{k\to\infty} \|u\chi_{\Omega^k} \mid X(\Omega)\| = 1$. Hence

$$\lim_{k\to\infty} \|c\chi_{\Omega^k} \mid X(\Omega)\| = 1.$$

Since $\Omega^{k+1} \subset \Omega^k$, it follows that for all $k \in \mathbb{N}$,

$$1 = \|c\chi_{\Omega^k} \mid X(\Omega)\|.$$

However, (4.2.16) and (4.2.19) show that

$$1 = \|c \mid X(\Omega)\| = \|c\chi_{\Omega^k} \mid X(\Omega)\|,$$

which leads to

$$\|\chi_\Omega \mid X(\Omega)\| = \|\chi_{\Omega^k} \mid X(\Omega)\|$$

for all $k \in \mathbb{N}$ and contradicts (P3). The proof is complete. □

We observe that (P3) is not always satisfied: think of $L_\infty(\Omega)$.

It is now convenient to make the following generalisation of the Poincaré inequality. Given any $F \in W(X,Y)^*$, we shall say that (F, X, Y) **supports the Poincaré inequality** if there is a positive constant K such that for all $u \in W(X,Y)$,

$$\|u \mid X(\Omega)\| \leq K\left\{|F(u)| + \|\nabla u \mid Y(\Omega)\|\right\}.$$

We see immediately from our work so far that we have

Theorem 4.2.6. *Suppose that (P2) and (P3) hold and let the functional* $F \in W(X(\Omega), Y(\Omega))^*$ *satisfy condition (iii) of Lemma 4.2.2 Then* (F, X, Y) *supports the Poincaré inequality if, and only if,* $A < 1$.

Corollary 4.2.7. *Suppose that (P2) and (P3) hold and let* $w \in \mathcal{W}(\Omega)$ *be such that (P1) holds. Then* (w, X, Y) *supports the Poincaré inequality if, and only if,* $A < 1$.

Use of Lemma 4.2.4 and Lemma 4.2.5 with $F = 0$ also gives

Theorem 4.2.8. *Suppose that (P2) holds and that* $\chi_\Omega \notin X(\Omega)$. *Then* $A < 1$ *if, and only if, there is a constant* K *such that for all* $u \in W(X(\Omega), Y(\Omega))$,

$$\|u \mid X(\Omega)\| \leq K \|\nabla u \mid Y(\Omega)\|.$$

Next, we establish other characterisations of the Poincaré inequality. Let $w \in \mathcal{W}(\Omega) \cap L_1(\Omega)$ and suppose that (P1) holds. Then the **weighted average** $u_{\Omega,w}$ of a function u defined on Ω is given by

$$u_{\Omega,w} = \frac{1}{w(\Omega)} \int_\Omega u(x)w(x)dx,$$

where $w(\Omega) = \int_\Omega w(x)dx$; it is defined for all $u \in W(X(\Omega), Y(\Omega))$. In the next Lemma we shall use condition (P1)*, which is more restrictive than (P1):

(P1)* $\quad F_w \in X(\Omega)^*$, where $F_w(u) = \int_\Omega uwdx$ for all $u \in X(\Omega)$.

Lemma 4.2.9. *Suppose that* $\chi_\Omega \in X(\Omega)$ *and that* $w \in \mathcal{W}(\Omega) \cap L_1(\Omega)$ *is such that (P1)* holds. Then the following statements are equivalent:*

(i) (w, X, Y) *supports the Poincaré inequality;*

(ii) there is a positive constant K_2 *such that for all* $u \in W(X(\Omega), Y(\Omega))$,

$$\|u - u_{\Omega,w} \mid X(\Omega)\| \leq K_2 \|\nabla u \mid Y(\Omega)\|;$$

(iii) there is a positive constant K_3 *such that for all* $u \in W(X(\Omega), Y(\Omega))$,

$$\inf_{c \in \mathbb{C}} \|u - c \mid X(\Omega)\| \leq K_3 \|\nabla u \mid Y(\Omega)\|.$$

This follows immediately from Lemma 4.2.11 below. In this more general result we employ the next Lemma, the simple proof of which is omitted.

Lemma 4.2.10. *Suppose that* $\chi_\Omega \in X(\Omega)$ *and let* $F \in X(\Omega)^*$ *be such that* $F(\chi_\Omega) \neq 0$. *Let*

$$Z(\Omega) = \{u \in X(\Omega) : u = c\chi_\Omega, c \in \mathbb{C}\}.$$

Then the map $L : W(X(\Omega), Y(\Omega)) \to X(\Omega)$ *given by*

$$L(u) = \frac{F(u)}{F(\chi_\Omega)}\chi_\Omega$$

is a projection onto $Z(\Omega)$ *such that for all* $u \in W(X(\Omega), Y(\Omega))$ *we have* $F(u) = F(L(u))$ *and* $\|L(u) \mid X(\Omega)\| = K_0 |F(u)|$, *where*

$$K_0 = \|\chi_\Omega \mid X(\Omega)\| / |F(\chi_\Omega)|.$$

Lemma 4.2.11. *Suppose that* $\chi_\Omega \in X(\Omega)$ *and let* $F \in X(\Omega)^*$ *be such that* $F(\chi_\Omega) \neq 0$. *Then the following statements are equivalent:*
(i) $(F, X(\Omega), Y(\Omega))$ *supports the Poincaré inequality;*
(ii) there is a positive constant K_2 *such that for all* $u \in W(X(\Omega), Y(\Omega))$,

$$\|u - L(u) \mid X(\Omega)\| \leq K_2 \|\nabla u \mid Y(\Omega)\|;$$

(iii) there is a positive constant K_3 *such that for all* $u \in W(X(\Omega), Y(\Omega))$,

$$\inf_{c \in \mathbb{C}} \|u - c \mid X(\Omega)\| \leq K_3 \|\nabla u \mid Y(\Omega)\|.$$

Proof. (i) $\Longrightarrow$ (ii) Let $u \in W(X(\Omega), Y(\Omega))$.Then

$$f := u - L(u) \in W(X(\Omega), Y(\Omega)), F(f) = 0$$

and, by (i),

$$\|u - L(u) \mid X(\Omega)\| \leq K_1 \|\nabla u \mid Y(\Omega)\|.$$

(ii) $\Longrightarrow$ (iii) This is clear.
(iii) $\Longrightarrow$ (ii) Since

$$\|u - L(u) \mid X(\Omega)\| \leq \|u - c \mid X(\Omega)\| + \|c\chi_\Omega - L(u) \mid X(\Omega)\|$$

and

$$\begin{aligned}\|c\chi_\Omega - L(u) \mid X(\Omega)\| &= \|L(c\chi_\Omega - u) \mid X(\Omega)\| \\ &= K_0 |F(u - c\chi_\Omega)| \leq K_0 \|F\| \|u - c \mid X(\Omega)\|,\end{aligned}$$

it follows that (ii) holds.
(ii) $\Longrightarrow$ (i) As

$$\|u \mid X(\Omega)\| \leq \|u - L(u) \mid X(\Omega)\| + \|L(u) \mid X(\Omega)\|$$

and

$$\|L(u) \mid X(\Omega)\| = K_0 |F(u)|,$$

(i) holds. □

Lemma 4.2.12. *Suppose that $\chi_\Omega \notin X(\Omega)$ and let $u \in X(\Omega)$. Then*

$$\inf_{c\in\mathbb{C}} \|u - c \mid X(\Omega)\| = \|u \mid X(\Omega)\| . \tag{4.2.20}$$

Proof. If $\|u - c \mid X(\Omega)\| < \infty$ for some $c \in \mathbb{C}$, then $c = u - (u - c) \in X(\Omega)$, and hence $c = 0$ since $\chi_\Omega \notin X(\Omega)$. Thus $\|u - c \mid X(\Omega)\| = \infty$ if $c \neq 0$ and the result follows. □

Theorem 4.2.13. *Suppose there is a function $w \in \mathcal{W}(\Omega) \cap L_1(\Omega)$ such that (P1)* holds; suppose also that (P2) and (P3) hold. Then the following statements are equivalent:*

(i) $A < 1$;

(ii) there is a positive constant K such that for all $u \in W(X(\Omega), Y(\Omega))$,

$$\inf_{c\in\mathbb{C}} \|u - c \mid X(\Omega)\| \leq K \|\nabla u \mid Y(\Omega)\| ; \tag{4.2.21}$$

(iii) the operator $T : W(X(\Omega), Y(\Omega)) \to (Y(\Omega))^n$ given by $Tf = \nabla f$ has closed range.

Proof. (i)⇔(ii) Let $\chi_\Omega \notin X(\Omega)$. Then by Lemma 4.2.12 , (4.2.20) holds for all $u \in W(X(\Omega), Y(\Omega))$. Then (4.2.21) holds for all $u \in W(X(\Omega), Y(\Omega))$ if, and only if, (4.2.13) also holds for all $u \in W(X(\Omega), Y(\Omega))$, which by Theorem 4.2.8 is equivalent to $A < 1$.

On the other hand, if $\chi_\Omega \in X(\Omega)$, then by Lemma 4.2.9, (4.2.21) holds for all $u \in W(X(\Omega), Y(\Omega))$ if, and only if, (w, X, Y) supports the Poincaré inequality; and by Corollary 4.2.7 this is equivalent to $A < 1$.

(ii)⇔(iii) By Theorem I.3.4 of [46], T has closed range if, and only if, the minimum modulus $\nu(T)$ of T , defined by

$$\nu(T) = \inf \{\|Tu\| / \text{dist}(u, \ker T) : u \in W(X(\Omega), Y(\Omega)) \setminus \ker T \},$$

where $\ker T = \{u \in W(X(\Omega), Y(\Omega)) : Tu = 0\}$, satisfies $\nu(T) > 0$. Since for all $u \in W(X(\Omega), Y(\Omega))$,

$$\inf_{c\in\mathbb{C}} \|u - c \mid W(X(\Omega), Y(\Omega))\| = \|\nabla u \mid Y(\Omega)\| + \inf_{c\in\mathbb{C}} \|u - c \mid X(\Omega)\| , \tag{4.2.22}$$

(4.2.21) can be rewritten as

$$\inf_{c\in\mathbb{C}} \|u - c \mid W(X(\Omega), Y(\Omega))\| \leq (K + 1) \|\nabla u \mid Y(\Omega)\|$$

or, equivalently,

$$1/(K + 1) \leq \|Tu\| / \inf_{c\in\mathbb{C}} \|u - c \mid W(X(\Omega), Y(\Omega))\| \tag{4.2.23}$$

for all $u \in W(X(\Omega), Y(\Omega))$ for which the denominator of the last quotient is non-zero. We now claim that for all $u \in W(X(\Omega), Y(\Omega))$,

$$\operatorname{dist}(u, \ker T) = \inf_{c \in \mathbb{C}} \|u - c \mid W(X(\Omega), Y(\Omega))\| \tag{4.2.24}$$

To prove this, note that if $\chi_\Omega \notin X(\Omega)$, then $\ker T = \{0\}$ and, by (4.2.22) and Lemma 4.2.12,

$$\operatorname{dist}(u, \ker T) = \|u \mid W(X(\Omega), Y(\Omega))\| = \inf_{c \in \mathbb{C}} \|u - c \mid W(X(\Omega), Y(\Omega))\|;$$

if $\chi_\Omega \in X(\Omega)$, then $\ker T = \mathbb{C}$ and our claim again holds.

The inequality (4.2.23) together with (4.2.24) imply that (4.2.21) is equivalent to

$$1/(K+1) \leq \|Tu\| / \operatorname{dist}(u, \ker T)$$

for all $u \in W(X(\Omega), Y(\Omega)) \setminus \ker(T)$, and as this is equivalent to $\nu(T) > 0$, the proof of the Theorem is complete. □

Remark 4.2.14. (i) The weight function w does not appear in the statements (i)-(iii) of the Theorem: our assumptions involving w represent a condition on the space $X(\Omega)$ only. The same is true of statement (iii) of Lemma 4.2.9

(ii) Let $X(\Omega)$ have absolutely continuous norm (see Definition 3.1.11) and let $\beta(id)$ be the ball measure of non-compactness of the embedding $id : W(X(\Omega), Y(\Omega)) \to X(\Omega)$ (see Section 1.3 and [46], I.2). For each $k \in \mathbb{N}$ let $id_k : W(X(\Omega), Y(\Omega)) \to X(\Omega)$ be defined by $id_k u = u\chi_{\Omega_k}$. Then it is easy to check that the maps $P_k : X(\Omega) \to X(\Omega)$ defined by $P_k f = f\chi_{\Omega_k}$ $(f \in X(\Omega))$ satisfy the conditions required in Proposition 3.1 of [27] and so

$$\beta(id) = \lim_{k \to \infty} \|id - id_k\| = \operatorname{dist}(id, \mathcal{K}(W, X)),$$

where $W = W(X(\Omega), Y(\Omega))$. Since

$$\begin{aligned} \lim_{k \to \infty} \|id - id_k\| &= \lim_{k \to \infty} \sup_{\|u|W\| \leq 1} \|(id - id_k)u \mid X(\Omega)\| \\ &= \lim_{k \to \infty} \sup_{\|u|W\| \leq 1} \|u\chi_{\Omega^k} \mid X(\Omega)\| = A, \end{aligned}$$

it follows that

$$A = \beta(id) = \operatorname{dist}(id, \mathcal{K}(W, X)).$$

Note that the only property of an absolutely continuous norm that is used in this argument is that for all $f \in X(\Omega)$, $\lim_{k \to \infty} \|u\chi_{\Omega^k} \mid X(\Omega)\| = 0$.Without this we would merely have

$$\beta(id) \leq A,$$

for $\beta(id) = \beta(id - id_k) \leq \|id - id_k\|$ for all $k \in \mathbb{N}$. Of course, this is good enough to ensure that $\beta(id) = 0$ if $A = 0$; that is, $A = 0$ implies that id is compact. Moreover, if $\beta(id) = 0$, then by Remark 4.2.3 (c) the inequality (4.2.7) holds, so that by Theorem 4.2.6 if we assume in addition that (P3) holds, and make none of the assumptions of absolute continuity, we see that $A < 1$ if $\beta(id) = 0$.

We now turn to the Friedrichs inequality. Let $X = X(\Omega)$ and $Y = Y(\Omega)$ be Banach function spaces. We say that (X, Y) **supports the Friedrichs inequality** if there is a positive constant K such that for all $u \in C_0^\infty(\Omega)$,

$$\|u \mid X(\Omega)\| \leq K \, \|\nabla u \mid Y(\Omega)\| \,. \tag{4.2.25}$$

For each $k \in \mathbb{N}$ we define

$$A_k^0 = \sup\{\|u\chi_{\Omega^k} \mid X(\Omega)\| : u \in C_0^\infty(\Omega), \|u \mid W(X,Y)\| \leq 1\}\,. \tag{4.2.26}$$

It is clear that

$$A_k^0 = \sup\{\|u\chi_{\Omega^k} \mid X(\Omega)\| : u \in W_0(X,Y), \|u \mid W(X,Y)\| \leq 1\}\,, \tag{4.2.27}$$

and since $0 \leq A_{k+1}^0 \leq A_k^0 \leq 1$, the limit

$$A^0 = \lim_{k\to\infty} A_k^0 \tag{4.2.28}$$

exists and $A^0 \in [0,1]$.

Lemma 4.2.15. *Assume that (P2) and the following condition hold:*
(F1) if $\chi_\Omega \in X(\Omega)$ and $\{u_j\}$ is a sequence in $C_0^\infty(\Omega)$ with $u_j \to c$ in $W(X,Y)$, where c is a constant function on Ω, then $c = 0$.
Suppose that $A^0 < 1$. Then (X, Y) supports the Friedrichs inequality.

Proof. Suppose that the Lemma is false. Then just as in the proof of Lemma 4.2.5 we see that there is a sequence $\{u_j\} \subset C_0^\infty(\Omega)$ such that $\|u_j \mid X(\Omega)\| = 1$ for all $j \in \mathbb{N}$, $\|\nabla u_j \mid Y(\Omega)\| \to 0$ and $\|u_j - c \mid X(\Omega)\| \to 0$ as $j \to \infty$. If $\chi_\Omega \notin X(\Omega)$, then the constant function c must be zero and we have an evident contradiction. However, if $\chi_\Omega \in X(\Omega)$, then $u_j \to c$ in $W(X,Y)$ and so by (F1) we must again have $c = 0$. The result follows. □

Remark 4.2.16. Condition (F1) is implied by
(F1)* if $\chi_\Omega \in X(\Omega)$, then there exist $x_0 \in \partial\Omega$ and $R, K_1, K_2 > 0$ such that if $\Omega(x_0,R) := \{x \in \Omega : |x - x_0| < R\} \in C^{0,1}$, then

$$\int_{\Omega(x_0,R)} |f| \, dx \leq K_1 \, \|f \mid X(\Omega)\| \text{ for all } f \in X(\Omega) \tag{4.2.29}$$

and

$$\int_{\Omega(x_0,R)} |g| \, dx \leq K_2 \, \|g \mid Y(\Omega)\| \text{ for all } g \in Y(\Omega). \tag{4.2.30}$$

To see this, observe that the assumptions in (F1)* together with standard Sobolev embedding theory imply that

$$W(X(\Omega), Y(\Omega)) \hookrightarrow W_1^1(\Omega(x_0,R)) \hookrightarrow L_1(\partial\Omega(x_0,R)).$$

Thus if $\{u_j\}$ is a sequence in $C_0^\infty(\Omega)$ with $u_j \to c$ in $W(X,Y)$, where c is a constant function on Ω, it follows that $u_j \to c$ in $L_1(\partial\Omega(x_0,R))$. As the restriction of u_j to $\partial\Omega \cap \overline{\Omega(x_0,R)}$ is zero for all $j \in \mathbb{N}$, we must have $c = 0$.

Lemma 4.2.17. *If (X, Y) suppports the Poincaré inequality, then $A^0 < 1$.*

Proof. If the result were false, there would be a sequence $\{u_j\} \subset C_0^\infty(\Omega)$ such that (4.2.16) and (4.2.18) hold. This would contradict (4.2.25). □

Theorem 4.2.18. *Suppose that (F1) and (P2) hold. The the following statements are equivalent:*
(i) $A^0 < 1$;
(ii) (X, Y) *supports the Friedrichs inequality;*
(iii) the operator $T : \mathring{W}(X(\Omega), Y(\Omega)) \to (Y(\Omega))^n$ *given by* $Tf = \nabla f$ *has closed range.*

Proof. (i)$\Longleftrightarrow$(ii) This follows from Lemmas 4.2.15 and 4.2.17.

(ii)$\Longleftrightarrow$(iii) Let $u \in \ker T$. Then $u \in \mathring{W}(X, Y)$ and $\nabla u = 0$: thus u is constant on Ω, say $u = c$. If $\chi_\Omega \notin X(\Omega)$, then since $u \in \mathring{W}(X, Y)$ we must have $c = 0$. On the other hand, if $\chi_\Omega \in X(\Omega)$, then as $c \in \mathring{W}(X, Y)$ there is a sequence $\{u_j\}$ in $C_0^\infty(\Omega)$ with $u_j \to c$ in $W(X, Y)$, so that in view of $(F1)$, $c = 0$. Hence $\ker T = \{0\}$. Since T is closed, it has closed range if, and only if,

$$0 < \nu(T) := \inf \left\{ \|Tu\| / \text{dist}(u, \ker T) : u \in \mathring{W}(X, Y) \setminus \ker(T) \right\},$$

and as $\ker T = \{0\}$ this is equivalent to

$$\|u \mid X(\Omega)\| \leq \|\nabla u \mid Y(\Omega)\| / \nu(T)$$

for all $u \in \mathring{W}(X, Y)$. The density of $C_0^\infty(\Omega)$ means that the last inequality holds if, and only if, (X, Y) supports the Friedrichs inequality. This completes the proof of the theorem. □

Remark 4.2.19. Suppose that (P2) holds and that for all $f \in X(\Omega)$,

$$\lim_{j \to \infty} \|f \chi_{\Omega^j} \mid X(\Omega)\| = 0.$$

Of course, this is the case, in particular, if $X(\Omega)$ has an absolutely continuous norm. Then as in Remark 4.2.14 it can be seen that

$$A^0 = \beta(id_0) = \text{dist}(id_0, \mathcal{K}(\mathring{W}, X)),$$

where $\mathring{W} = \mathring{W}(X(\Omega), Y(\Omega))$ and $\beta(id_0)$ is the (ball) measure of non-compactness of the embedding $id_0 : \mathring{W} \to X(\Omega)$.

4.2.2 Examples

We give several examples of Banach function spaces $X(\Omega)$ and $Y(\Omega)$ for which the important conditions of the last subsection hold.

Weighted Lebesgue spaces

Let $p \in [1, \infty)$ and $w \in \mathcal{W}(\Omega)$. We recall that the **weighted Lebesgue space** $L_p(\Omega, w)$ (sometimes also denoted by $L_p(\Omega, wdx)$ is the set of all $u \in \mathcal{M}(\Omega)$ with finite norm

$$\|u \mid L_p(\Omega, w)\| = \left(\int_\Omega |u(x)|^p \, w(x)dx \right)^{1/p},$$

that it is complete when given this norm, and that it is a Banach function space if $w, w^{-1/p'} \in L_{1,loc}(\Omega)$. If we take $v_0, v_1 \in \mathcal{W}(\Omega)$, assume that

$$v_0, v_1, v_0^{-1/p'}, v_1^{-1/p'} \in L_{1,loc}(\Omega)$$

and put

$$X(\Omega) = L_p(\Omega, v_0), \quad Y(\Omega) = L_p(\Omega, v_1),$$

then the abstract Sobolev space $W(X(\Omega(, Y(\Omega))$ is just the weighted Sobolev space usually denoted by $W_p^1(\Omega, v_0, v_1)$ and with norm written in the form $\|\cdot \mid W_p^1(\Omega, v_0, v_1)\|$: see, for example, [147]. The space $\overset{\circ}{W}(X(\Omega), Y(\Omega))$ is then represented as $\overset{\circ}{W}{}_p^1(\Omega, v_0, v_1)$. If $v_0 = v_1 = 1$ this space is denoted by $\overset{\circ}{W}{}_p^1(\Omega)$. We now consider the various conditions imposed in the last subsection in connection with the Poincaré inequality. With $w = v_0$, Hölder's inequality shows that (P1)* (and hence also (P1)) holds if

$$v_0(\Omega) := \int_\Omega v_0 dx < \infty. \tag{4.2.31}$$

Suppose that the sequence $\{\Omega_k\}$ is such that $\Omega_k \in C^{0,1}$ for all $k \in \mathbb{N}$ (see [46], Definition V.4.1) and assume that

$$v_0, v_1 \in \mathcal{W}_c(\Omega),$$

where $\mathcal{W}_c(\Omega)$ is the subset of $\mathcal{W}(\Omega)$ consisting of those functions which are locally bounded on Ω from above and below by positive constants. Then the standard (unweighted) embedding theory shows that (P2) holds. Condition (P3) means that $v_0(\Omega^k) < v_0(\Omega)$ for at least one $k \in \mathbb{N}$ provided that (4.2.31) holds. This is plainly the case for any $v_0 \in \mathcal{W}(\Omega)$.

To take stock of the situation, we have from the last subsection the following assertions. Suppose that $v_0, v_1 \in \mathcal{W}_c(\Omega)$ and that

$$\Omega_k \in C^{0,1} \quad \text{for all } k \in \mathbb{N}. \tag{4.2.32}$$

Suppose that $F \in W_p^1(\Omega, v_0, v_1)^*$ is such that $F(\chi_\Omega) \neq 0$ if $v_0(\Omega) < \infty$. Then $(F, L_p(\Omega, v_0), L_p(\Omega, v_1))$ supports the Poincaré inequality if, and only if, $A < 1$, where A is given by (4.2.6).

Assume additionally that (4.2.31) holds. Then $(v_0, L_p(\Omega, v_0), L_p(\Omega, v_1))$ supports the Poincaré inequality if, and only if, $A < 1$.

Lastly, suppose that (4.2.31) is replaced by the condition $v_0(\Omega) = \infty$. Then $A < 1$ if, and only if, there is a positive constant K such that for all $u \in W_p^1(\Omega, v_0, v_1)$,

$$\|u \mid L_p(\Omega, v_0)\| \leq K \|\nabla u \mid L_p(\Omega, v_1)\| .$$

Turning to the Friedrichs inequality, we observe that condition (F1) (and *a fortiori* (F1)*) holds if we assume

(F1)$_1^*$ If $v_0(\Omega) < \infty$, then there exist $x_0 \in \partial\Omega$ and $R > 0$ such that

$$\Omega(x_0, R) = \{x \in \Omega : |x - x_0| < R\} \in C^{0,1}$$

and

$$v_0^{-1/p}, v_1^{-1/p} \in L_{p'}(\Omega(x_0, R)).$$

The results of the last subsection imply the following. Suppose that (4.2.32) holds and that Ω and $v_0, v_1 \in \mathcal{W}_c(\Omega)$ are such that (F1)$_1^*$ is satisfied. Then $(L_p(\Omega, v_0), L_p(\Omega, v_1))$ supports the Friedrichs inequality if, and only if, $A^0 < 1$, where A^0 is given by (4.2.28).

Since $L_p(\Omega; v_0)$ has an absolutely continuous norm for all $v_0 \in \mathcal{W}(\Omega)$ (here the assumption that p is finite is crucial), it follows from our earlier remarks that A and A^0 are the measures of non-compactness of the embedding maps from $W_p^1(\Omega, v_0, v_1)$ and $\mathring{W}_p^1(\Omega, v_0, v_1)$ respectively to $L_p(\Omega, v_0)$.

Orlicz spaces

Here we pass from spaces of Lebesgue type to the broader class of Orlicz spaces. We recall that a Young function is a map $\phi : [0, \infty) \to [0, \infty)$ which is continuous, increasing and convex, with $\phi(0) = 0$ and

$$\lim_{t \to 0} \phi(t)/t = \lim_{t \to \infty} t/\phi(t) = 0.$$

Its complementary function $\widetilde{\phi}$ is defined by

$$\widetilde{\phi}(t) = \sup_{s > 0}(st - \phi(s)),$$

and has the same properties as those of ϕ mentioned above.

Let $w \in \mathcal{W}(\Omega)$ satisfy

$$w(\Omega_k) < \infty \quad \text{for all } k \in \mathbb{N} \tag{4.2.33}$$

and let ϕ be a Young function. The **weighted Orlicz space** $L_\phi(\Omega, w)$ is defined to be the set of all $f \in \mathcal{M}(\Omega)$ such that for some $\alpha > 0$ (depending on f),

$$\int_{\Omega} \phi(\alpha |f(x)|) w(x) dx < \infty.$$

This space may be equipped with the **Orlicz norm**

$$\|f \mid L_{\phi}(\Omega, w)\|_0 = \sup \left\{ \int_{\Omega} fgw dx : \int_{\Omega} \widetilde{\phi}(g) w dx \leq 1 \right\}, \tag{4.2.34}$$

or with the **Luxemburg norm**

$$\|f \mid L_{\phi}(\Omega, w)\| = \inf \left\{ \lambda > 0 : \int_{\Omega} \phi(|f(x)| / \lambda) w(x) dx \leq 1 \right\}. \tag{4.2.35}$$

Just as in [144] it can be shown that these norms are equivalent: in fact,

$$\|f \mid L_{\phi}(\Omega, w)\| \leq \|f \mid L_{\phi}(\Omega, w)\|_0 \leq 2 \|f \mid L_{\phi}(\Omega, w)\|$$

for all $f \in L_{\phi}(\Omega, w)$. We also have Hölder's inequality:

$$\left| \int_{\Omega} fgw dx \right| \leq \|f \mid L_{\phi}(\Omega, w)\| \, \|g \mid L_{\phi}(\Omega, w)\|_0 . \tag{4.2.36}$$

With either of these norms, $L_{\phi}(\Omega, w)$ is complete. Moreover, for any $Q \subset \Omega$ with $0 < w(Q) < \infty$ we have

$$\|\chi_Q \mid L_{\phi}(\Omega, w)\| = 1/\phi^{-1}(1/w(Q)) \tag{4.2.37}$$

and

$$\|\chi_Q \mid L_{\phi}(\Omega; w)\|_0 = w(Q)(\widetilde{\phi})^{-1}(1/w(Q)). \tag{4.2.38}$$

For all this we refer to [162] and [202].

It is easy to check that $X(\Omega) = L_{\phi}(\Omega, w)$ is a Banach function space provided that

$$\left\| 1/w \mid L_{\widetilde{\phi}}(\Omega_k, w) \right\| < \infty \text{ for all } k \in \mathbb{N}.$$

Thus if we take $v_0, v_1 \in \mathcal{W}(\Omega)$, assume that

$$v_0(\Omega_k), v_1(\Omega_k) < \infty \text{ for all } k \in \mathbb{N}, \tag{4.2.39}$$

$$\left\| 1/v_i \mid L_{\widetilde{\phi}}(\Omega_k, v_i) \right\| < \infty \text{ for all } k \in \mathbb{N},\ i = 0, 1, \tag{4.2.40}$$

and set

$$X(\Omega) = L_{\phi}(\Omega, v_0),\ Y(\Omega) = L_{\phi}(\Omega, v_1), \tag{4.2.41}$$

the abstract Sobolev space $W(X(\Omega), Y(\Omega))$ is a weighted Orlicz-Sobolev space, denoted by $W^1 L_{\phi}(\Omega, v_0, v_1)$, and with norm written as

$$\left\|\cdot \mid W^1 L_\phi(\Omega, v_0, v_1)\right\|_0 \text{ or } \left\|\cdot \mid W^1 L_\phi(\Omega, v_0, v_1)\right\|$$

depending on whether the Orlicz or the Luxemburg norms are taken in $X(\Omega)$ and $Y(\Omega)$. The space $\mathring{W}(X(\Omega), Y(\Omega))$ is then written as $\mathring{W}^1 L_\phi(\Omega, v_0, v_1)$. If $v_0 = v_1 = 1$, $W^1 L_\phi(\Omega, v_0, v_1)$ is the unweighted Orlicz-Sobolev space $W^1 L_\phi(\Omega)$ with norms $\left\|\cdot \mid W^1 L_\phi(\Omega)\right\|_0$ or $\left\|\cdot \mid W^1 L_\phi(\Omega)\right\|$ (see [148]). In the same way, if $w = 1$, then $L_\phi(\Omega, w) := L_\phi(\Omega)$.

To verify the conditions imposed (in connection with the Poincaré and Friedrichs inequalities) in the context of Orlicz spaces it is convenient to establish some Lemmas. First, however, we recall that given Young functions ϕ_1 and ϕ_2 we say that ϕ_1 **increases essentially more slowly than** ϕ_2 (near infinity) if for every $\lambda > 0$,

$$\lim_{t\to\infty} \phi_2(\lambda t)/\phi_1(t) = \infty :$$

we write this as $\phi_1 \ll \phi_2$.

Lemma 4.2.20. *Let ϕ_1 and ϕ_2 be Young functions. Then $\phi_1 \ll \phi_2$ if, and only if,*

$$\lim_{t\to\infty} \phi_2^{-1}(\lambda t)/\phi_1^{-1}(t) = 0 \text{ for all } \lambda > 0.$$

Proof. Just argue by contradiction. □

Now let ϕ be a Young function and put

$$g_\phi(t) = \phi^{-1}(t)/t^{1+1/n}, \; t > 0. \tag{4.2.42}$$

The **Sobolev conjugate** ϕ^* of ϕ is defined by

$$(\phi^*)^{-1}(t) = \int_0^t g_\phi(s)ds. \tag{4.2.43}$$

Note that with no loss of generality we can always assume that

$$\int_0^1 g_\phi(s)ds < \infty$$

(see [148]).

Lemma 4.2.21. *For any Young function ϕ, we have $\phi \ll \phi^*$.*

Proof. Lemma 4.2.20 together with the fact that ϕ^{-1} is increasing mean that it is enough to show that for all $\lambda > 1$,

$$\lim_{t\to\infty} (\phi^*)^{-1}(\lambda t)/\phi^{-1}(t) = 0.$$

Let $\lambda > 1$. Since ϕ^{-1} is concave and $\phi^{-1}(0) = 0$ it follows that $\phi^{-1}(\lambda r) \leq \lambda\phi^{-1}(r)$ for all $r > 0$. Hence

$$\frac{(\phi^*)^{-1}(\lambda t)}{\phi^{-1}(t)} = \frac{1}{\phi^{-1}(t)}\int_0^{\lambda t}\frac{\phi^{-1}(s)}{s^{1+1/n}}ds = \frac{\lambda}{\phi^{-1}(t)}\int_0^t \frac{\phi^{-1}(\lambda r)}{(\lambda r)^{1+1/n}}dr$$

$$\leq \frac{\lambda^{1-1/n}}{\phi^{-1}(t)}\int_0^t \frac{\phi^{-1}(\lambda r)}{r^{1+1/n}}dr = \lambda^{1-1/n}(\phi^*)^{-1}(t)/\phi^{-1}(t).$$

It is therefore enough to show that

$$\lim_{t\to\infty}(\phi^*)^{-1}(t)/\phi^{-1}(t) = 0. \tag{4.2.44}$$

To do this, note that

$$\lim_{t\to\infty}\phi^{-1}(t)/t = 0, \tag{4.2.45}$$

and so $\phi^{-1}(t) < t$ for large enough t. Put

$$(\phi^*)^{-1}(t)/\phi^{-1}(t) = I_1(t) + I_2(t) + I_3(t),$$

where

$$I_1(t) = (\phi^*)^{-1}(1)/\phi^{-1}(t),\ I_2(t) = \frac{1}{\phi^{-1}(t)}\int_1^{\phi^{-1}(t)}\frac{\phi^{-1}(s)}{s^{1+1/n}}ds$$

and

$$I_3(t) = \frac{1}{\phi^{-1}(t)}\int_{\phi^{-1}(t)}^{t}\frac{\phi^{-1}(s)}{s^{1+1/n}}ds.$$

Since $\phi^{-1}(t)\to\infty$ as $t\to\infty$, we see that $I_1(t)\to 0$ as $t\to\infty$. Moreover, since

$$\lim_{t\to\infty}\phi^{-1}(\phi^{-1}(t))/\phi^{-1}(t) = 0$$

and

$$I_2(t) \leq \frac{\phi^{-1}(\phi^{-1}(t))}{\phi^{-1}(t)}\int_1^{\phi^{-1}(t)}\frac{1}{s^{1+1/n}}ds \leq \frac{\phi^{-1}(\phi^{-1}(t))}{\phi^{-1}(t)}\int_1^{\infty}\frac{1}{s^{1+1/n}}ds$$

$$= n\phi^{-1}(\phi^{-1}(t))/\phi^{-1}(t),$$

we have $I_2(t)\to 0$ as $t\to\infty$. Also

$$I_3(t) \leq \int_{\phi^{-1}(t)}^{t}\frac{\phi^{-1}(s)}{s^{1+1/n}}ds \leq n\left\{\phi^{-1}(t)\right\}^{-1/n}\to 0 \text{ as } t\to\infty.$$

This completes the proof. □

We are now able to prove the embedding result which we need.

Lemma 4.2.22. *If Θ is a bounded domain in $\mathbb{R}^n$ with boundary of class $C^{0,1}$, then*

$$W^1L_\phi(\Theta) \hookrightarrow\hookrightarrow L_\phi(\Theta). \tag{4.2.46}$$

Proof. Suppose that $n \geq 2$. Assume that

$$\int_1^\infty g_\phi(s)ds < \infty.$$

By Theorem 7.2.5 of [148],

$$W^1L_\phi(\Theta) \hookrightarrow L_\infty(\Theta).$$

Let ϕ_1 be a Young function such that $\phi \ll \phi_1$: for example, we could take

$$\phi_1(t) = \begin{cases} \phi(t) & if\ 0 \leq t \leq \phi^{-1}(1), \\ (\phi(t))^2\ if & t > \phi^{-1}(1). \end{cases}$$

As Θ is bounded, $L_\infty(\Theta) \hookrightarrow L_{\phi_1}(\Theta)$ and so $W^1L_\phi(\Theta) \hookrightarrow L_{\phi_1}(\Theta)$. In view of Lemma 7.4.1 of [148], we thus have (4.2.46). On the other hand, if

$$\int_1^\infty g_\phi(s)ds = \infty,$$

then by Theorem 7.2.3 of [148],

$$W^1L_\phi(\Theta) \hookrightarrow L_{\phi^*}(\Theta).$$

By Lemma 4.2.21, $\phi \ll \phi^*$; the result now follows again from Lemma 7.4.1 of [148].

To complete the proof we have to deal with the case in which $n = 1$, so that Θ is a bounded interval (a, b) in $\mathbb{R}$. Let

$$B = \{u \in W^1L_\phi(\Theta) : \|u|W^1L_\phi(\Theta)\| \leq 1\}$$

and let $U \in B$. Since, by Hölder's inequality, $W^1L_\phi(\Theta) \hookrightarrow W_1^1(\Theta)$ and every element of $W_1^1(\Theta)$ may be identified with an absolutely continuous function on $\overline{\Theta}$, we may suppose that u is absolutely continuous on $[a, b]$. Thus for all $x, y \in \Theta$ we have

$$|u(x) - u(y)| = \left|\int_y^x u'(s)ds\right| \leq |x-y|\,(\widetilde{\phi})^{-1}\left(|x-y|^{-1}\right)\|u'|L_\phi(\Theta)\|$$

$$\leq |x-y|\,(\widetilde{\phi})^{-1}\left(|x-y|^{-1}\right).$$

In view of the behaviour of the Young function $\widetilde{\phi}$ at infinity, it follows that B is an equicontinuous subset of $C(\overline{\Theta})$. Moreover,

$$u(x) = \int_a^x u'(s)ds + u(a)$$

and, by Fubini's theorem,

$$\int_a^b u(x)dx = (b-a)u(a) + \int_a^b u'(s)(b-s)ds.$$

Hence

$$|u(a)| = (b-a)^{-1} \left| \int_a^b u(x)dx - \int_a^b u'(s)(b-s)ds \right| \leq K_1 \left\| u \mid W_1^1(\Theta) \right\|$$

and

$$|u(x)| \leq \left| \int_a^x u'(s)ds \right| + |u(a)| \leq K_2 \left\| u \mid W_1^1(\Theta) \right\| .$$

Hence B is relatively compact in $C(\overline{\Theta})$ and so $W^1L_\phi(\Theta) \hookrightarrow\hookrightarrow C(\overline{\Theta})$.Thus

$$W^1L_\phi(\Theta) \hookrightarrow\hookrightarrow L_\infty(\Theta) \hookrightarrow L_\phi(\Theta),$$

and the proof is complete. □

We are now able to discuss conditions (P1), (P1)*, (P2) and (P3) in the context of Orlicz spaces. Let $v_0, v_1 \in \mathcal{W}_c(\Omega)$ be weights satisfying (4.2.39) and (4.2.40), and let $X(\Omega), Y(\Omega)$ be given by (4.2.41). Use of the Hölder inequality (4.2.36) shows that (P1)* (and *a fortiori* (P1)) holds if $\left\| \chi_\Omega \mid L_{\tilde{\phi}}(\Omega, v_0) \right\|$ is finite, which is equivalent to requiring that

$$v_0(\Omega) < \infty. \tag{4.2.47}$$

We also choose the sequence $\{\Omega_k\}$, which is naturally assumed to satisfy (4.2.1) and (4.2.2), so that for all $k \in \mathbb{N}$, $\partial\Omega_k$ is of class $C^{0,1}$. By Lemma 4.2.22 this ensures that (P2) holds.

Turning to (P3), assume that $X(\Omega)$, $Y(\Omega)$ are given the Luxemburg norms. Then by (4.2.37), (P3) means that

$$1/\phi^{-1}(1/v_0(\Omega^k)) < 1/\phi^{-1}(1/v_0(\Omega)) \tag{4.2.48}$$

for at least one $k \in \mathbb{N}$ provided that (4.2.47) holds, which is certainly the case since it is equivalent to $v_0(\Omega^k) < v_0(\Omega)$.

If $X(\Omega)$ and $Y(\Omega)$ are given the Orlicz norms then, in view of (4.2.38), (P3) means that for at least one $k \in \mathbb{N}$,

$$v_0(\Omega^k) \left(\widetilde{\phi}\right)^{-1} (1/v_0(\Omega^k)) < v_0(\Omega) \left(\widetilde{\phi}\right)^{-1} (1/v_0(\Omega)),$$

or equivalently,

$$\left(\widetilde{\phi}\right)^{-1}(\tau_k)/\tau_k < \left(\widetilde{\phi}\right)^{-1}(\tau)/\tau$$

provided that (4.2.47) holds, where $\tau_k = 1/v_0(\Omega^k)$, $\tau = 1/v_0(\Omega)$. Since $\tau_k \to \infty$ and $\lim_{t\to\infty}\left(\widetilde{\phi}\right)^{-1}(t)/t = 0$, (P3) is satisfied.

As for the Friedrichs inequality, we observe that (F1)* (and *a fortiori* (F1)) holds if we assume:

$(\mathrm{F1})_2^*$ If $v_0(\Omega) < \infty$, then there exist $x_0 \in \partial\Omega$ and $R > 0$ such that

$$\Omega(x_0, R) := \{x \in \Omega : |x - x_0| < R\}$$

has boundary of class $C^{0,1}$ and

$$\left\|1/v_i \mid L_{\widetilde{\phi}}\left(\Omega(x_0, R), v_i\right)\right\| < \infty \quad \text{for } i = 0, 1.$$

By [202], if $v_0(\Omega_k) < \infty$ for all $k \in \mathbb{N}$, the space $X(\Omega) = L_\phi(\Omega, v_0)$ has absolutely continuous norm if, and only if, the Young function ϕ satisfies the Δ_2-condition. We recall that by this is meant the following:
(i) If $v_0(\Omega) < \infty$, then ϕ is said to satisfy the Δ_2-condition if there are positive constants M and N such that $\phi(2v)/\phi(v) < M$ if $v > N$.
(ii) If $v_0(\Omega) = \infty$, then ϕ is said to satisfy the Δ_2-condition if there is a positive constant M such that $\phi(2v)/\phi(v) < M$ for all $v > 0$.

Thus if $v_0(\Omega_k) < \infty$ for all $k \in \mathbb{N}$ and ϕ satisfies the Δ_2-condition, we have $A = \beta(id)$, where id is the natural embedding of $W^1L_\phi(\Omega, v_0, v_1)$ in $L_\phi(\Omega, v_0)$.

We may now specialise the results obtained earlier in this section to give various theorems concerning the Poincaré and Friedrichs inequalities in Orlicz-Sobolev spaces. For example, Theorem 4.2.13 and the remarks above give

Theorem 4.2.23. *Let $v_0, v_1 \in \mathcal{W}_c(\Omega)$ and suppose that $v_0(\Omega) < \infty$. Then $A < 1$ if, and only if, there is a positive constant K such that for all $u \in W^1L_\phi(\Omega, v_0, v_1)$,*

$$\inf_{c\in\mathbb{C}} \|u - c \mid L_\phi(\Omega, v_0)\| \le K \|\nabla u \mid L_\phi(\Omega, v_1)\|.$$

If the Young function ϕ satisfies the Δ_2-condition, then $A = \beta(id)$, where id is the natural embedding of $W^1L_\phi(\Omega, v_0, v_1)$ in $L_\phi(\Omega, v_0)$.

In the same way, Theorem 4.2.18 gives the next result about the Friedrichs inequality:

Theorem 4.2.24. *Let $v_0, v_1 \in \mathcal{W}_c(\Omega)$ and suppose that $(F1)_2^*$ holds. Then $A^0 < 1$ if, and only if, there is a positive constant K such that for all $u \in C_0^\infty(\Omega)$,*

$$\|u \mid L_\phi(\Omega, v_0)\| \le K \|\nabla u \mid L_\phi(\Omega, v_1)\|.$$

If ϕ satisfies the Δ_2-condition, then $A^0 = \beta(id_0)$, where id_0 is the natural embedding of $\overset{\circ}{W}{}^1L_\phi(\Omega, v_0, v_1)$ in $L_\phi(\Omega, v_0)$.

We now provide concrete examples of Orlicz-Sobolev spaces in which the Poincaré inequality holds.

Let Ω have boundary of class $C^{0,1}$, let $n > 1$ and let ϕ be one of the following Young functions:

(a) $$\phi(x) = \begin{cases} \left(\frac{e}{\alpha+1}\right)^{\alpha+1} x^{\alpha+1} & \text{for } 0 \le x < \alpha+1, \\ e^x & \text{for } x \ge \alpha+1, \end{cases}$$

where $0 < \alpha < n-1$;

(b) $$\phi(x) = \begin{cases} \beta^{-1}e^{\beta}x^{\beta} & \text{for } 0 \le x < 1, \\ \beta^{-1}e^{\beta x} & \text{for } x \ge 1, \end{cases}$$

where $1 < \beta \le n$;

(c) $$\phi(x) = \begin{cases} \alpha^{-1}(x/e)^{\alpha+1} & \text{for } 0 \le x < e^{1+1/\alpha}, \\ x(logx - 1) & \text{for } x \ge e^{1+1/\alpha}, \end{cases}$$

where $0 < \alpha < n-1$.

Then $\int_0^1 t^{-1-1/n}\phi^{-1}(t)dt < \infty$. From Lemma 4.2.22 and Remark 4.2.3 (c), it follows that there is a positive constant K such that for all $u \in W^1L_\phi(\Omega)$,

$$\|u \mid L_\phi(\Omega)\| \le K\left\{\left|\int_\Omega u dx\right| + \|\nabla u \mid L_\phi(\Omega)\|\right\}.$$

4.2.3 Higher-order cases

Here we outline how the general results given above may be extended to include abstract Sobolev spaces of higher order. In a slight and convenient variation of Definition 3.6.1, we suppose that $X = X(\Omega)$ and $Y = Y(\Omega)$ are Banach function spaces on a domain Ω in $\mathbb{R}^n$ equipped with Lebesgue $n-$measure, that $k \in \mathbb{N}$ and that $W^k(X,Y)$ is the abstract Sobolev space defined by

$$W^k(X,Y) = \{f \in X : D^\alpha f \in Y \text{ for all } \alpha \in \mathbb{N}_0^n \text{ with } |\alpha| = k\}$$

and equipped with the norm

$$\left\|f \mid W^k(X,Y)\right\| = \|f \mid X\| + \left\|\nabla^k f \mid Y\right\|,$$

where by $\left\|\nabla^k f \mid Y\right\|$ we mean $\sum_{|\alpha|=k} \|D^\alpha f \mid Y\|$. (In Definition 3.6.1 all the derivatives up to and including those of order k were taken, not merely those of order k.) It is easy to see that $W^k(X,Y)$ is complete. The closure of $C_0^\infty(\Omega)$ in $W^k(X,Y)$ is denoted by $\mathring{W}^k(X,Y)$.With reference to the nested sequence of domains (Ω_m) familiar from the earlier discussions, and with $\Omega^m = \Omega\backslash\Omega_m$, we write

$$A_m = \sup_{\|f|W^k(X,Y)\|\le 1} \|f\chi_{\Omega^m} \mid X\|, \; A = \lim_{m\to\infty} A_m \in [0,1].$$

If X has absolutely continuous norm, $A = 0$ if, and only if, $W^k(X,Y)$ is compactly embedded in X.

By an inequality of Poincaré type we shall mean one of the form

$$\|f \mid X\| \leq K\left\{|F(u)| + \left\|\nabla^k f \mid Y\right\|\right\}, \ f \in W^k(X,Y). \tag{4.2.49}$$

Here F is a continuous functional such that $F(\lambda f) = \lambda F(f)$ for all $\lambda > 0$ and all $f \in W^k(X,Y)$, and $F(f) = 0 \Rightarrow f = 0$ if f belongs to a suitable subspace of $\mathcal{P}_{k-1} \cap W^k(X,Y)$, where $\mathcal{P}_{k-1}$ is the family of polynomials on $\mathbb{R}^n$ of degree at most $k-1$. A typical result, which can be obtained by following the same line of argument as in the case $k = 1$ already discussed, is the following.

Theorem 4.2.25. *Let X have absolutely continuous norm, let $F \in W^k(X,Y)^*$ be such that $F(f) = 0 \Rightarrow f = 0$ if $f \in \mathcal{P}_{k-1} \cap W^k(X,Y)$, and suppose that for each $m \in \mathbb{N}$, $W^k(X(\Omega_m), Y(\Omega_m))$ is compactly embedded in $X(\Omega_m)$. Then the Poincaré inequality (4.2.49) holds on $W^k(X,Y)$ if, and only if, $A < 1$.*

For a proof of this, and for diverse related results, some even in the setting of anisotropic spaces, see [73].

The main thrust of the arguments developed so far in this section and in the previous one is the connection between the measure of non-compactness and the Poincaré and Friedrichs inequalities. An alternative method of establishing inequalities of Poincaré type in the context of $W^k(X,Y)$, which does not rely on knowledge of the measure of non-compactness, is provided by the work of [24]. This is based on the following result.

Theorem 4.2.26. *Let W stand for either $W^k(X,Y)$ or $\mathring{W}^k(X,Y)$, and let F be a continuous functional on W such that $F(\lambda f) = \lambda F(f)$ for all $\lambda > 0$ and all $f \in W$, and $F(f) = 0 \Rightarrow f = 0$ if $f \in \mathcal{P}_{k-1} \cap W$. Suppose there are a Banach function space Z in which W is compactly embedded; and a functional G defined on W, with $G(0) = 0$, which is continuous at 0 with respect to the norm*

$$\|f\|_{Z,Y} := \|f \mid Z\| + \left\|\nabla^k f \mid Y\right\|.$$

Then a sufficient condition for the validity of the Poincaré inequality (4.2.49) on W is that the inequality

$$\|u \mid X\| \leq K(\|u\|_{Z,Y} + |G(0)|) \tag{4.2.50}$$

should hold for all $u \in W$.

Proof. Suppose that (4.2.49) is false. Then for each $m \in \mathbb{N}$ there exists $f_m \in W$, $f_m \neq 0$, such that

$$\|f_m \mid X\| \geq m\left(|F(f_m)| + \left\|\nabla^k f_m \mid Y\right\|\right).$$

Set $u_m = f_m / \|f_m \mid X\|$: the homogeneity of the norms and of the functional F implies that $\|u_m \mid X\| = 1, F(u_m) \to 0$ and $\left\|\nabla^k u_m \mid Y\right\| \to 0$ as $m \to$

∞. Hence (u_m) is bounded in W; as W is compactly embedded in Z, this sequence has a subsequence (u_{m_i}) which is a Cauchy sequence in Z and so also with respect to the norm $\|\cdot\|_{Z,Y}$. By (4.2.50) and the definition of G, (u_{m_i}) is a Cauchy sequence in X and thus in W. Since W is complete, (u_{m_i}) converges to a point u in W. It follows that if $|\alpha| = m$, $D^\alpha u_{m_i} \to D^\alpha u$ in Y : plainly $D^\alpha u = 0$. Hence u is a polynomial of degree at most $k-1$. Moreover, $\|u \mid X\| = 1$. But by the continuity of F, $F(u) = 0$, and so by the assumed property of F on polynomials, $u = 0$, giving a contradiction. □

Remark 4.2.27. If $G = 0$, to show that the Poincaré inequality on $W^k(X,Y)$ holds, all we have to do is to find a Banach function space Z such that $W^k(X,Y)$ is compactly embedded in Z and $W^k(Z,Y) \hookrightarrow X$; a similar statement holds for $\overset{\circ}{W}{}^k(X,Y)$. Details of how this procedure may be implemented in diverse situations may be found in [24]. Of course, if $W^k(X,Y)$ is compactly embedded in X, Theorem 4.2.25 and the fact that $A = 0$ show immediately that the Poincaré inequality holds.

4.3 Concrete spaces

In this section we deal with inequalities of classical Sobolev or Poincaré type when the underlying space domain may have an irregular boundary. If this boundary is of class $C^{0,1}$ or satisfies some type of cone condition, these inequalities are very familiar (see, for example, [46], Chapter V), but now they are available under much less stringent restrictions. We begin with a description of some of the conditions which have been placed on the boundary in recent years with the derivation of such inequalities at least partly in mind.

4.3.1 Classes of domains

Definition 4.3.1. *Let $0 < \alpha \le \beta < \infty$. A domain Ω in $\mathbb{R}^n$ is said to be an (α,β)–**John domain**, written $\Omega \in J(\alpha,\beta)$, if there is a point $x_0 \in \Omega$ (called a central point) such that given any $x \in \Omega$, there is a rectifiable path $\gamma : [0,d] \to \Omega$ which is parametrised by arc-length, such that $\gamma(0) = x$, $\gamma(d) = x_0$, $d \le \beta$ and*

$$\operatorname{dist}(\gamma(t), \partial\Omega) \ge \frac{\alpha}{d} t$$

*for every $t \in [0,d]$. If $c \in (0,1]$, we say that Ω is a c–**John domain** if it is an (α,β)–John domain for some α and β with $c = \alpha/\beta$. A domain is called a **John domain** if it belongs to $J(\alpha,\beta)$ for some α and β; the class of all John domains is denoted by J.*

These domains were first introduced by Fritz John in [132], although naturally he was not responsible for the terminology, which is due to Martio and Sarvas [167].

Remark 4.3.2. (a) If $\Omega \in J(\alpha,\beta)$, then diam $(\Omega) \leq 2\beta$. Thus John domains are bounded. An extension of the notion of John domains to unbounded domains is given by Väisälä [224], who in [225] showed that these unbounded domains had the convenient property that they could be represented as the union of an increasing sequence of c–John domains in the sense given above.
(b) Martio and Sarvas [167] show that J coincides with the family of all bounded domains Ω with the property that there are $\delta \in (0,1]$ and $x_0 \in \Omega$ such that given any $x \in \Omega$ there is a path $\gamma : [0,1] \to \Omega$ with $\gamma(0) = x, \gamma(1) = x_0$ and

$$\gamma([0,t]) \subset B\left(\gamma(t), \delta^{-1}\text{dist}\,(\gamma(t), \partial\Omega)\right)$$

for all $t \in [0,1]$; thus

$$\text{dist}\,(\gamma(t), \partial\Omega) \geq \delta\,|x - \gamma(t)|$$

for all $t \in [0,1]$.
(c) Another characterisation of John domains was given by Martio in [166] by means of bilipschitz mappings: we recall that if $A \subset \mathbb{R}^n$ and $L \geq 1$, then a map $f : A \to \mathbb{R}^n$ is said to be L–bilipschitz if

$$|x - y|\,/L \leq |f(x) - f(y)| \leq L\,|x - y|$$

for all $x, y \in A$. He shows that if $\Omega \in J(\alpha,\beta)$ has a central point x_0, then given any $x \in \Omega$, there is an L–bilipschitzian map $T : B(0,\alpha) \to \Omega$, with $L = (\beta/\alpha)^4 c(n)$ and $x_0 = T(0)$, such that $x \in T(B(0,\alpha))$. Conversely, suppose Ω is a domain in $\mathbb{R}^n$ with the property that there are a point $x_0 \in \Omega$ and $L \geq 1$ such that given any $x \in \Omega$, there is an L–bilipschitzian map $T : B(0,\alpha) \to \Omega$ with $x_0 = T(0)$ and $x \in T(B(0,\alpha))$. Then $\Omega \in J(\alpha/L^3, \alpha L)$.

Examples

(i) Every bounded domain with a Lipschitz boundary is a John domain. In particular, every bounded convex domain is a John domain.
(ii) Let Ω be a bounded domain which has the interior cone property. We recall that this means that there is a cone

$$V = \left\{x \in \mathbb{R}^n : x_n \geq 0,\ x_1^2 + ... + x_{n-1}^2 \leq b^2 x_n^2,\ \ |x| \leq a\right\},$$

with $a, b > 0$, such that every point of $\overline{\Omega}$ is the vertex of a cone $V_x \subset \overline{\Omega}$ congruent to V. Then Ω is a John domain: to see this, extend the central lines of the cones so that they end at a fixed point x_0 as far from the boundary as possible. Crudely speaking, the difference between a domain with the interior cone condition and a John domain is that the cone is replaced by a 'twisted cone'. However, as we shall see in some of the following examples, this replacement can make a very big difference.
(iii) The Koch snowflake is a John domain: see [167]. This shows that the boundary of a plane John domain may have infinite length.

(iv) The set $B(0,1)\backslash\left\{x \in \mathbb{R}^n : |x_n| \leq |x|^2\right\}$ is a John domain, and so are the annulus $B(0,1)\backslash\overline{B(0,r)}$, where $0 < r < 1$, and the punctured ball $B(0,1)\backslash\{0\}$.
(v) In the plane, the domain $B((1,0),1)\backslash\{(x,y) \in \mathbb{R}^2 : |y| \geq x^2\}$ is not a John domain. Another example of a plane domain which is not a John domain is given by the 'comb'

$$\{(x,y) : 0 < x < 1, 1/2 < y < 1\} \cup \bigcup_{j=1}^{\infty} \{(x,y) : 2^{-j} < x < 2^{-j+1}, 0 < y < 1/2\}.$$

(vi) An important class of John domains arises from the **uniform domains** which were introduced by Martio and Sarvas [167]. Given positive real numbers α and β, with $\alpha \leq \beta$, a domain $\Omega \subset \mathbb{R}^n$ is said to be $(\alpha,\beta)-$**uniform** if for each pair of points x, y in Ω, with $x \neq y$, there is a domain $\Theta \in J(\alpha|x-y|, \beta|x-y|)$ such that $x, y \in \Theta \subset \Omega$. The family of all $(\alpha,\beta)-$uniform domains is denoted by $U(\alpha,\beta)$; a domain $\Omega \subset \mathbb{R}^n$ is called **uniform**, written $\Omega \in U$, if $\Omega \in U(\alpha,\beta)$ for some α and β. For equivalent versions of this definition we refer to [166] and [223]: thus a domain $\Omega \subset \mathbb{R}^n$ is a uniform domain if there are positive numbers a and b such that each pair of points $x_1, x_2 \in \Omega$ can be joined by a rectifiable path γ in Ω, of length

$$l(\gamma) \leq a\,|x_1 - x_2|, \tag{4.3.1}$$

such that for all points x on γ,

$$\min_{j=1,2} l(\gamma(x_j, x)) \leq b\,\mathrm{dist}(x, \partial\Omega), \tag{4.3.2}$$

where $\gamma(x_j, x)$ is that part of γ which is between x_j and x. A local version of this class of domains is also available: a domain $\Omega \subset \mathbb{R}^n$ is **locally uniform** if there exists a positive number r such that each pair of points $x_1, x_2 \in \Omega$ with $|x_1 - x_2| < r$ can be joined by a rectifiable path γ in Ω such that (4.3.1) and (4.3.2) hold, for some a and b. Every uniform domain is locally uniform; and every bounded locally uniform domain is uniform. In fact, the locally uniform domains are just the $(\varepsilon,\delta)-$domains(with $\varepsilon,\delta \in (0,\infty)$) introduced by Jones [133], the uniform domains being his $(\varepsilon,\infty)-$domains. We recall that an $(\varepsilon,\delta)-$domain Ω is defined by the property that given any $x, y \in \Omega$ with $0 < |x-y| < \delta$, there is a rectifiable path γ in Ω joining x to y, with length $l(\gamma) \leq |x-y|/\varepsilon$, satisfying

$$\mathrm{dist}\,(z, \partial\Omega) \geq \varepsilon\,|x-z|\,|y-z|\,/\,|x-y|$$

for all z on γ.

It is shown in [97] that every bounded uniform domain is a John domain. However, there are John domains which are not uniform: a simple example is given by $B(0,1)\backslash\{x \in \mathbb{R}^n : x_n = 0, x_1 \geq 0\}$, which contains points whose

Euclidean distance apart is very small while the length of any path in the set joining the points is large. Examples of uniform domains are provided by the Koch snowflake, any convex John domain and any domain which satisfies.the interior cone condition.

Another interesting class of domains is formed by those that satisfy a plumpness condition.

Definition 4.3.3. *Let $b \in (0,1]$. A domain Ω in $\mathbb{R}^n$ is said to be $b-$**plump** if there is a positive number σ such that for every $y \in \partial\Omega$ and for all $t \in (0,\sigma]$ there is a point $x \in \Omega \cap \overline{B(y,t)}$ with $\operatorname{dist}(x,\partial\Omega) \geq bt$. If Ω is $b-$plump for some b, we say that Ω is plump.*

This definition is due to Martio and Väisälä [168]; the hypothesis in it is called a **corkscrew condition** by Jerison and Kenig [131]. There is a connection with the **exterior regular domains** of Triebel and Winkelvoss (see [74], p. 59); these are the bounded domains Ω such that (a) Ω is the interior of its closure; (b) there is a positive number c such that for any cube Q centred on $\partial\Omega$ with side length $l \leq 1$, there is a subcube Q^e with side length cl and $Q^e \subset Q \cap (\mathbb{R}^n \backslash \overline{\Omega})$. If Ω coincides with the interior of its closure, then Ω is exterior regular if, and only if, the interior of the complement of Ω is plump.

Every $(\alpha,\beta)-$John domain is $(\alpha/\beta)-$plump: for this result we refer to Martio and Vuorinen [169], Lemma 6.3.

Yet another extension of the class of John domains is provided by domains which satisfy a so-called quasi-hyperbolic boundary condition. Introduced by Gehring and Osgood [98], this relies on the quasi-hyperbolic distance between arbitrary points x_1, x_2 in the domain Ω. This is defined by

$$k_\Omega(x_1,x_2) = \inf_\gamma \int_0^l \frac{1}{\operatorname{dist}(\gamma(s),\partial\Omega)} ds,$$

where the infimum is taken over all rectifiable paths $\gamma : [0,l] \to \Omega$, parameterised by arc-length, with $\gamma(0) = x_1$, $\gamma(l) = x_2$. It can be shown that k_Ω is a metric on Ω. Moreover, in [98] it is proved that the infimum is attained; a path winning this competition is called a **quasi-hyperbolic geodesic**.

Definition 4.3.4. *A domain Ω in $\mathbb{R}^n$ is said to be a **quasi-hyperbolic boundary condition domain**, written $Q \in QHBC$, if there are a point $x_0 \in \Omega$ and positive constants a and b such that for all $x \in \Omega$,*

$$k_\Omega(x,x_0) \leq a \log\left(\frac{1}{d(x)}\right) + b.$$

Here, as usual, $d(x)$ stands for the distance of x from $\partial\Omega$. If Ω is bounded, an equivalent condition is that there should exist a constant $\alpha \geq 1$ such that for all $x \in \Omega$,

$$k_{\Omega}(x, x_0) \leq \alpha \log\left(\frac{|x - x_0|}{\min\{d(x_0), d(x)\}}\right).$$

In fact, these domains were first introduced by Herron and Vuorinen [125] by means of (a slightly more general version of) this equivalent condition. The equivalence was established by Herron [124]. Using the first definition given above, Gehring and Martio [97] showed that every QHBC domain is bounded, with

$$\operatorname{diam} \Omega \leq 2ae^{b/a},$$

and that every John domain is a QHBC domain. An example of a QHBC domain which is not plump, and is therefore not a John domain, is given by Hurri in [129]. Her example is the 'tunnel' in $\mathbb{R}^2$ defined by

$$\Omega = \operatorname{int}\left(\bigcup_{j=1}^{\infty} \overline{\Omega}_j\right),$$

where $\Omega_j = (d_j + \varepsilon_j \log \varepsilon_j, d_j) \times (-\varepsilon_j, \varepsilon_j)$; (ε_j) is the sequence defined by $\varepsilon_1 = 1/2$, $\varepsilon_j = \varepsilon_{j-1}^2$ $(j > 1)$, and

$$d_k = -\sum_{j=1}^{k} \varepsilon_j \log \varepsilon_j \quad (k \in \mathbb{N}).$$

The lack of plumpness stems from the point at which the sections of the tunnel accumulate. For an example of a domain which is not of QHBC type we also refer to [129], Remark 7.18.

Next, we introduce the Whitney cube #-condition of Martio and Vuorinen [169]. To do this we first recall that by the Whitney covering $\mathcal{W}$ of a bounded domain Ω is meant a family of closed cubes Q, each with sides parallel to the coordinate axes and with side length $l_Q = 2^{-k}$ and diameter $d_Q = 2^{-k}\sqrt{n}$ for some $k \in \mathbb{N}$, such that

(i) $\Omega = \bigcup_{Q \in \mathcal{W}} Q$;
(ii) the interiors of distinct cubes are disjoint;
(iii) for all $Q \in \mathcal{W}$, $1 \leq \operatorname{dist}(Q, \partial\Omega)/d_Q \leq 4$.

Such a covering always exists: see, for example, [211]. With d as the distance function as above, (iii) implies that for all $x \in Q \in \mathcal{W}$,

$$l_Q\sqrt{n} \leq d(x) \leq 5l_Q\sqrt{n}.$$

Now write

$$\mathcal{W}_k = \left\{Q \in \mathcal{W} : l_Q = 2^{-k}\right\}, \quad k \in \mathbb{N},$$

and let $n(k)$ be the number of cubes in $\mathcal{W}_k$. The method of construction of $\mathcal{W}$ shows that each $Q \in \mathcal{W}_k$ satisfies

$$Q \subset \left\{x \in \Omega : \sqrt{n}2^{-k} \leq d(x) \leq 5\sqrt{n}2^{-k}\right\}. \tag{4.3.3}$$

Definition 4.3.5. *Let $h : (0,\infty) \to (0,\infty)$ be continuous and increasing. A domain Ω is said to satisfy a* ***Whitney cube #-condition with function*** *h if there exists $k_0 \in \mathbb{N}$ such that*

$$n(k) \leq h(k) \text{ for all } k \geq k_0, k \in \mathbb{N}.$$

If there exist $M > 0$ and $\lambda \in (0,n)$ such that

$$n(k) \leq 2^{\lambda k} M \text{ for all } k \in \mathbb{N},$$

we simply say that Ω satisfies a ***Whitney cube #-condition.***

Note that every bounded domain Ω satisfies a Whitney cube #-condition with function h when $h(t) = 2^{nt} M$ for some $M > 0$, since obviously

$$n(k) = \# \left\{ Q \in \mathcal{W} : |Q| = 2^{-kn} \right\} \leq 2^{kn} |\Omega| .$$

Remark 4.3.6. It is shown in [169] that every plump domain satisfies a Whitney cube #-condition. However, thanks to the work of Smith and Stegenga [208] it is now known that every QHBC-domain satisfies such a condition.

There is a connection between domains which satisfy such a condition and the Minkowski dimension of the boundary. To be more precise, we let $0 < \lambda \leq n$ and $r > 0$, and put

$$M_{\Omega}^{\lambda}(\partial\Omega, r) = r^{-(n-\lambda)} \left| (\partial\Omega + B(0,r)) \cap \Omega \right| ,$$

$$M_{\Omega}^{\lambda}(\partial\Omega) = \limsup_{r \to 0+} M_{\Omega}^{\lambda}(\partial\Omega, r)$$

and

$$\dim_{M,\Omega}(\partial\Omega) = \inf \left\{ \lambda : M_{\Omega}^{\lambda}(\partial\Omega) < \infty \right\} .$$

This last quantity is called the **inner Minkowski dimension** of Ω. The corresponding quantities obtained by replacement of $|(\partial\Omega + B(0,r)) \cap \Omega|$ by $|\partial\Omega + B(0,r)|$ are denoted by $M^{\lambda}(\partial\Omega, r)$, $M^{\lambda}(\partial\Omega)$ and $\dim_M(\partial\Omega)$, the last of these being the Minkowski dimension of $\partial\Omega$ which was introduced in Definition 1.1.4. Plainly $M_{\Omega}^{\lambda}(\partial\Omega) \leq M^{\lambda}(\partial\Omega)$ and $\dim_{M,\Omega}(\partial\Omega) \leq \dim_M(\partial\Omega)$.

The connection to which we referred above is the following.

Lemma 4.3.7. *Let Ω be a bounded domain in $\mathbb{R}^n$ and let $0 < \lambda \leq n$. Then $M_{\Omega}^{\lambda}(\partial\Omega) < \infty$ if, and only if, there are positive constants K and k_0 such that $n(k) \leq 2^{\lambda k} K$ for all $k \geq k_0, k \in \mathbb{N}$.*

Proof. First suppose that $M_{\Omega}^{\lambda}(\partial\Omega) < \infty$. Then there exist $K, r_0 > 0$ such that

$$\left| (\partial\Omega + B(0,r)) \cap \Omega \right| \leq K r^{n-\lambda}$$

for all $r \in (0, r_0]$. Take $k \in \mathbb{N}$, $k \geq (\log 2)^{-1} \log(12\sqrt{n}/r_0)$ and set $r = 6\sqrt{n} 2^{-k}$. Then $2r \leq r_0$. By a standard covering theorem (see, for example,

[46], Theorem XI.5.3), there are points $x_1, ..., x_m \in \partial\Omega$ and a positive constant C, depending only on n, such that

$$\partial\Omega \subset \bigcup_{j=1}^{m} B(x_j, r), \quad \sum_{j=1}^{m} \chi_{B(x_j,r)} \leq C.$$

Every cube $Q \in \mathcal{W}_k$ is contained in at least one of the balls $B(x_j, 2r)$, $j = 1, ..., m$. For given $x \in Q$, choose $y \in \partial\Omega$ so that $d(x) = |x - y| : y \in B(x_j, r)$ for some $j \in \{1, ..., m\}$. Then for every $z \in Q$ we have

$$|z - x_j| \leq |z - x| + |x - y| + |y - x_j| = 12\sqrt{n}2^{-k} = 2r.$$

Let $n_j(k)$ be the number of cubes $Q \in \mathcal{W}_k$ which are contained in $B(x_j, 2r)$. Then plainly

$$\begin{aligned} n(k) &\leq \sum_{j=1}^{k} n_j(k) \leq \sum_{j=1}^{k} |B(x_j, 2r) \cap \Omega| \,/\, |Q| \\ &\leq C2^{nk} \left|(\partial\Omega + B(0, 2r)) \cap \Omega\right| \\ &\leq CK(12\sqrt{n})^{n-\lambda}2^{\lambda k}. \end{aligned}$$

Conversely, suppose that $n(k) \leq K2^{\lambda k}$ for all $k \geq k_0$, $k \in \mathbb{N}$. We may suppose that $\lambda < n$ since

$$\limsup_{r \to 0} \left|(\partial\Omega + B(0, r)) \cap \Omega\right| = |\partial\Omega| < \infty.$$

Fix $r > 0$ with $r \leq \sqrt{n}2^{-k_0}$ and choose $k' \geq k_0$ such that

$$\sqrt{n}2^{-k'-1} \leq r < \sqrt{n}2^{-k'}.$$

Then from the method of construction of the Whitney decomposition (see (4.3.3)),

$$(\partial\Omega + B(0, r)) \cap \Omega \subset \bigcup_{k \geq k'} \mathcal{W}_k.$$

Hence

$$\left|(\partial\Omega + B(0, r)) \cap \Omega\right| \leq \sum_{k=k'}^{\infty} K2^{k(\lambda-n)} = \frac{K2^{k'(\lambda-n)}}{1 - 2^{\lambda-n}}.$$

Thus

$$r^{-(n-\lambda)} \left|(\partial\Omega + B(0, r)) \cap \Omega\right| \leq \frac{K2^{k'(\lambda-n)}}{(1 - 2^{\lambda-n})(\sqrt{n}2^{-k'-1})^{n-\lambda}}$$

$$\leq \frac{K2^{n-\lambda}}{(1 - 2^{\lambda-n})n^{(n-\lambda)/2}}$$

and the result follows. □

This result is given in [24] ; the proof closely follows that of [169], Theorem 3.11 and Lemma 3.4, in which the Minkowski dimension, rather than the inner Minkowski dimension, is used.

Remark 4.3.8. In [169], Corollary 6.4, it is shown that if Ω is an (α, β)–John domain, then

$$\dim_M(\partial\Omega) \leq c < n,$$

where c depends only on β/α and n. For an example of a set Ω such that $M_\Omega^\lambda(\partial\Omega) = \infty$ for all $\lambda \in (0, n)$, see [129], Remark 7.18.

Following naturally from this connection with the Whitney cube #-condition, there is a relationship, given in [24], between the inner Minkowski dimension of the boundary and the distance function.

Theorem 4.3.9. *Let Ω be a bounded domain in $\mathbb{R}^n$. Then the following conditions are equivalent:*
(i) $\dim_{M,\Omega}(\partial\Omega) < n$;
(ii) there exists $\mu \in (0, n)$ such that $\int_\Omega d(x)^{-\mu} dx < \infty$.

Proof. First suppose that (i) holds. Let $\mathcal{W}$ be a Whitney covering of Ω and put $\lambda = \dim_{M,\Omega}(\partial\Omega)$. Then if $\mu > 0$,

$$\int_\Omega d(x)^{-\mu} dx = \sum_{Q \in \mathcal{W}} \int_Q d(x)^{-\mu} dx = \sum_{k=1}^{\infty} \sum_{Q \in \mathcal{W}_k} \int_Q d(x)^{-\mu} dx$$

$$\approx \sum_{k=1}^{\infty} n(k) 2^{-kn} (2^{-k})^{-\mu} \lesssim \sum_{k=1}^{\infty} 2^{(\lambda - n + \mu)k}.$$

Since $\lambda < n$ the last sum is finite for a suitable $\mu < n - \lambda$.

Conversely, suppose that (ii) holds and that $\dim_{M,\Omega}(\partial\Omega) = n$. Then by Lemma 4.3.7, no matter what $K > 0$ and $\lambda \in (0, n)$ are chosen, Ω does not satisfy the Whitney cube #-condition with the function $h(t) = K2^{\lambda t}$. Thus if we take $\lambda = n - \mu$, there is a sequence of natural numbers $k_j = k_j(\lambda)$ such that $n(k_j) > 2^{\lambda k_j}$. Then

$$\int_\Omega d(x)^{-\mu} dx = \sum_{k=1}^{\infty} \sum_{Q \in \mathcal{W}_k} \int_Q d(x)^{-\mu} dx \gtrsim \sum_{k=1}^{\infty} n(k) (2^{-k})^{-\mu} |Q|$$

$$\geq \sum_{j=1}^{\infty} n(k_j) (2^{-k_j})^{-\mu} 2^{-k_j n} > \sum_{j=1}^{\infty} 2^{k_j(\lambda + \mu - n)} = \infty.$$

This contradiction completes the proof. □

Remark 4.3.10. In view of Remark 4.3.6 and Lemma 4.3.7, we know that condition (i) is satisfied by all plump domains and therefore also by every John domain. The integrability property of the distance function is accordingly available in this wide family of domains. We refer to [24] for details of how this integrability property may be used to obtain inequalities of Poincaré type.

It may be convenient for the reader to have a crude summary of the more important relationships between the various classes of domains that have been introduced. For simplicity we confine ourselves here to bounded domains in $\mathbb{R}^n$: with this understanding, let $C^{0,1}$ stand for the class of all Lipschitz domains, IC for the domains satisfying an interior cone condition, Jo for the Jones (ε, ∞) domains, J for the John domains, P for the plump domains, $QHBC$ for the quasi-hyperbolic boundary condition domains and W for the domains satisfying a Whitney cube #-condition. Then

$$C^{0,1}, IC, Jo \subsetneq J \subsetneq P \subset W \text{ and } J \subsetneq QHBC \subset W.$$

4.3.2 Sobolev and Poincaré inequalities

Our main concern here is to show how inequalities of Poincaré type may be obtained in domains with possibly unpleasant boundaries. We focus principally on the John domains mentioned in the last subsection.

To begin with, we establish a result which illustrates how, in certain circumstances, weak estimates can imply strong ones. The underlying idea is that of double truncation of functions, which is due to Maz'ya [171]. Given a scalar-valued function u on a set X and real numbers s, t with $0 < s < t < \infty$, we put for each $x \in X$,

$$\widetilde{u}_s^t(x) = \begin{cases} t, & \text{if } \ |u(x)| \geq t, \\ |u(x)|, & \text{if } s \leq |u(x)| < t, \\ s, & \text{if } \ |u(x)| < s, \end{cases}$$

and

$$u_s^t(x) = \widetilde{u}_s^t(x) - s.$$

This simply means that the function $|u|$ is truncated above and below, at the levels t and s respectively, and is then lowered to the coordinate axis.

Theorem 4.3.11. *Let μ and ν be positive measures on the same $\sigma-$algebra in a set R, let $0 < p \leq q < \infty$ and suppose that $u \in L_{q,\infty}(R, \mu)$ and $v \in L_p(R, \nu)$. Let $\{A_t : t > 0\}$ be a family of measurable subsets of R with the property that if $0 < s < t < \infty$, then $A_t \subset A_s$ and*

$$\left\| u_s^t \mid L_{q,\infty}(R, \mu) \right\| \leq \left\| v\chi_{A_t \setminus A_s} \mid L_p(R, \nu) \right\|.$$

Then $u \in L_q(R, \mu)$ and

$$\| u \mid L_q(R, \mu) \| \leq 4 \| v \mid L_p(R, \nu) \|.$$

Proof. Plainly

$$\int_R |u|^q \, d\mu \le \sum_{k\in\mathbb{Z}} 2^{kq}\mu\left(\{x \in R : 2^{k-1} < |u(x)| \le 2^k\}\right)$$

$$\le \sum_{k\in\mathbb{Z}} 2^{kq}\mu\left(\{x \in R : |u(x)| \ge 2^{k-1}\}\right)$$

$$= \sum_{k\in\mathbb{Z}} 2^{kq}\mu\left(\left\{x \in R : \left|u_{2^{2-2}}^{2^{k-1}}(x)\right| \ge 2^{k-2}\right\}\right)$$

$$\le \sum_{k\in\mathbb{Z}} 2^{kq}2^{-(k-2)q}\left(\int_{A_{2^{k-2}}\setminus A_{2^{k-1}}} |v|^p \, d\nu\right)^{q/p}.$$

Since $q/p \ge 1$, this last expression can be estimated from above by

$$2^{2q}\left(\sum_{k\in\mathbb{Z}} \int_{A_{2^{k-2}}\setminus A_{2^{k-1}}} |v|^p \, d\nu\right)^{q/p} \le 4^q \left\| v \mid L_p(R,\nu)\right\|^q,$$

which completes the proof. □

As an immediate application we have the following equivalence between certain inequalities of strong and weak type.

Theorem 4.3.12. *Let μ and ν be positive Borel measures on an open set Ω in $\mathbb{R}^n$, with ν absolutely continuous with respect to Lebesgue measure; let $0 < p \le q < \infty$. Then the following statements are equivalent:*
(i) There exists a constant C_1 such that for every Lipschitz function u with compact support in Ω,

$$\left\| u \mid L_{q,\infty}(\Omega,\mu)\right\| \le C_1 \left\| |\nabla u| \mid L_p(\Omega,\nu)\right\|.$$

(ii) There exists a constant C_2 such that for every Lipschitz function u with compact support in Ω,

$$\left\| u \mid L_q(\Omega,\mu)\right\| \le C_2 \left\| |\nabla u| \mid L_p(\Omega,\nu)\right\|.$$

Proof. Suppose that (i) holds. We use the fact that if w is a Lipschitz function which is constant on a measurable subset E of Ω, then $\nabla w = 0$ a.e. in Ω. Put $v = |\nabla u|$, $A_t = \{x \in \Omega : |u(x)| > t\}$ and observe that $|\nabla u|\,\chi_{A_t\setminus A_s} = |\nabla u_s^t|$ a.e. with respect to Lebesgue measure and therefore also ν-a.e. Then (ii) follows immediately from Theorem 4.3.11. The reverse implication is trivial. □

We now give another result with the same flavour, this time concerning the so-called Sobolev-Poincaré inequality.

Theorem 4.3.13. *Let μ and ν be positive Borel measures on an open set Ω in $\mathbb{R}^n$, with ν absolutely continuous with respect to Lebesgue measure and $\mu(\Omega) < \infty$; let $0 < p \leq q < \infty$. Then the following statements are equivalent:*
(i) There is a constant C_1 such that for every locally Lipschitz function u on Ω,

$$\inf_{c\in\mathbb{R}} \|u - c \mid L_{q,\infty}(\Omega,\mu)\| \leq C_1 \||\nabla u| \mid L_p(\Omega,\nu)\| .$$

(ii) There is a constant C_2 such that for every locally Lipschitz function u on Ω,

$$\inf_{c\in\mathbb{R}} \|u - c \mid L_q(\Omega,\mu)\| \leq C_2 \||\nabla u| \mid L_p(\Omega,\nu)\| .$$

Proof. We simply have to show that (i) implies (ii). Choose $b \in \mathbb{R}$ so that

$$\mu(\{x \in \Omega : u(x) \geq b\}) \geq \mu(\Omega)/2 \text{ and } \mu(\{x \in \Omega : u(x) \leq b\}) \geq \mu(\Omega)/2.$$

Let $v_+ = \max\{u - b, 0\}$ and $v_- = -\min\{u - b, 0\} : |u - b| = v_+ + v_-$. It is thus enough to show that

$$\|v_\pm \mid L_q(\Omega,\mu)\| \leq C \||\nabla u| \mid L_p(\Omega,\nu)\| .$$

Let v stand for either v_+ or v_- : the argument that follows is the same for both cases. We use the easily proved result that if γ is a positive measure on R with $\gamma(R) < \infty$, and w is a non-negative measurable function such that $\gamma(\{x \in R : w(x) = 0\}) \geq \gamma(R)/2$, then for every $t > 0$,

$$\gamma(\{x \in R : w(x) > t\}) \leq 2 \inf_{c\in\mathbb{R}} \gamma(\{x \in R : |w(x) - c| > t/2\}) .$$

We note that if $0 < t_1 < t_2 < \infty$, the truncated function $v_{t_1}^{t_2}$ is locally Lipschitz and $\mu\left(\{x \in \Omega : v_{t_1}^{t_2}(x) = 0\}\right) \geq |\Omega|/2$. Application of the result just quoted plus (i) then shows that

$$\begin{aligned}\sup_{t>0} \mu\left(\{x \in \Omega : v_{t_1}^{t_2}(x) > t\}\right) &\leq C \inf_{c\in\mathbb{R}} \sup_{t>0} \mu\left(\{x : |v_{t_1}^{t_2}(x) - c| > t/2\}\right) (\frac{t}{2})^q \\ &\leq C \left\||\nabla v_{t_1}^{t_2}| \mid L_p(\Omega,\nu)\right\|^q \\ &\leq C \left\||\nabla u| \chi_{\{x:t_1<|u-b|\leq t_2\}} \mid L_p(\Omega,\nu)\right\|^q .\end{aligned}$$

The result now follows from Theorem 4.3.11. □

Following immediately from the proof we have

Corollary 4.3.14. *Under the assumptions of Theorem 4.3.13 and with b as defined in the proof of that Theorem, there are two additional conditions equivalent to the conditions (i) and (ii):*
(iii) There is a constant C_3 such that for every locally Lipschitz function u on Ω,

$$\|u - b \mid L_{q,\infty}(\Omega, \mu)\| \leq C_3 \||\nabla u| \mid L_p(\Omega, \nu)\| .$$

(iv) There is a constant C_4 such that for every locally Lipschitz function u on Ω,

$$\|u - b \mid L_q(\Omega, \mu)\| \leq C_4 \||\nabla u| \mid L_p(\Omega, \nu)\| .$$

Theorem 4.3.12 has a direct application to the classical Sobolev embedding theorem. The starting point is the familiar inequality

$$|u(x)| \leq C(n) \int_{\mathbb{R}^n} \frac{|\nabla u(z)|}{|x-z|^{n-1}} dz \text{ for a.a } x \text{ in } \mathbb{R}^n, \tag{4.3.4}$$

which holds for all Lipschitz functions u with compact support in $\mathbb{R}^n$: see, for example, [46], Lemma V.3.15 or [231]. We now use

Lemma 4.3.15. *Let Ω be an open subset of $\mathbb{R}^n$, let $g \in L_1(\Omega)$ and put*

$$I_1 g(x) = \int_\Omega \frac{g(z)}{|x-z|^{n-1}} dz, \ x \in \Omega.$$

Then

$$\|I_1 g \mid L_{n',\infty}(\Omega)\| \leq C(n) \|g \mid L_1(\Omega)\| ,$$

where n' is the conjugate of n, $n' = n/(n-1)$.

Proof. Let E be a measurable subset of $\mathbb{R}^n$ with positive, finite n-measure $|E|$ and let $B = B(x,r)$ be a ball in $\mathbb{R}^n$ with $|B| = |E|$. Then (see, for example, [231], p. 58)

$$\int_E \frac{dz}{|x-z|^{n-1}} \leq \int_B \frac{dz}{|x-z|^{n-1}} = Cr = C_1 |E|^{1/n} .$$

Take $E = \{x \in \Omega : I_1 g(x) > \lambda\}$. Then

$$\lambda |E| \leq \int_E I_1 g(x) dx = \int_\Omega \int_E \frac{g(z)}{|x-z|^{n-1}} dx dz \leq C_1 |E|^{1/n} \int_\Omega |g(z)| \, dz.$$

The result follows immediately. □

After this preparation it is a simple matter to prove the basic form of the Sobolev embedding theorem.

Theorem 4.3.16. *There is a constant C, depending only on n, such that for all compactly supported Lipschitz functions u on $\mathbb{R}^n$,*

$$\|u \mid L_{n'}(\mathbb{R}^n)\| \leq C \||\nabla u| \mid L_1(\mathbb{R}^n)\| .$$

Proof. By Lemma 4.3.15,

$$\|I_1(|\nabla u|) \mid L_{n',\infty}(\mathbb{R}^n)\| \le C \,\||\nabla u| \mid L_1(\mathbb{R}^n)\|.$$

The result now follows immediately from (4.3.4) and Theorem 4.3.12. □

Remark 4.3.17. This inequality is, of course, just the case $p = 1$ of the classical Sobolev inequality

$$\|u \mid L_{p^*}(\mathbb{R}^n)\| \le C(n,p) \,\||\nabla u| \mid L_p(\mathbb{R}^n)\|,$$

valid when $1 \le p < n$ and $p^* = \frac{np}{n-p}$. The argument just given, which follows the presentation of Hajłasz [115], can be regarded as a substitution for the traditional Gagliardo-Nirenberg approach (see, for example, [46], Theorem V.3.6). The extension from $p = 1$ to $p \in [1, n)$ is routine.

With inequalities of Poincaré type in mind, we need the following familiar lemma.

Lemma 4.3.18. *Let $x_0 \in \mathbb{R}^n$, $r > 0$ and $u \in C^1(\mathbb{R}^n)$; put $B = B(x_0, r)$. Then for all $x \in B$,*

$$|u(x) - u_B| \le C \int_B \frac{|\nabla u(y)|}{|x-y|^{n-1}} dy, \tag{4.3.5}$$

where $u_B = |B|^{-1} \int_B u(y)dy$ and C is a constant which depends only on n.

The proof is well known: see, for example [46], p. 240 or [99], p. 155. In fact, this result holds for all $u \in W^1_p(\mathbb{R}^n)$ if $1 < p < \infty$. This follows easily from an inequality of Hedberg [122] which involves the maximal function. The form of the (Hardy-Littlewood) maximal function which we shall consider is, for convenience here, slightly different from that in Remark 3.5.2 and is given by

$$\mathbb{M}(u)(x) = \sup_{r>0} \frac{1}{|B(x.r)|} \int_{B(x,r)} |u(y)|\, dy, \; x \in \mathbb{R}^n, \; u \in L_{1,loc}(\mathbb{R}^n).$$

Hedberg's result is the following.

Lemma 4.3.19. *Let $u \in L_{1,loc}(\mathbb{R}^n)$, $x_0 \in \mathbb{R}^n$, $r > 0$ and $B = B(x_0, r)$. Then there is a constant C, depending only on n, such that for all $x \in B$,*

$$\int_B \frac{|u(y)|}{|x-y|^{n-1}} dy \le Cr\mathbb{M}(u)(x). \tag{4.3.6}$$

Proof. Take $x \in B$ and put $A_i = \left(B(x, 2^{-i+1}r)\backslash B(x, 2^{-i}r)\right) \cap B$, $i \in \mathbb{N}_0$. Then

$$\int_B \frac{|u(y)|}{|x-y|^{n-1}} dy = \sum_{i=0}^{\infty} \int_{A_i} \frac{|u(y)|}{|x-y|^{n-1}} dy \le \sum_{i=0}^{\infty} (2^{-i}r)^{1-n} \int_{B(x,2^{1-i}r)} |u(y)|\, dy$$

$$= \sum_{i=0}^{\infty} 2^{i(n-1)} \frac{|B(0,1)|\, 2^{(1-i)n}}{|B(x,2^{1-i}r)|} \int_{B(x,2^{1-i}r)} |u(y)|\, dy$$

$$= 2^n r\, |B(0,1)| \sum_{i=0}^{\infty} 2^{-i} \int_{B(x,2^{1-i}r)} |u(y)|\, dy / \left|B(x,2^{1-i}r)\right|$$

$$\leq Cr\mathbb{M}(u)(x).$$

□

Corollary 4.3.20. *If $1 < p < \infty$ and $u \in W_p^1(\mathbb{R}^n)$, then (4.3.5) holds for a.e. $x \in B = B(x_0, r)$.*

Proof. There is a sequence (u_i) in $C^1(\mathbb{R}^n) \cap W_p^1(\mathbb{R}^n)$ such that $u_i \to u$ in $L_p(\mathbb{R}^n)$, $\nabla u_i \to \nabla u$ in $L_p(\mathbb{R}^n)$ and $\nabla u_i(x) \to \nabla u(x)$ a.e. We observe that by Lemma 4.3.19,

$$\left| \int_B \frac{|\nabla u_i(y)|}{|x-y|^{n-1}} dy - \int_B \frac{|\nabla u(y)|}{|x-y|^{n-1}} dy \right| \leq \int_B \frac{|\nabla u_i(y) - \nabla u(y)|}{|x-y|^{n-1}} dy$$

$$\leq Cr\mathbb{M}(|\nabla u_i - \nabla u|)(x) \to 0$$

for a.e. $x \in B$. Now let $i \to \infty$ in (4.3.5) applied to u_i. □

This extension of Lemma 4.3.18 enables us to obtain the following interesting result (see Hajłasz [114]).

Theorem 4.3.21. *Let $1 < p < \infty$ and $u \in W_p^1(\mathbb{R}^n)$. Then there is a constant C, depending only on n, such that for a.e. $x, y \in \mathbb{R}^n$,*

$$|u(x) - u(y)| \leq C\, |x-y|\, \{\mathbb{M}(|\nabla u|)(x) + \mathbb{M}(|\nabla u|)(y)\}. \tag{4.3.7}$$

Proof. Corollary 4.3.20 and Lemma 4.3.19 show that for a.e. $x, y \in \mathbb{R}^n$ we have, with $B = B(\frac{x+y}{2}, |x-y|)$,

$$|u(x) - u(y)| \leq |u(x) - u_B| + |u_B - u(y)|$$

$$\leq C \left\{ \int_B \frac{|\nabla u(z)|}{|x-z|^{n-1}} dz + \int_B \frac{|\nabla u(z)|}{|y-z|^{n-1}} dz \right\}$$

$$\leq C\, |x-y|\, \{\mathbb{M}(|\nabla u|)(x) + \mathbb{M}(|\nabla u|)(y)\}.$$

□

Remark 4.3.22. If $1 < p < \infty$ and $u \in L_p(\mathbb{R}^n)$ is such that there is a function $g \in L_p(\mathbb{R}^n)$ with

$$|u(x) - u(y)| \leq |x - y| \{g(x) + g(y)\}$$

for a.e. $x, y \in \mathbb{R}^n$, then Hajłasz [114] has shown that $u \in W_p^1(\mathbb{R}^n)$. Since the maximal operator $\mathbb{M}$ maps $L_p(\mathbb{R}^n)$ boundedly into itself, this means that (4.3.7) is necessary and sufficient for u to belong to $W_p^1(\mathbb{R}^n)$. This is the starting point of the development of the theory of Sobolev spaces on a metric space which is given in [114] (see also [123]).

The connection of the maximal operator with the Sobolev space $W_p^1(\mathbb{R}^n)$ which has just been discussed makes it natural to wonder about the action of $\mathbb{M}$ on $W_p^1(\mathbb{R}^n)$. The matter is settled by work of Kinnunen [137] which shows that $\mathbb{M}$ maps $W_p^1(\mathbb{R}^n)$ boundedly into itself. To explain this some preparation is desirable.

Lemma 4.3.23. *Let $1 < p < \infty$ and suppose that (u_i) is a bounded sequence in $W_p^1(\mathbb{R}^n)$ such that $u_i(x) \to u(x)$ a.e. Then $u \in W_p^1(\mathbb{R}^n)$ and $u_i \to u$, $\nabla u_i \to \nabla u$ weakly in $L_p(\mathbb{R}^n)$.*

Proof. Since (u_i) is bounded in $L_p(\mathbb{R}^n)$ and $u_i \to u$ a.e., it follows from [126], Theorem 13.44 that $u_i \to u$ weakly in $L_p(\mathbb{R}^n)$. Moreover, as $(D_k u_i)$ is bounded in $L_p(\mathbb{R}^n)$, there are a subsequence $(D_k u_{i(j)})$ and a $v \in L_p(\mathbb{R}^n)$ such that $D_k u_{i(j)} \to v$ weakly in $L_p(\mathbb{R}^n)$. For all $\phi \in C_0^\infty(\mathbb{R}^n)$,

$$\int_{\mathbb{R}^n} v\phi dx = \lim_{j\to\infty} \int_{\mathbb{R}^n} \phi D_k u_{i(j)} dx = -\lim_{j\to\infty} \int_{\mathbb{R}^n} u_{i(j)} D_k \phi dx = -\int_{\mathbb{R}^n} u D_k \phi dx.$$

It follows that v is the weak derivative $D_k u$, and so $u \in W_p^1(\mathbb{R}^n)$. That the whole sequence $(D_k u_i)$ converges weakly to $D_k u$ in $L_p(\mathbb{R}^n)$ follows from a standard contradiction argument. □

Lemma 4.3.24. *Let $1 < p < \infty$, suppose that $u \in W_p^1(\mathbb{R}^n)$ and let $f \in L_\infty(\mathbb{R}^n)$ have compact support. Then $u * f \in W_p^1(\mathbb{R}^n)$ and $D_k(u * f) = D_k u * f$, $k = 1, ..., n$.*

Proof. Given any $\phi \in C_0^\infty(\mathbb{R}^n)$,

$$\int_{\mathbb{R}^n} \phi(x)(D_k u * f)(x) dx = \int_{\mathbb{R}^n} f(y) \int_{\mathbb{R}^n} D_k u(x - y)\phi(x) dx dy$$

$$= -\int_{\mathbb{R}^n} (u * f) D_k \phi dx.$$

Since $u * f$ and $D_k u * f$ belong to $L_p(\mathbb{R}^n)$, it follows that $D_k u * f$ is the weak derivative of $u * f$. □

The promised result is contained in the next theorem.

Theorem 4.3.25. *Let* $1 < p < \infty$ *and* $u \in W_p^1(\mathbb{R}^n)$. *Then* $\mathbb{M}(u) \in W_p^1(\mathbb{R}^n)$,

$$|D_k \mathbb{M}(u)| \leq \mathbb{M}(D_k |u|) \text{ a.e., } k = 1, ..., n,$$

and there is a constant c, *depending only on* n *and* p, *such that*

$$\|\mathbb{M}(u) \mid L_p(\mathbb{R}^n)\| \leq c \, \|u \mid L_p(\mathbb{R}^n)\| \, .$$

Proof. For each $r > 0$ write $\chi_r = |B(0,r)|^{-1} \chi_{B(0,r)}$. Then for each $x \in \mathbb{R}^n$,

$$|B(x,r)|^{-1} \int_{B(x,r)} |u(y)| \, dy = |B(0,r)|^{-1} \int_{B(0,r)} |u(x-y)| \, dy$$

$$= (|u| * \chi_r)(x) = u_r(x), \text{ say.} \tag{4.3.8}$$

We use the crucial fact that $|u| \in W_p^1(\mathbb{R}^n)$: see, for example, [46], Proposition V.2.6. By Lemma 4.3.24, $D_k u_r = D_k |u| * \chi_r$, and by (4.3.8),

$$\mathbb{M}(u)(x) = \sup_j u_{r_j}(x),$$

where the supremum is taken over all rational positive r_j. Put $u_j = u_{r_j}$ and

$$v_i(x) = \max \{u_j(x) : 1 \leq j \leq i\}, \; i \in \mathbb{N}.$$

Then $v_i \leq \mathbb{M}(u)$ a.e. and

$$|D_k v_i| \leq \max_{1 \leq j \leq i} |D_k u_j| = \max_{1 \leq j \leq i} |D_k |u| * \chi_{r_j}| \leq \max_{1 \leq j \leq i} \chi_{r_j} * |D_k |u||$$

$$\leq \sup_j \chi_{r_j} * |D_k |u|| = \mathbb{M}(D_k |u|) \tag{4.3.9}$$

almost everywhere. Since $\mathbb{M}$ maps $L_p(\mathbb{R}^n)$ boundedly into itself, (4.3.9) shows that

$$\|v_i \mid W_p^1(\mathbb{R}^n)\|^p = \|v_i \mid L_p(\mathbb{R}^n)\|^p + \|\nabla v_i \mid L_p(\mathbb{R}^n)\|^p$$

$$\leq \|\mathbb{M}(u) \mid L_p(\mathbb{R}^n)\|^p + \left\| \left(\sum_{k=1}^n \mathbb{M}(D_k |u|)^2 \right)^{1/2} \mid L_p(\mathbb{R}^n) \right\|^p$$

$$\leq C \left(\|u \mid L_p(\mathbb{R}^n)\|^p + \|\nabla u \mid L_p(\mathbb{R}^n)\|^p \right)$$

$$= C \left\| u \mid W_p^1(\mathbb{R}^n) \right\|^p.$$

Hence (v_i) is a bounded sequence in $W_p^1(\mathbb{R}^n)$ and $v_i \to \mathbb{M}(u)$ a.e. By Lemma 4.3.23, $\mathbb{M}(u) \in W_p^1(\mathbb{R}^n)$ and the rest is an immediate consequence of (4.3.9). □

In fact, a local version of the mapping property of the maximal operator is also available, thanks to Kinnunen and Lindqvist [138]. To explain this, let Ω be a proper open subset of $\mathbb{R}^n$ and define the **local Hardy-Littlewood maximal operator** $\mathbb{M}_\Omega$ by

$$\mathbb{M}_\Omega(u)(x) = \sup |B(x,r)|^{-1} \int_{B(x,r)} |u(y)|\, dy, \;\; u \in L_{1,loc}(\Omega), \;\; x \in \Omega,$$

where the supremum is taken over all $r \in (0, d(x))$, with $d(x) = \text{dist}\,(x, \partial\Omega)$. For each $t > 0$ put

$$u_t(x) = |B(x, td(x))|^{-1} \int_{B(x,td(x))} u(y)dy.$$

Lemma 4.3.26. *Let $1 < p \le \infty$, $u \in W_p^1(\Omega)$ and $0 < t < 1$. Then $u_t \in W_p^1(\Omega)$ and for almost all $x \in \Omega, |\nabla u_t(x)| \le 2\mathbb{M}_\Omega(|\nabla u|)(x)$.*

Proof. First suppose that $u \in C^\infty(\Omega)$; let $t \in (0,1)$. Then

$$\begin{aligned} D_i u_t(x) =& D_i \left(\frac{1}{\omega_n \, (td(x))^n} \right) \int_{B(x,td(x))} u(y)dy \\ &+ \frac{1}{\omega_n \, (td(x))^n} D_i \int_{B(x,td(x))} u(y)dy, \end{aligned}$$

$i = 1, ..., n$, for a.e. x. As

$$\frac{\partial}{\partial r} \int_{B(x,r)} u(y)dy = \int_{\partial B(x,r)} u(y) d\mathcal{H}_{n-1}(y),$$

we see that for a.e. $x \in \Omega$ and all $i \in \{1, ..., n\}$,

$$\begin{aligned} D_i \int_{B(x,td(x))} u(y)dy = & \int_{B(x,td(x))} D_i u(y)dy \\ & + t \int_{\partial B(x,td(x))} u(y) dH^{n-1}(y) \cdot D_i d(x). \end{aligned}$$

Hence for a.e. $x \in \Omega$,

$$\nabla u_t(x) = \frac{n \nabla d(x)}{d(x)} \left\{ \frac{1}{|\partial B|} \int_{\partial B} u(y) dH^{n-1}(y) - \frac{1}{|B|} \int_B u(y)dy \right\}$$

$$+\frac{1}{|B|}\int_B \nabla u(y)dy,$$

where for shortness we have written B instead of $B(x,td(x))$, and ∂B for $\partial B(x,td(x))$. Now use Green's first identity

$$\int_{\partial B(x,R)} u(y)\frac{\partial v}{\partial \nu}(y)dH^{n-1}(y)=\int_{B(x,R)}\{u(y)\Delta v(y)+\nabla u(y)\cdot\nabla v(y)\}\,dy,$$

where $B(x,R)\subset\Omega$, $\nu(y)=(y-x)/R$ is the unit outer normal of $B(x,R)$ and $v(y)=\frac{1}{2}|y-x|^2$. This gives

$$\left|\frac{1}{|\partial B(x,R)|}\int_{\partial B(x,R)} u(y)dH^{n-1}(y)-\frac{1}{|B(x,R)|}\int_{B(x,R)} u(y)dy\right|$$

$$=n^{-1}|B(x,R)|^{-1}\left|\int_{B(x,R)}\nabla u(y)\cdot(y-x)dy\right|$$

$$\leq Rn^{-1}\mathbb{M}_\Omega(|\nabla u|)(x).$$

Using this with $R=td(x)$, it follows that if $e\in\mathbb{R}^n$, $|e|=1$, then since $|\nabla d(x)|=1$ a.e. (see, for example, Theorem 5.1.5 below),

$$|e\cdot\nabla u_t(x)|\leq\frac{n\,|e\cdot\nabla d(x)|}{d(x)}\cdot\frac{td(x)}{n}\mathbb{M}_\Omega(|\nabla u|)(x)+$$

$$|B(x,td(x))|^{-1}\left|\int_{B(x,td(x))} e\cdot\nabla u(y)dy\right|$$

$$\leq t\mathbb{M}_\Omega(|\nabla u|)(x)+|B(x,td(x))|^{-1}\int_{B(x,td(x))}|\nabla u(y)|\,dy$$

$$\leq(t+1)\mathbb{M}_\Omega(|\nabla u|)(x)$$

for a.e. x $\in\Omega$. This gives the desired inequality for a smooth function u.

Now suppose that $u\in W_p^1(\Omega)$ and $1<p<\infty$. Let (ϕ_j) be a sequence of functions in $W_p^1(\Omega)\cap C^\infty(\Omega)$ such that $\|\phi_j-u\mid W_p^1(\Omega)\|\to 0$ as $j\to\infty$. Let $t\in(0,1)$. Then for all $x\in\Omega$,

$$u_t(x)=\lim_{j\to\infty}(\phi_j)_t(x)$$

and

$$|(\phi_j)_t(x)|\leq|B(x,td(x))|^{-1}\int_{B(x,td(x))}|\phi_j(y)|\,dy\leq\mathbb{M}_\Omega(\phi_j)(x)\quad(j\in\mathbb{N}).$$

Application of what we have already proved gives

$$|\nabla((\phi_j)_t)(x)| \leq 2\mathrm{M}_\Omega(|\nabla\phi_j|)(x)$$

for a.e. x $\in \Omega$. Together with the mapping properties of M_Ω in $L_p(\Omega)$ this shows that

$$\left\|(\phi_j)_t \mid W_p^1(\Omega)\right\| \leq c(n,p) \left\|\phi_j \mid W_p^1(\Omega)\right\|.$$

Hence $((\phi_j)_t)$ is a bounded sequence in $W_p^1(\Omega)$ which converges to u_t pointwise: by Lemma 4.3.23 it follows that $u_t \in W_p^1(\Omega)$, while $(\phi_j)_t$ and $\nabla(\phi_j)_t$ converge weakly in $L_p(\Omega)$ to u_t and ∇u_t respectively. Moreover, for all $x \in \Omega$,

$$|\mathrm{M}_\Omega(|\nabla\phi_j|) - \mathrm{M}_\Omega(|\nabla u|)| \leq \mathrm{M}_\Omega(|\nabla\phi_j| - |\nabla u|).$$

Thus

$$\begin{aligned}
\|\mathrm{M}_\Omega(|\nabla\phi_j|) - \mathrm{M}_\Omega(|\nabla u|) \mid L_p(\Omega)\| &\leq \|\mathrm{M}_\Omega(|\nabla\phi_j| - |\nabla u|) \mid L_p(\Omega)\| \\
&\leq c(n,p) \||\nabla\phi_j| - |\nabla u| \mid L_p(\Omega)\| \\
&\to 0 \text{ as } j \to \infty.
\end{aligned}$$

Passage to the limit in

$$|\nabla((\phi_j)_t(x)| \leq 2\mathrm{M}_\Omega(|\nabla\phi_j|)(x)$$

now gives the claimed inequality. This complete the proof when $1 < p < \infty$.

It remains to deal with the case $p = \infty$. Here it follows, much as above, that for all $p \in (1,\infty)$, $u_t \in W_{p,loc}^1(\Omega)$ and

$$|\nabla u_t(x)| \leq 2\mathrm{M}_\Omega(|\nabla u|)(x)$$

for a.e. $x \in \Omega$. Since $\mathrm{M}_\Omega(|\nabla u|) \in L_\infty(\Omega)$, we see that $u_t \in W_\infty^1(\Omega)$ and the proof is complete. □

We can now give the counterpart of Theorem 4.3.25 for functions in $W_p^1(\Omega)$.

Theorem 4.3.27. *Let $1 < p \leq \infty$ and $u \in W_p^1(\Omega)$. Then $\mathrm{M}_\Omega(u) \in W_p^1(\Omega)$,*

$$|\nabla\mathrm{M}_\Omega(u)(x)| \leq 2\mathrm{M}_\Omega(|\nabla u|)(x)$$

for a.e. $x \in \Omega$, and there is a constant c, depending only on n and p,such that

$$\left\|\mathrm{M}_\Omega(u) \mid W_p^1(\Omega)\right\| \leq c(n,p) \left\|u \mid W_p^1(\Omega)\right\|.$$

Proof. First suppose that $1 < p < \infty$. As in the proof of Theorem 4.3.25, we use the fact that $|u| \in W_p^1(\Omega)$. Enumerate the rationals in $(0,1)$ by t_j $(j \in \mathbb{N})$ and put $u_j = |u|_{t_j}$. By Lemma 4.3.26, $u_j \in W_p^1(\Omega)$ and

$$|\nabla u_j(x)| \leq 2\mathrm{M}_\Omega(|\nabla u|)(x)$$

for $a.e.\ x \in \Omega$. Define

$$v_k(x) = \max_{1\leq j\leq k} u_j(x),\ k \in \mathbb{N},\ x \in \Omega.$$

Then (v_k) is an increasing sequence of functions in $W_p^1(\Omega)$ which converges to $\mathrm{M}_\Omega(u)$ pointwise, and

$$|\nabla v_k(x)| = \left|\nabla \max_{1\leq j\leq k} u_j(x)\right| \leq \max_{1\leq j\leq k} |\nabla u_j(x)| \leq 2\mathrm{M}_\Omega(|\nabla u|)(x)$$

for all $k \in \mathbb{N}$ and a.e. $x \in \Omega$. Moreover, $v_k(x) \leq \mathrm{M}_\Omega(u)(x)$ for all $k \in \mathbb{N}$ and all $x \in \Omega$. The proof may now be completed just as in the proof of Theorem 4.3.25.

When $p = \infty$, we argue as above to show that $\mathrm{M}_\Omega(u) \in W^1_{p,loc}(\Omega)$ for all $p \in (1, \infty)$. Together with the mapping property of M_Ω on the Lebesgue spaces this gives the result. □

Corollary 4.3.28. *Let* $1 < p < \infty$ *and* $u \in \mathring{W}_p^1(\Omega)$. *Then* $\mathrm{M}_\Omega(u) \in \mathring{W}_p^1(\Omega)$.

Proof. Let (ϕ_j) be a sequence of functions in $C_0^\infty(\Omega)$ such that

$$\left\|\phi_j - u \mid W_p^1(\Omega)\right\| \to 0 \quad \text{as } j \to \infty.$$

By Theorem 4.3.27, $\mathrm{M}_\Omega(\phi_j) \in W_p^1(\Omega)$ for all $j \in \mathbb{N}$. Plainly $\mathrm{M}_\Omega(\phi_j)(x) = 0$ if $d(x) < \frac{1}{2}\mathrm{dist}(\mathrm{supp}\ \phi_j, \partial\Omega)$. Hence $\mathrm{M}_\Omega(\phi_j) \in \mathring{W}\,{}_p^1(\Omega)$. As M_Ω is sublinear,

$$|\mathrm{M}_\Omega(\phi_j)(x) - \mathrm{M}_\Omega(u)(x)| \leq \mathrm{M}_\Omega(\phi_j - u)(x)$$

for all $x \in \Omega$, and so

$$\|\mathrm{M}_\Omega(\phi_j) - \mathrm{M}_\Omega(u) \mid L_p(\Omega)\| \leq c(n,p)\, \|\phi_j - u \mid L_p(\Omega)\|\,.$$

In addition, by Theorem 4.3.27 again,

$$\|\nabla\mathrm{M}_\Omega(\phi_j) \mid L_p(\Omega)\| \leq 2\, \|\mathrm{M}_\Omega(|\nabla\phi_j|) \mid L_p(\Omega)\| \leq c(n,p)\, \|\nabla\phi_j \mid L_p(\Omega)\|\,,$$

and so $(\mathrm{M}_\Omega(\phi_j))$ is a bounded sequence in $\mathring{W}\,{}_p^1(\Omega)$ which converges to $\mathrm{M}_\Omega(u)$ in $L_p(\Omega)$. A standard weak compactness argument now shows that $\mathrm{M}_\Omega(u) \in \mathring{W}\,{}_p^1(\Omega)$. □

After these diversions, we return to the matter of Sobolev and Poincaré inequalities in domains with possibly bad boundaries, beginning with an extension of Lemma 4.3.18 to John domains. We follow the treatment of Hajłasz [115].

Theorem 4.3.29. *Let Ω be a John domain in $\mathbb{R}^n$. Then there is a constant C, depending only on n and the parameter δ of Remark 4.3.2 (b), such that for every locally Lipschitz function u on Ω and all $x \in \Omega$,*

$$|u(x) - u_\Omega| \leq C \int_\Omega \frac{|\nabla u(z)|}{|x-z|^{n-1}} dz.$$

Proof. Let x_0 be a central point of Ω and put $B_0 = B(x_0, \frac{1}{4}d(x_0))$. We claim that there is a constant K, depending only on n and δ, such that given any $x \in \Omega$ there is a sequence (or chain) of balls $B_i = B(x_i, r_i) \subset \Omega$ ($i \in \mathbb{N}_0$) with the following properties: (i) For each $i \in \mathbb{N}_0$, $|B_i \cup B_{i+1}| \leq K\ |B_i \cap B_{i+1}|$; (ii) dist $(x, B_i) \leq M r_i$, $r_i \to 0$, $x_i \to x$ as $i \to \infty$; (iii) no point of Ω belongs to more than K balls.

To justify this claim, suppose first that $x \in \Omega \backslash 2B_0$. Let γ be a John curve that joins x_0 to x. We now construct a chain of balls B_i, each centred on γ, and starting with B_0, in the following inductive way. Given that $B_0, ..., B_i$ have been defined, start from the centre x_i of B_i and follow γ towards x until B_i is left for the last time, at x_{i+1}, say: define $B_{i+1} = B(x_{i+1}, \frac{1}{4}|x - x_{i+1}|\delta)$. It follows that each B_i constructed in this way is contained in Ω; we must verify that the sequence (B_i) has properties (i)-(iii). Note that consecutive balls have comparable radii, these radii being comparable to the distance of the centres of the balls from x : this gives (i) and the inequality in (ii). As for (iii), suppose that $y \in B_{i_1} \cap ... \cap B_{i_k}$: the radii of the balls $B_{i_1}, ..., B_{i_k}$ are comparable to $|x-y|$. If $i < j$, the method of construction of the balls shows that the centre of B_j does not belong to B_i : thus the distances between the centres of the balls B_{i_j} are comparable to $|x-y|$. However, the number of points in $\mathbb{R}^n$ with pairwise comparable distances is bounded: if $z_1, ... z_N$ satisfy $c^{-1}r < |z_i - z_j| < cr$ for $i \neq j$, then $N \leq C(c,n)$. Hence k is bounded by a constant depending only on n and δ, and (iii) follows. Thus $r_i \to 0$ and consequently $x_i \to x$ as $i \to \infty$. This completes the proof of (ii). When $x \in 2B_0$ the analysis is much easier and is accordingly omitted.

Now let u be a locally Lipschitz function on Ω. Since $u_{B_i} = |B_i|^{-1} \int_{B_i} u dy \to u(x)$ as $i \to \infty$, we see that

$$\begin{aligned}
|u(x) - u_{B_0}| &\leq \sum_{i=0}^{\infty} |u_{B_i} - u_{B_{i+1}}| \\
&\leq \sum_{i=0}^{\infty} \left\{ |u_{B_i} - u_{B_i \cap B_{i+1}}| + |u_{B_{i+1}} - u_{B_i \cap B_{i+1}}| \right\} \\
&\leq \sum_{i=0}^{\infty} \frac{1}{|B_i \cap B_{i+1}|} \left\{ \int_{B_i} |u - u_{B_i}|\, dy + \int_{B_{i+1}} |u - u_{B_{i+1}}|\, dy \right\} \\
&\leq C \sum_{i=0}^{\infty} \frac{1}{|B_i|} \int_{B_i} |u - u_{B_i}|\, dy
\end{aligned}$$

$$\leq C \sum_{i=0}^{\infty} \int_{B_i} \frac{|\nabla u(z)|}{r_i^{n-1}} dz,$$

the last two inequalities following from (i) and the standard Poincaré inequality in a ball, respectively. By (ii), $|x - z| \leq Cr_i$ if $z \in B_i$, so that with the help of (iii) we have

$$|u(x) - u_{B_0}| \leq C \sum_{i=0}^{\infty} \int_{B_i} \frac{|\nabla u(z)|}{|x - z|^{n-1}} dz \leq C \int_{\Omega} \frac{|\nabla u(z)|}{|x - z|^{n-1}} dz. \tag{4.3.10}$$

Since

$$|u(x) - u_{\Omega}| \leq |u(x) - u_{B_0}| + |u_{B_0} - u_{\Omega}|,$$

all that is left to do is to estimate $|u_{B_0} - u_{\Omega}|$. To do this, observe that

$$|u_{B_0} - u_{\Omega}| \leq |\Omega|^{-1} \int_{\Omega} |u(x) - u_{B_0}| \, dx \leq C |\Omega|^{-1} \int_{\Omega} \int_{\Omega} \frac{|\nabla u(z)|}{|x - z|^{n-1}} dx dz$$

$$\leq C |\Omega|^{-(n-1)/n} \int_{\Omega} |\nabla u(z)| \, dz,$$

the final inequality following as in the proof of Lemma 4.3.15. Since Ω is a John domain,

$$C |\Omega|^{1/n} \geq d(x_0) \geq \delta |x - x_0|,$$

so that for all $z \in \Omega$,

$$\text{diam } \Omega \leq C(n, \delta) |\Omega|^{1/n},$$

from which we have

$$|\Omega|^{-(n-1)/n} \leq \frac{C}{|x - z|^{n-1}}$$

for all $z \in \Omega$. Hence

$$|u_{B_0} - u_{\Omega}| \leq C \int_{\Omega} \frac{|\nabla u(z)|}{|x - z|^{n-1}} dz,$$

and the proof is complete. □

Now that we have this theorem at our disposal we can establish the Sobolev-Poincaré inequality for John domains.

Theorem 4.3.30. *Let Ω be a John domain and $1 \leq p < n$. Then there is a constant C, depending only on n, p and δ, such that for all locally Lipschitz functions u on Ω,*

$$\left(\int_{\Omega} |u - u_{\Omega}|^{p^*} dx \right)^{1/p^*} \leq C \left(\int_{\Omega} |\nabla u|^p \, dx \right)^{1/p}, \tag{4.3.11}$$

where $p^* = np/(n - p)$.

Proof. First suppose that $p = 1$. From Lemma 4.3.15 and Theorem 4.3.29 we have

$$\|u - u_\Omega \mid L_{n',\infty}(\Omega)\| \leq C(n, \delta) \, \| |\nabla u| \mid L_1(\Omega)\| \, .$$

Together with Theorem 4.3.12 this gives the desired result.

Now suppose that $1 < p < n$. As in the proof of Theorem 4.3.13, choose $b \in \mathbb{R}$ such that

$$|\Omega_+| \geq |\Omega| \, /2, \ \ |\Omega_-| \geq |\Omega| \, /2,$$

where

$$\Omega_+ = \{x \in \Omega : u(x) \geq b\} \text{ and } \Omega_- = \{x \in \Omega : u(x) \leq b\} \, .$$

By Corollary 4.3.14,

$$\|u - b \mid L_{n'}(\Omega)\| \leq C \, \| |\nabla u| \mid L_1(\Omega)\| \, . \tag{4.3.12}$$

Put $\alpha = p(n-1)/(n-p) > 1$ and

$$v = \begin{cases} |u - b|^\alpha & \text{on } \Omega_+ \\ -|u - b|^\alpha & \text{on } \Omega_-. \end{cases}$$

Then v is locally Lipschitz, $|\{x \in \Omega : v(x) \geq 0\}| \geq |\Omega| \, /2$ and $|\{x \in \Omega : v(x) \leq 0\}| \geq |\Omega| \, /2$. Application of Corollary 4.3.14 to v gives, with the aid of Hölder's inequality,

$$\begin{aligned} \left(\int_\Omega |u - b|^{p^*} dx \right)^{1/n'} &= \left(\int_\Omega |v|^{n'} dx \right)^{1/n'} \leq C \int_\Omega |\nabla v| \, dx \\ &\leq C' \left(\int_\Omega |u - b|^{p^*} dx \right)^{1/p'} \left(\int_\Omega |\nabla v|^p \, dx \right)^{1/p}, \end{aligned}$$

and the proof is complete. □

This result was proved independently by Martio [166] and Reshetnyak [204] when $1 < p < n$; the case $p = 1$ was covered by Boyarski [20]. The proof given here is that of Hajłasz [115]. When Ω is a bounded, simply-connected domain in the plane and $1 \leq p < 2$, Buckley and Koskela [25] have established the remarkable result that the Sobolev-Poincaré inequality holds if, and only if, Ω is a John domain.

4.4 Hardy inequalities

Here we shall discuss the higher-dimensional analogue of the Hardy inequality treated in Chapter 2. To be more precise, let Ω be a domain in $\mathbb{R}^n$, $\Omega \neq \mathbb{R}^n$, let $p \in (1, \infty)$ and let d be the distance function defined by $d(x) = \mathrm{dist}(x, \partial\Omega)$, $x \in \Omega$. The Hardy inequality is

$$\|f/d \mid L_p(\Omega)\| \le C \|\nabla f \mid L_p(\Omega)\|, \; f \in \mathring{W}_p^1(\Omega). \tag{4.4.1}$$

The constant C is supposed to be independent of f. It is known that (4.4.1) holds if Ω is bounded and has Lipschitz boundary: see, for example, [191]. Our initial concern is to extend this result to a more general class of domains.

If $n < p < \infty$, it has been shown by Lewis [156] that the Hardy inequality holds with no conditions on Ω. We give an account of this using the approach of Kinnunen and Martio [139], which proceeds via the fractional maximal function; we also deal with the case $p \le n$, when restrictions on Ω are needed. We start with the fractional maximal function and establish natural analogues of results given in 4.3 for the Hardy-Littlewood maximal function.

Given $\alpha \in [0, n]$, the **fractional maximal function** of a function $f \in L_{1,loc}(\mathbb{R}^n)$ is defined by

$$\mathbb{M}_\alpha f(x) = \sup_{r>0} r^{\alpha-n} \int_{B(x,r)} |f(y)| \, dy; \tag{4.4.2}$$

we also put

$$\mathbb{M}_{\alpha,R} f(x) = \sup_{0<r<R} r^{\alpha-n} \int_{B(x,r)} |f(y)| dy.$$

Note that when $\alpha = 0$, the fractional maximal function is just a constant multiple of the maximal function defined in Section 4.3.

Lemma 4.4.1. *Let $\gamma > \lambda \ge 0$. Then there is a constant C such that for all cubes Q in $\mathbb{R}^n$, all $g \in L_1(Q)$ and all $x \in Q$,*

$$\int_Q \frac{|g(z)|}{|x-z|^{n-\gamma}} dz \le C \, |Q|^{(\gamma-\lambda)/n} \, \mathbb{M}_{\lambda, \mathrm{diam} Q} g(x).$$

Proof. Let $\delta = \mathrm{diam}\, Q$ and write the integral above as a sum of integrals over the sets $Q_k := Q \cap \left(B(x, \delta/2^k) \setminus B(x, \delta/2^{k+1})\right)$, $k \in \mathbb{N}_0$. In each Q_k, $|x-z|^{n-\gamma} \approx (\delta/2^k)^{n-\gamma}$; hence

$$\int_{Q_k} \frac{|g(z)|}{|x-z|^{n-\gamma}} dz \le C(\delta/2^k)^{-(n-\gamma)} \int_{B(x, 2^{-k}\delta)} |g(z)| \, dz,$$

from which the result follows easily. □

Lemma 4.4.2. *Let $u \in C^\infty(\mathbb{R}^n), 1 \le p < \infty$ and $0 \le \alpha < p$. Then there is a constant c, depending only on n, such that for every $x, y \in \mathbb{R}^n$,*

$$|u(x) - u(y)| \le c\,|x-y|^{1-\alpha/p} \left\{ \mathbb{M}_{\alpha/p, 2|x-y|}(|\nabla u|\,\chi)(x) + \mathbb{M}_{\alpha/p, 2|x-y|}(|\nabla u|\,\chi)(y) \right\}. \tag{4.4.3}$$

Proof. Let $x, y \in \mathbb{R}^n, x \neq y$, and let Q be a cube centred at $(x+y)/2$, containing x and y and with diameter $= 2|x-y|$. Then by Theorem 4.3.29 (or a straightforward adaptation of Lemma 4.3.18),

$$|u(x) - u(y)| \leq |u(x) - u_Q| + |u(y) - u_Q|$$

$$\leq C \left\{ \int_Q \frac{|\nabla u(z)|}{|x-z|^{n-1}} dz + \int_Q \frac{|\nabla u(z)|}{|y-z|^{n-1}} dz \right\}.$$

The desired inequality now follows from Lemma 4.4.1 with $\gamma = 1$ and $\lambda = \alpha/p$. □

Since smooth functions are dense in $W_p^1(\mathbb{R}^n)$, it follows that if $u \in W_p^1(\mathbb{R}^n)$, then (4.4.3) holds for almost all $x, y \in \mathbb{R}^n$. Note also that by Hölder's inequality,

$$\mathbb{M}_{\alpha/p}(|\nabla u|)(x) \leq c(n,p) \left(\mathbb{M}_\alpha(|\nabla u|^p)(x)\right)^{1/p}. \qquad (4.4.4)$$

Thus if $n < p < \infty$, we may take $\alpha = n$ in this and obtain

$$\mathbb{M}_{n/p}(|\nabla u|)(x) \leq c(n,p) \left\| \nabla u \mid L_p(\mathbb{R}^n) \right\|, \; x \in \mathbb{R}^n.$$

Together with Lemma 4.4.2 this shows that, after possible redefinition on a set of zero measure, $u \in C^{1-n/p}(\mathbb{R}^n)$. This approach goes back to work of Hedberg [122]; see also [231]. For details of further results obtainable by this strategy see also [139].

We are now in a position to deal with the Hardy inequality.

Theorem 4.4.3. *Let Ω be an open subset of $\mathbb{R}^n$, $\Omega \neq \mathbb{R}^n$, and let $n < p < \infty$. Then there is a constant C such that for all $u \in \overset{\circ}{W}{}_p^1(\Omega)$,*

$$\left\| u/d \mid L_p(\Omega) \right\| \leq C \left\| \nabla u \mid L_p(\Omega) \right\|.$$

Proof. Choose any $q \in (n,p)$, fix $x \in \Omega$ and let $x_0 \in \partial\Omega$ be such that $|x - x_0| = \mathrm{dist}(x, \partial\Omega) = R$. Then (4.4.3) and (4.4.4) show that for all $u \in C_0^\infty(\Omega)$ (supposed extended to all of $\mathbb{R}^n$ by 0),

$$|u(x)| \leq c(n,q) \, |x - x_0|^{1-n/q} \left(\int_{B(x,4R)} |\nabla u(y)|^q \, dy \right)^{1/q}$$

$$= c(n,q) R \left(R^{-n} \int_{B(x,4R)} |\nabla u(y)|^q \, dy \right)^{1/q}$$

$$\leq c(n,q) d(x) \left(\mathbb{M}(|\nabla u|^q)(x)\right)^{1/q}. \qquad (4.4.5)$$

This, plus the boundedness of the Hardy-Littlewood operator $\mathbb{M}$, shows that

$$\int_{\Omega} (|u(x)| / d(x))^p \, dx \leq c \int_{\Omega} (\mathbb{M}(|\nabla u|^q)(x))^{p/q} dx \leq c \int_{\Omega} |\nabla u(x)|^p \, dx,$$

where c depends only on n, p and q. A routine approximation argument now gives this inequality for all $u \in \overset{\circ}{W}{}^1_p (\Omega)$. □

This shows that when $n < p < \infty$, Hardy's inequality holds for all open sets apart from $\mathbb{R}^n$: no conditions on the boundary are needed. The situation when $1 < p \leq n$ is more complicated, as conditions on Ω are then necessary: see [156], Theorem 3. Various conditions on Ω, sufficient to ensure that Hardy's inequality holds, are known. We mention one due to Lewis [156] which involves the capacity density of the complement of Ω. To explain this, recall that given $p \in (1, \infty)$ and a compact set $K \subset \Omega$, the p-capacity of K relative to Ω is

$$C_p(K, \Omega) = \inf \left\{ \int_{\Omega} |\nabla u(x)|^p \, dx : u \in C_0^{\infty}(\Omega), \ u(x) \geq 1 \text{ for all } x \in K \right\}.$$

A closed set $E \subset \mathbb{R}^n$ is said to be **uniformly p-fat** if there is a constant $\gamma > 0$ such that for all $x \in E$ and all $r > 0$,

$$C_p \left(E \cap \overline{B(x,r)}, B(x, 2r) \right) \geq \gamma C_p \left(\overline{B(x,r)}, B(x, 2r) \right) = \gamma c(n,p) r^{n-p}.$$

Some examples may help to put this definition in perspective:
(i) If $n < p < \infty$, every non-empty closed set is uniformly p-fat.
(ii) Every closed set satisfying the interior cone condition is uniformly p-fat for every $p \in (1, \infty)$.
(iii) The complement of a Lipschitz domain is uniformly p-fat for every $p \in (1, \infty)$.
(iv) If there is a constant $\gamma > 0$ such that such that for all $x \in E$ and all $r > 0$,

$$|E \cap B(x,r)| \geq \gamma |B(x,r)|,$$

then E is uniformly p-fat for every $p \in (1, \infty)$.

The basic property of uniformly fat sets is that if E is uniformly p-fat, then it is also uniformly q-fat for some $q \in (1, p)$: this was shown by Lewis [156].

We shall now give a proof of Lewis's result concerning the Hardy inequality, using the approach of Kinnunen and Martio [139]. A lemma of Poincaré type is needed.

Lemma 4.4.4. *Let $p \in (1, \infty)$, $x_0 \in \mathbb{R}^n$, $R > 0$, let $u \in C^{\infty}(\mathbb{R}^n)$ and put $Z = \{x \in \mathbb{R}^n : u(x) = 0\}$. Then there is a constant c, depending only on n and p, such that*

$$\left(\frac{1}{|B(x_0, 2R)|} \int_{B(x_0,R)} |u(y)|^p \, dy \right)^{1/p} \leq c \left(C_p \left(Z \cap \overline{B(x_0, R)}, B(x_0, 2R) \right) \right)^{-1}$$

$$\times \int_{B(x_0,2R)} |\nabla u(y)|^p \, dy)^{1/p}.$$

The proof may be found in [171], Chapter 10.

Theorem 4.4.5. *Let $1 < p \leq n$, $0 \leq \alpha < p$, let $\Omega \subset \mathbb{R}^n$ be an open set such that $\mathbb{R}^n \backslash \Omega$ is uniformly $p-$fat and suppose that $u \in C_0^\infty(\Omega)$. Then there is a constant c, depending only on n and p, such that for all $x \in \Omega$,*

$$|u(x)| \leq c d(x)^{1-\alpha/p} (\mathbb{M}_\alpha (|\nabla u|^p)(x))^{1/p}. \tag{4.4.6}$$

Proof. Let $x \in \Omega$ and let $x_0 \in \partial\Omega$ be such that $|x - x_0| = d(x) = R$. Let $u \in C_0^\infty(\Omega)$, and extend u and ∇u by defining them to be zero in $\mathbb{R}^n \backslash \Omega$. Then as in the proof of Lemma 4.4.2 and with the help of (4.4.4) we have for all $x \in B(x_0, 2R)$,

$$\left| u(x) - u_{B(x_0,2R)} \right| \leq c(n,p) R^{1-\alpha/p} \left(\mathbb{M}_\alpha |\nabla u|^p \right)(x))^{1/p}.$$

Thus

$$\begin{aligned} |u(x)| &\leq \left| u(x) - u_{B(x_0,2R)} \right| + \left| u_{B(x_0,2R)} \right| \\ &\leq c R^{1-\alpha/p} \left(\mathbb{M}_\alpha |\nabla u|^p \right)(x))^{1/p} + \left| u_{B(x_0,2R)} \right|. \end{aligned}$$

Now Hölder's inequality and Lemma 4.4.4 show that, with $I_a = \int_{B(x_0,aR)} |\nabla u(y)|^p \, dy$ and $B_a = B(x_0, aR)$,

$$\begin{aligned} \frac{1}{|B(x_0,2R)|} \int_{B(x_0,2R)} |u(y)| \, dy &\leq c \left((C_p(Z \cap \overline{B_2}, B_4))^{-1} I_4 \right)^{1/p} \\ &\leq c \left(\left(C_p \left((\mathbb{R}^n \backslash \Omega) \cap \overline{B_2}, B_4 \right) \right)^{-1} I_4 \right)^{1/p} \\ &\leq c \left(R^{p-n} \int_{B(x_0,8R)} |\nabla u(y)|^p \, dy \right)^{1/p} \\ &\leq c R^{1-\alpha/p} \left(\mathbb{M}_\alpha |\nabla u|^p \right)(x))^{1/p}. \end{aligned}$$

The proof is complete. □

The promised theorem due to Lewis is the following.

Theorem 4.4.6. *Let $1 < p \leq n$ and suppose that Ω is an open set in $\mathbb{R}^n$ such that $\mathbb{R}^n \backslash \Omega$ is uniformly $p-$fat. Then there is a constant c, depending only on n and p, such that for all $u \in \overset{\circ}{W}{}_p^1(\Omega)$,*

$$\int_\Omega \left(\frac{|u(x)|}{d(x)} \right)^p dx \leq c \int_\Omega |\nabla u(x)|^p \, dx.$$

Proof. By the basic property of uniformly $p-$fat sets, $\mathbb{R}^n \backslash \Omega$ is uniformly $q-$fat for some $q \in (1, p)$. From (4.4.6) with $\alpha = 0$ we have for all $u \in C_0^\infty(\Omega)$,

$$|u(x)| \leq c(n, q) d(x) \left(\mathbb{M}_\alpha |\nabla u|^q \right)(x) \right)^{1/q}, \ x \in \Omega.$$

From now on the proof is the same as for Theorem 4.4.3. □

Note that in fact Lewis [156] proved more when $p = n$, for he showed that Hardy's inequality when $p = n$ holds if, and only if, $\mathbb{R}^n \backslash \Omega$ is uniformly $n-$fat.

In the opposite direction from the last theorem we have the result (see [46], p.223) that if $u \in W_p^1(\Omega)$ and $u/d \in L_p(\Omega)$, then $u \in \overset{\circ}{W}{}_p^1(\Omega)$. This naturally holds if Hardy's inequality holds. However, Kinnunen and Martio [139] have shown that the requirement that u/d should belong to $L_p(\Omega)$ can be weakened to $u/d \in L_{p,\infty}(\Omega)$, thus providing another example of how weak conditions can serve as well as strong ones. We give their proof below.

Theorem 4.4.7. *Let Ω be a proper open subset of $\mathbb{R}^n$ and suppose that $u \in W_p^1(\Omega)$ is such that $u/d \in L_{p,\infty}(\Omega)$. Then $u \in \overset{\circ}{W}{}_p^1(\Omega)$.*

Proof. Define u and ∇u to be zero in $\mathbb{R}^n \backslash \Omega$, let N be the exceptional set of zero measure on which (4.4.3) fails to hold and for every $\lambda > 0$ put

$$F_\lambda = \{x \in \Omega \backslash N : |u(x)| \leq \lambda, \mathbb{M} \left(|\nabla u|^p \right)(x) \leq \lambda^p, |u(x)| / d(x) \leq \lambda \} \cup (\mathbb{R}^n \backslash \Omega).$$

We claim that $u \mid_{F_\lambda}$ is Lipschitz-continuous. Let $x, y \in \Omega \cap F_\lambda$. Then by (4.4.3),

$$\begin{aligned} |u(x) - u(y)| &\leq c(n, p) |x - y| \left\{ \left(\mathbb{M} \left(|\nabla u|^p \right)(x) \right)^{1/p} + \left(\mathbb{M} \left(|\nabla u|^p \right)(y) \right)^{1/p} \right\} \\ &\leq c(n, p) \lambda |x - y| ; \end{aligned}$$

without loss of generality we may assume that $c(n, p) \geq 1$. If $x \in \Omega \cap F_\lambda$ and $y \in \mathbb{R}^n \backslash \Omega$, then

$$|u(x) - u(y)| = |u(x)| \leq \lambda d(x) \leq \lambda |x - y| .$$

If $x, y \in \mathbb{R}^n \backslash \Omega$ the claim is obvious. It follows that $u \mid_{F_\lambda}$ is Lipschitz-continuous with constant $c(n, p)\lambda$.

The next step is to extend $u \mid_{F_\lambda}$ to a Lipschitz-continuous function on $\widetilde{u}_\lambda$ on $\mathbb{R}^n$ with the same Lipschitz constant by defining

$$\widetilde{u}_\lambda(x) = \inf_{y \in F_\lambda} \{u(y) + c(n,p)\lambda\,|x-y|\};$$

then we put $u_\lambda(x) = \operatorname{sgn} u(x) \min\{|\widetilde{u}_\lambda(x)|, \lambda\}$. This procedure is a modification of that of McShane [173]. The function u_λ has the properties: (i) $u_\lambda(x) = u(x)$ for all $x \in F_\lambda$; (ii) $|u_\lambda(x)| \le \lambda$ and $|\nabla u_\lambda(x)| \le c(n,p)\lambda$ for all $x \in \mathbb{R}^n$; (iii) $u_\lambda(x) = 0$ for all $x \in \mathbb{R}^n \backslash \Omega$; (iv) $\nabla u_\lambda(x) = \nabla u(x)$ for a.e. $x \in F_\lambda$. It is convenient to introduce the following sets:

$$F_\lambda^1 = \{x \in \Omega : |u(x)| > \lambda\},\ F_\lambda^2 = \{x \in \Omega : M(|\nabla u|^p)(x) > \lambda^p\},$$
$$F_\lambda^3 = \{x \in \Omega : |u(x)|/d(x) > \lambda\}.$$

Then

$$\int_\Omega |\nabla u_\lambda(x)|^p\,dx \le \int_{F_\lambda \cap \Omega} |\nabla u_\lambda(x)|^p\,dx + \sum_{j=1}^{3} \int_{F_\lambda^j} |\nabla u_\lambda(x)|^p\,dx.$$

Since $|\nabla u_\lambda(x)| \le c(n,p)\lambda$ we have

$$\int_{F_\lambda^1} |\nabla u_\lambda(x)|^p\,dx \le c(n,p)\lambda^p\,|\{x \in \Omega : |u(x)| > \lambda\}| \le c(n,p)\int_\Omega |u(x)|^p\,dx.$$

By the familiar property of the maximal function on $L_{1,\infty}$ (see, for example, [70]),

$$\int_{F_\lambda^2} |\nabla u_\lambda(x)|^p\,dx \le c(n,p)\lambda^p\,|\{x \in \Omega : \mathbb{M}(|\nabla u|^p)(x) > \lambda^p\}|$$

$$\le c(n,p)\int_\Omega |\nabla u(x)|^p\,dx.$$

Moreover, since $u/d \in L_{p,\infty}(\Omega)$,

$$\int_{F_\lambda^3} |\nabla u_\lambda(x)|^p\,dx \le c(n,p)\lambda^p\,|\{x \in \Omega : |u(x)|/d(x) > \lambda\}| \le c(n,p).$$

Putting all these estimates together we see that

$$\int_\Omega |\nabla u_\lambda(x)|^p\,dx \le c(n,p)\left\{\int_\Omega |u(x)|^p\,dx + \int_\Omega |\nabla u(x)|^p\,dx + 1\right\}$$

for all $\lambda > 0$. In the same way it can be shown that

$$\int_\Omega |u_\lambda(x)|^p\,dx \le c(n,p)\left\{\int_\Omega |u(x)|^p\,dx + \int_\Omega |\nabla u(x)|^p\,dx + 1\right\}$$

for all $\lambda > 0$. As u_λ is Lipschitz-continuous and is zero on $\mathbb{R}^n \backslash \Omega$, it follows that $u_\lambda \in \overset{\circ}{W}{}_p^1(\Omega)$. Hence $\{u_\lambda : \lambda > 0\}$ is a bounded subset of $\overset{\circ}{W}{}_p^1(\Omega)$. Since $|\Omega \backslash F_\lambda| \to 0$ as $\lambda \to \infty$ and u_λ coincides with u in F_λ, we have $u_\lambda \to u$ a.e. in Ω. Thus just as in Lemma 4.3.23 we conclude that $u \in \overset{\circ}{W}{}_p^1(\Omega)$. The proof is complete. □

Corollary 4.4.8. *Let $\Omega \neq \mathbb{R}^n$. If $1 < p \leq n$ and $\mathbb{R}^n \backslash \Omega$, then $u \in W_p^1(\Omega)$ belongs to $\mathring{W}_p^1(\Omega)$ if, and only if, $u/d \in L_{p,\infty}(\Omega)$. If $n < p < \infty$ and $u \in W_p^1(\Omega)$, then $u \in \mathring{W}_p^1(\Omega)$ if, and only if, u satisfies Hardy's inequality (4.4.1).*

Remark 4.4.9. Inequalities similar to the classical Hardy inequality can be obtained under quite mild conditions on Ω. For example, in [60] (see also [226] and [227]) it is shown that inequalities of the type

$$\int_\Omega \left(\frac{|u(x)|}{d^\alpha(x)}\right)^p dx \leq c \int_\Omega \left(\frac{|\nabla u(x)|}{d^\beta(x)}\right)^p dx, \ u \in C_0^\infty(\Omega), \tag{4.4.7}$$

hold under various restrictions. Thus (4.4.7) is true if Ω is $b-$plump and has finite measure, $0 \leq \alpha < \min\{\varepsilon/p, \beta+1\}$, $-\varepsilon/p' < \beta \leq 0$, where

$$1 < p < \infty \text{ and } \varepsilon = \frac{\log\left(1+(b/24)^n\right)}{\log(120/b)} \in (0,1).$$

The argument relies on the fact that a plump domain satisfies a Whitney cube #-condition. Moreover, if Ω is $b-$plump and has infinite measure, $1 < p < \infty$, ε is as defined above, $l \in \mathbb{N}$ and $-\varepsilon/p' < \beta \leq 0$, then a logarithmic form of the Hardy inequality holds, namely

$$\int_\Omega |u(x)|^p \left(\log \frac{1}{d(x)}\right)^l dx \leq c \int_\Omega \left(\frac{|\nabla u(x)|}{d^\beta(x)}\right)^p dx, \ u \in C_0^\infty(\Omega).$$

For other logarithmic Hardy inequalities see [191] and [210].

There are domains in which the classical Hardy inequality fails but a modification in the form of (4.4.7) holds. We illustrate this by means of two examples:

(i) Let $\Omega = B(0,1)\backslash\{0\} \subset \mathbb{R}^n$. The complement of Ω in $\mathbb{R}^n$ is not $n-$uniformly fat and hence, by Theorem 3 of [156], the classical Hardy inequality with $p = n$ does not hold. However, Ω is a John domain and so a Hardy inequality of the form of (4.4.7) holds, with $p = n$.

(ii) Lewis [156] constructed a set F of Cantor type such that the classical Hardy inequality fails in $B(0,1)\backslash F$ whenever $1 < p < n$. Since $B(0,1)\backslash F$ satisfies a Whitney cube #-condition, a Hardy inequality of the form of (4.4.7) holds for $1 < p < n$, with certain values of α and β.

To conclude this section, we turn from the question of when the Hardy inequality (4.4.1) holds to that of finding the best constant C in it, given that the inequality is true. Naturally associated with (4.4.1) is the variational problem of determining

$$\mu_p(\Omega) = \inf\left\{\frac{\int_\Omega |\nabla f|^p \, dx}{\int_\Omega |f/d|^p \, dx} : f \in \mathring{W}_p^1(\Omega), \ f \neq 0\right\}. \tag{4.4.8}$$

Plainly (4.4.1) holds if, and only if, $\mu_p(\Omega) > 0$; and if $\mu_p(\Omega) > 0$, then the best constant C is $1/\mu_p(\Omega)$. When $n = 1$ we know that $\mu_p(\Omega) = (1 - 1/p)^p$.

When $n > 1$ the situation is more complicated, but thanks to Matskewich and Sobolevskii [170], who dealt with the case $n = 2$, and Marcus, Mizel and Pinchover [165], who gave a proof for the general case, it turns out that $\mu_p(\Omega) = (1 - 1/p)^p$ if Ω is convex and $\partial\Omega$ is smooth in the neighbourhood of one of its points. We give this result below, together with a sketch of its proof [165].

Theorem 4.4.10. *Let $n \geq 2$, let Ω be a convex domain in $\mathbb{R}^n$ and suppose there exists $x_0 \in \partial\Omega$ such that $\partial\Omega$ is of class C^2 in some neighbourhood of x_0. Then $\mu_p(\Omega) = (1 - 1/p)^p$.*

Proof. First suppose that Ω is bounded. Since Ω is convex, there is a sequence $\{\Omega_k\}$ of domains, each of which is a bounded convex polytope which contains Ω, such that for all $x \in \Omega$, $\lim_{k\to\infty} d_{\Omega_k}(x) = d(x)$: see [77], Theorem 33. Here d_{Ω_k} is the distance function for the set Ω_k and d, as usual, is the distance function for Ω. Let $u \in C_0^\infty(\Omega)$: then $u \in C_0^\infty(\Omega_k)$ for all k and $R_{\Omega_k}(u) := \int_{\Omega_k} |\nabla u|^p\, dx / \int_{\Omega_k} |u/d|^p\, dx \to R_\Omega(u) := \int_\Omega |\nabla u|^p\, dx / \int_\Omega |u/d|^p\, dx$ as $k \to \infty$. Hence

$$\limsup_{k\to\infty} \mu_p(\Omega_k) \leq \mu_p(\Omega). \tag{4.4.9}$$

We now claim that

$$\limsup_{k\to\infty} \mu_p(\Omega_k) = (1 - 1/p)^p \Longrightarrow \mu_p(\Omega) = (1 - 1/p)^p. \tag{4.4.10}$$

To establish this, we first write any point $x \in \mathbb{R}^n$ in the form $x = (x', x_n)$, where $x' = (x_1, ..., x_{n-1})$. The assumption concerning x_0 means that there is a tangent hyperplane Π at x_0 : without loss of generality we shall assume that $x_0 = 0$, that $\Pi = \{x \in \mathbb{R}^n : x_n = 0\}$ and that Ω contains a segment $\{(0, x_n) : 0 < x_n < b\}$ for some $b > 0$. Let $H = \{x \in \mathbb{R}^n : x_n > 0\}$ and $\varepsilon \in (0, 1)$. Since $\mu_p(H) = (1 - 1/p)^p$, there exists $\phi \in C_0^\infty(H)$ such that $|R_H(\phi) - (1 - 1/p)^p| < \varepsilon$. Moreover, there exists $A > 0$ such that

$$\text{supp}\, \phi \subset K = \{x \in \mathbb{R}^n : x_n > 0,\ |x'| < Ax_n\},$$

and there is a neighbourhood U of 0 such that for all $x \in U \cap \Omega$,

$$|\text{dist}\,(x, \Pi) - d(x)| \leq o(1)\ |x|.$$

Since R_H and K are invariant with respect to transformations of the form $x \longmapsto \alpha x$ with $\alpha > 0$, we may assume that

$$\text{supp}\, \phi \subset U \cap \Omega \text{ and } d(x) < (1 + \varepsilon)x_n \text{ for all } x \in \text{supp}\, \phi.$$

Putting all this together we see that

$$R_\Omega(\phi) \leq (1 + \varepsilon)R_H(\phi) \leq (1 + \varepsilon)\left((1 - 1/p)^p + \varepsilon\right),$$

and as ε may be chosen arbitrarily small, it follows that

$$\mu_p(\Omega) \le (1 - 1/p)^p, \tag{4.4.11}$$

which justifies our claim.

Next, let D be a domain in $\mathbb{R}^{n-1}$ and let S be a bounded domain in $\mathbb{R}^n$ of the form $\{(x', x_n); x' \in D, 0 < x_n < A(x')\}$. We claim that if $u \in C^1(\overline{S})$ and $u = 0$ on $x_n = 0$, then

$$\int_S |\nabla u|^p \, dx \ge (1 - 1/p)^p \int_S |u/x_n|^p \, dx. \tag{4.4.12}$$

To see this, note that if $(0, A)$ is a bounded interval, $f \in W_p^1(A)$ and $f(0) = 0$, then

$$\int_0^A |f'|^p \, dt \ge (1 - 1/p)^p \int_0^A |f(t)/t|^p \, dt.$$

Hence

$$\int_S |\nabla u|^p \, dx \ge \int_D \int_0^{A(x')} \left| \frac{\partial u}{\partial x_n} \right|^p dx_n dx' \ge (1 - 1/p)^p \int_D \int_0^{A(x')} |u/x_n|^p \, dx_n dx'$$

$$= (1 - 1/p)^p \int_S |u/x_n|^p \, dx.$$

Now we prove the theorem when Ω is a bounded convex polytope S with open faces $\Gamma_1, ..., \Gamma_q$. We put

$$\Gamma_j' = \partial S \backslash \Gamma_j, \quad S_j = \left\{ x \in S : \text{dist } (x, \Gamma_j) < \text{dist } (x, \Gamma_j') \right\}$$

and $S^* = \cup_{j=1}^q S_j$. If $u \in C_0^\infty(S)$, then by (4.4.12),

$$\int_{S_j} |\nabla u|^p \, dx \ge (1 - 1/p)^p \int_{S_j} |u/d_S(x)|^p \, dx. \tag{4.4.13}$$

Since S is convex, $S \backslash S^*$ has zero measure. The theorem for S is now a consequence of (4.4.11) and (4.4.13).

For the case when Ω is simply a bounded set satisfying the conditions of the theorem, we consider the sequence $\{\Omega_k\}$ mentioned at the beginning of the proof, observe that for each k, $\mu_p(\Omega_k) = (1 - 1/p)^p$ and appeal to (4.4.10). Finally, if Ω is unbounded, approximate it by the increasing sequence $\{\Theta_k\}$, where $\Theta_k = \Omega \cap B(0, k)$. If k is large enough, $x_0 \in \partial\Theta_k$ and there is a neighbourhood U of x_0 such that $U \cap \partial\Omega = U \cap \partial\Theta_k$. The result now follows quite easily, using (4.4.10) and (4.4.11). □

Remark 4.4.11. (i) In [165] it is also shown that when Ω is bounded and satisfies all the conditions of the theorem except that of convexity, then $\mu_p(\Omega) \le (1 - 1/p)^p$, with equality if the variational problem (4.4.8) has no minimiser; if $p = 2$ there is equality if, and only if, there is no minimiser.

(ii) It is clearly of interest to have some idea of how $\mu_p(\Omega)$ behaves when the last theorem does not apply; a reasonable amount of information about this is now available. We list below some results in this direction:

(a) It follows from [150] (see also [191]) that $\mu_p(\mathbb{R}^n \setminus \{0\}) = \left|\frac{n-p}{p}\right|^p$.

(b) Let Ω be a Lipschitz domain with $0 \in \Omega$ and let $\Omega^* = \Omega \setminus \{0\}$. Then, by scale invariance and the possibility of approximating $\mu_p(\mathbb{R}^n \setminus \{0\})$ by means of test functions with supports in arbitrarily small neighbourhoods of 0, we see that

$$\mu_p(\Omega^*) \leq \mu_p(\mathbb{R}^n \setminus \{0\}) = \left|\frac{n-p}{p}\right|^p.$$

It follows that $\mu_n(\Omega^*) = 0$. If $p < n$, then $\mathring{W}_p^1(\Omega) = \mathring{W}_p^1(\Omega^*)$, and since $d(x) \geq d_{\Omega^*}(x)$ we have $\mu_p(\Omega^*) \leq \mu_p(\Omega)$.

(c) Let Ω_1, Ω_2 be bounded Lipschitz domains in $\mathbb{R}^n$ with $\overline{\Omega_1} \subset \Omega_2$; put $\Omega_0 = \mathbb{R}^n \setminus \Omega_1$, $\Omega = \Omega_0 \cap \Omega_2$ and $\Omega^k = \Omega_0 \cap (k\Omega_2)$, $k \in \mathbb{N}$. Set $\mu_{p,i} = \mu_p(\Omega_i)$. Then in [165] it is shown that

$$\frac{\mu_{p,0}\mu_{p,2}}{\mu_{p,0} + \mu_{p,2}} \leq \mu_p(\Omega).$$

If $0 \in \Omega_1$, it is also shown that

$$\lim_{k\to\infty} \mu_p(\Omega_k) = \mu_p(\Omega_0) \leq \min\left\{(1 - 1/p)^p, \left|\frac{n-p}{p}\right|^p\right\}.$$

In particular, $\lim_{k\to\infty} \mu_n(\Omega_k) = 0$. Thus for annular plane domains μ_2 may be arbitrarily small. This is in contrast to the result of Davies [42] that for simply connected plane domains $\mu_2(\Omega) \geq 1/16$.

(d) For radially symmetric domains greater precision is possible as explicit solutions may be constructed. For example, if Ω is either $\mathbb{R}^n \setminus B(0,R)$, $B(0,R) \setminus B(0,r)$ where $0 < r < R$, or $B(0,R) \setminus \{0\}$, then (see [165])

$$\mu_2(\Omega) = \begin{cases} 0 & \text{if } n = 2, \\ 1/4 & \text{if } n \geq 3. \end{cases}$$

4.5 Notes

4.2. The connection between the Poincaré inequality and the measure of non-compactness of an embedding map has its roots in the work of Amick [6]; see also [45], [80] and the account given in [46], V.5. The material given here, relating to first-order spaces, is based on [69] and [72]; for further details of the higher-order situation see [73] and [24]. Chapter 4 of [231] also gives an abstract approach to Poincaré inequalities which leads easily to concrete inequalities.

4.3. There is now a considerable literature on inequalities, such as those of Poincaré and Sobolev, on domains with irregular boundaries. An excellent account of these inequalities, and the connection between them, is given in [116], which deals with Sobolev spaces on metric spaces, a topic also covered in [123]; the paper [129] should be consulted for Poincaré inequalities. See [58] for a connection between Poincaré inequalities and the Minkowski content of the boundary. The mapping property of the Hardy-Littlewood maximal operator that is given here and relating to Sobolev spaces on the whole of $\mathbb{R}^n$ is due to Kinnunen [137]; the corresponding results on open subsets of $\mathbb{R}^n$ were established by Kinnunen and Lindqvist [138]. For an extension of the classical Rellich theorem concerning the compactness of the embedding $id : W_p^1(\Omega) \to L_p(\Omega)$ we refer to [207], where it is shown that if Ω is a $QHBC$ domain and $p \geq n$, then id is compact; in [62] it is proved that for the same class of domains this conclusion is also true if $p \in (p_0, n)$, for a certain $p_0 < n$. For embeddings into spaces of exponential type see [209] and [61]. The book [172] by Maz'ya and Poborchi gives an interesting account of various aspects of the theory of Sobolev spaces on domains with irregular boundaries.

4.4. For regularity properties of the fractional maximal function, see [140], where analogues are given of the mapping properties of the classical Hardy-Littlewood operator which were described in 4.3.

Another version of the Hardy inequality is provided by the mean distance function m of Davies [41], who shows that for all domains $\Omega \subset \mathbb{R}^n$ and all $f \in C_0^\infty(\Omega)$,

$$\int_\Omega \left| \frac{f(x)}{m(x)} \right|^2 dx \leq \frac{4}{n} \int_\Omega |\nabla f(x)|^2 \, dx.$$

He also shows that m is equivalent to the usual distance function d if Ω satisfies a cone condition; see [46], X.6 for a discussion of this and applications to spectral theory. Further remarks on inequalities of Hardy type are contained in [59].

5

Generalised ridged domains

5.1 Introduction

The need for considering Sobolev spaces on scales of general function spaces to fine-tune classical results was demonstrated in Chapter 3. In Chapter 4 it was made clear that the nature of the boundary of the underlying domain Ω has considerable significance in many problems, and therefore a precise description of the boundary is essential for a detailed investigation. The main goal of this chapter is to study embeddings $E = E(\Omega) : W(X,Y)(\Omega) \to Z(\Omega)$, where $X = X(\Omega), Y = Y(\Omega)$ and $Z = Z(\Omega)$ are Banach function spaces defined on a domain $\Omega \subset \mathbb{R}^n, W(X,Y)(\Omega)$ denotes the Sobolev space defined in Section 3.6.1 (with $k = 1$), and the domain Ω is what we call a generalised ridged domain, GRD for short. We shall be mainly concerned with embedding theorems and the measure of non-compactness of $E(\Omega)$. The analysis in Section 4.2 indicated that the measure of non-compactness of the embedding map $E(\Omega)$ is determined by the nature of the boundary of Ω. For GRDs we can go a step further and highlight specific subsets of the boundary of Ω which are responsible for any non-compactness. Furthermore, two-sided estimates can be derived for the measure of non-compactness $\alpha(E)$ which determine whether or not $\alpha(E)$ is 0 (and hence E is compact) or is less than 1. To obtain such estimates which are also manageable in examples was a guiding principle behind the introduction of the GRDs in [80], [82]. It is a very wide class of domains which includes horns, spirals, "rooms and passages" and ones with fractal boundaries like the Koch snowflake. A characteristic feature of a GRD is a central axis called a generalised ridge. This was introduced in [80] and was modelled on the ridge, central set or skeleton of an open subset of $\mathbb{R}^n$. This section will be concerned with the definitions and properties of these sets.

5.1.1 Ridges and skeletons

Let $d_F(x) := \text{dist}(x, F)$ denote the distance from a point $x \in \mathbb{R}^n$ to a closed, non-empty proper subset F of $\mathbb{R}^n$. We shall call the set $N_F(x) := \{y \in F : |y - x| = d_F(x)\}$ the *near set* of x on F, and we shall denote its diameter by $w_F(x)$. The following lemma is an easy consequence of the definition.

Lemma 5.1.1. *(i) For each $x \in F^c := \mathbb{R}^n \setminus F, N_F(x)$ is a compact set.*
(ii) If B is a bounded subset of F^c, then $\cup_{x \in B} N_F(x)$ is bounded.
(iii) Let $\{x_n\}$ be a sequence in F^c which converges to $x \in F^c$. If $y_n \in N_F(x_n)$ for all n and $y_n \to y$, then $y \in N_F(x)$.

Lemma 5.1.2. *The functions w_F and $\theta_F := w_F/d_F$ are upper-semicontinuous on F^c.*

Proof. Since d_F is continuous and positive on F^c it is enough to prove that w_F is upper-semi-continuous.

If $x \in F^c$, there exists a sequence $\{x_n\}$ in F^c which converges to x and is such that

$$w_F(x_n) \to \limsup_{u \to x} w_F(u).$$

By Lemma 5.1.1(i) there exist $y_n^1, y_n^2 \in N_F(x_n)$ such that $w_F(x_n) = |y_n^1 - y_n^2|$. Moreover, by Lemma 5.1.1(i) and (ii), we may suppose that $\{y_n^1\}, \{y_n^2\}$ converge, say to y^1, y^2 respectively. Then, by Lemma 5.1.1(iii), $y^1, y^2 \in N_F(x)$ and we have

$$\limsup_{u \to x} w_F(u) = \lim_{n \to \infty} |y_n^1 - y_n^2| = |y^1 - y^2| \leq w_F(x)$$

The lemma is therefore proved. □

Lemma 5.1.3. *If $x \in F^c, y \in N_F(x)$ and $u = tx + (1-t)y$, where $0 < t < 1$, then $N_F(u) = \{y\}$.*

Proof. Suppose to the contrary that there exists a $y' \in B(u, |y-u|) \cap N_F(u)$. Then

$$\begin{aligned} d(x, y') &\leq d(x, u) + d(u, y') \\ &< d(x, u) + |y - u| \\ &= d(x, y), \end{aligned}$$

which is a contradiction. □

An immediate consequence of Lemma 5.1.3 is

Corollary 5.1.4. *For $x \in F^c$ and $y \in N_F(x)$ let*

$$\lambda := \sup\{t \in (0, \infty) : y \in N_F(y + t[x - y])\}.$$

Then, for all $t \in (0, \lambda), N_F(y + t[x - y]) = \{y\}$.

Theorem 5.1.5. *The function d_F is differentiable at $x \in F^c$ if and only if* card $N_F(x) = 1$, *i.e.* $N_F(x)$ *contains only one element. If d_F is differentiable at x then $\nabla d_F(x) = (x-y)/|x-y|$ where $\{y\} = N_F(x)$; ∇d_F is continuous on its domain of definition.*

Proof. Suppose $N_F(x) = \{y\}$ and $y + k \in N_F(x+h)$. Then

$$\begin{aligned} d_F(x+h)^2 - d_F(x)^2 &= |y+k-x-h|^2 - |y-x|^2 \\ &= 2(x-y)\cdot h + 2(y-x)\cdot k - 2h\cdot k + |h|^2 + |k|^2 \\ &= 2(x-y)\cdot h + \eta, \end{aligned}$$

where $\eta = 2(y-x)\cdot k - 2h\cdot k + |h|^2 + |k|^2$ and $\cdot$ denotes the usual scalar product on $\mathbf{R}^n$. But $|y-x|^2 \le |y+k-x|^2$ and $|y+k-x-h|^2 \le |y-x-h|^2$ so that

$$0 \le 2(y-x)\cdot k + |k|^2$$

and

$$2(y-x-h)\cdot k + |k|^2 \le 0.$$

Therefore

$$-2h\cdot k + |h|^2 \le \eta \le |h|^2,$$

and as $h \to 0, \eta = o(h)$, since $k \to 0$ by Lemma 5.1.1(iii). Thus d_F^2 is differentiable at x with gradient $2(x-y)$ and so d_F is differentiable with

$$\nabla d_F(x) = (x-y)/|x-y|.$$

Conversely, suppose d_F is differentiable at x and that $y \in N_F(x)$. Let $u = tx + (1-t)y$, where $0 < t < 1$. Then $y \in N_F(u)$ by Lemma 5.1.3, and as $|u-x| \to 0$, we have

$$-|u-x| = |u-y| - |x-y| = d_F(u) - d_F(x) = \nabla d_F(x)\cdot(u-x) + o(|u-x|).$$

On dividing through by $1-t$ and letting $t \to 1$ from below, we obtain

$$-|y-x| = \nabla d_F(x)\cdot(y-x).$$

Now $|d_F(x+h) - d_F(x)| \le |h|$ so that $|\nabla d_F(x)| \le 1$. It follows that

$$\nabla d_F(x) = (x-y)/|x-y|$$

and hence

$$y = x - d_F(x)\nabla d_F(x);$$

y is therefore unique. The continuity of ∇d_F on its domain of definition follows from Lemma 5.1.1(iii). □

We now set $F^c = \Omega$, where Ω *contains no half-space.* Unless there is a risk of ambiguity we omit the subscript $F(= \Omega^c)$ in $N_F(x), d_F, w_F$, and θ_F.

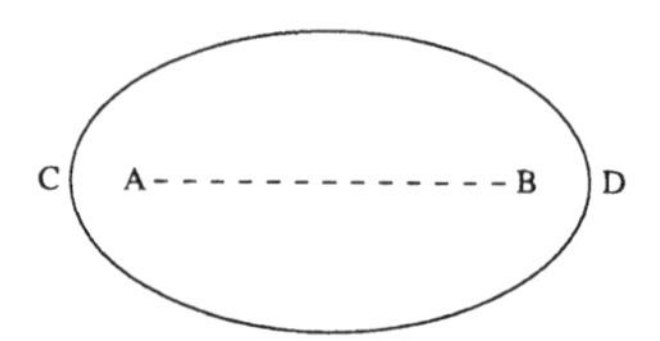

Fig. 5.1. The ridge of an ellipse

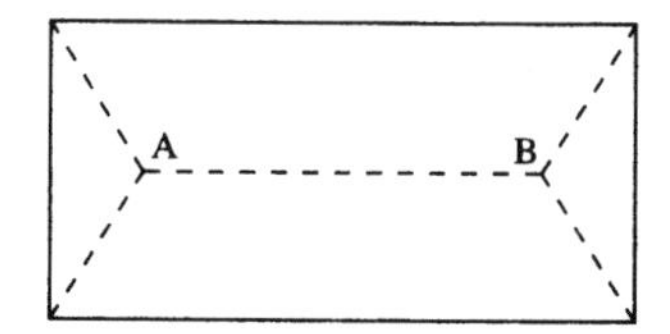

Fig. 5.2. The ridge of a rectangle

Definition 5.1.6. *For $x \in \Omega$ and $y \in N(x)$, the point $p(x) := y + \lambda(x - y)$ in Corollary 5.1.4 is called the* **ridge point** *of x in Ω and the set $\mathcal{R}(\Omega) := \{p(x) : x \in \Omega\}$ is called the* **ridge**, *or* **central set** *of Ω.*

The point of assuming that Ω contains no half-space is to ensure that $p(x)$ is defined for each $x \in \Omega$. A consequence of Lemma 5.1.3 is that card $N(x) = 1$ if x is not a ridge point. However, the converse is false since it is possible to have a ridge point $p(x)$ with card $N(p(x)) = 1$. The ridge of the ellipse in Figure 5.1 provides a simple counterexample. In Figure 5.1 the ridge is the closed line segment AB between the centres of curvature A, B of C, D respectively; $N(A) = \{C\}$, $N(B) = \{D\}$ and card $N(p(x)) = 2$ for $p(x) \neq A, B$. For the rectangle in Figure 5.2, card $N(p(x)) = 3$ for $p(x) = A$ or B and 2 otherwise.

Definition 5.1.7. *The* **skeleton** *of Ω is the set*

$$\mathcal{S}(\Omega) := \{x \in \Omega : \text{card } N(x) > 1\}.$$

Therefore, in view of Theorem 5.1.5, the skeleton of Ω coincides with the set of points in Ω at which the function $d : x \mapsto \text{dist}(x, \partial\Omega)$ is not differentiable. We see from Figure 5.1 that the inclusion $\mathcal{S}(\Omega) \subset \mathcal{R}(\Omega)$ is strict in general; in fact the ridge can be substantially larger than the skeleton, as is shown in [95]. In [95] it is proved that $\mathcal{S}(\Omega) \subset \mathcal{R}(\Omega) \subset \overline{\mathcal{S}(\Omega)}$ for any open set Ω. We note in particular the following properties of a domain Ω from the many interesting results established in [95]:-

1. the sets $\mathcal{S}(\Omega)$ and $\mathcal{R}(\Omega)$ are connected,
2. if Ω is a bounded subset of $\mathbb{R}^2, \mathcal{S}(\Omega)$ is rectifiable,
3. the Hausdorff and Minkowski (and indeed, any usual) dimensions of $\mathcal{S}(\Omega)$ are at most $n - 1$.

Some basic questions remain unresolved. While Rademacher's Theorem 1.2.1 implies that the skeleton, which by Theorem 5.1.5 is the set of points at which the Lipschitz function d ceases to be differentiable, has Lebesgue measure zero, is this also true for the ridge ? Can the ridge have Hausdorff dimension greater than $n-1$?

We now analyse properties of the functions p and $r = d \circ p$ associated with the ridge which help to motivate the definition of a generalised ridged domain (GRD) in Section 5.2 below. Note that

$$\mathcal{R}(\Omega) = \{x \in \Omega : p(x) = x\} = \{x \in \Omega : d(x) = r(x)\}.$$

Lemma 5.1.8. *The function r is upper-semicontinuous on Ω.*

Proof. We prove that, if $\{x_n\}$ is a sequence of points in Ω which converges to $x \in \Omega$ and is such that $\{r(x_n)\}$ tends to a limit, or ∞, then $\lim_{n\to\infty} r(x_n) \le r(x)$. Let $y_n \in N(x_n)$ and

$$\lambda_n = \min\{r(x_n)/d(x_n), 2r(x)/d(x)\},$$

so that $1 \le \lambda_n \le 2r(x)/d(x)$. Then $p(x_n) = y_n + [r(x_n)/d(x_n)](x_n - y_n)$. By Lemma 5.1.1(ii), $\{y_n\}$ is bounded and hence there exists a subsequence $\{y_{n(k)}\}$ converging to y say and such that $\{\lambda_{n(k)}\}$ converges to some $\lambda \ge 1$. Since

$$y_n \in N(y_n + \lambda_n[x_n - y_n])$$

it follows from Lemma 5.1.1(iii) that $y \in N(y + \lambda[x - y])$ and hence $\lambda \le r(x)/d(x)$. Therefore

$$\lim_{n\to\infty} r(x_n) = \lambda d(x) \le r(x)$$

and the lemma is proved. □

For the next theorem we recall that $w(x)$ denotes the diameter of the near set $N(x)$ of $x \in \Omega$ and $\theta(x) = w(x)/d(x)$.

Theorem 5.1.9. *The functions r and p are continuous at $x_0 \in \Omega$ if there exists a neighbourhood of x_0 on which $w \circ p$ (or equivalently $\theta \circ p$) is bounded away from zero.*

Proof. Since r is finite and upper semi-continuous, it is bounded above on compact subsets of Ω and so is $|p|$. Therefore, to prove that p, and consequently $r = d \circ p$, is continuous at x_0 it is sufficient to show that if the sequence $\{x_n\}$ converges to x_0 in Ω and $\{p(x_n)\}$ converges to u then $u = p(x_0)$.

Let $y_n \in N(x_n)$. Then, by Lemma 5.1.1(ii) and (iii), $\{y_n\}$ contains a subsequence which converges to a limit y say, and $y \in N(x_0)$. Since $y_n \in N(p(x_n))$ we also obtain from Lemma 5.1.1(iii) that $y \in N(u)$. But, by Lemma 5.1.2,

$$w(u) \ge \limsup_{n\to\infty} w \circ p(x_n) > 0$$

and so, by Lemma 5.1.3, $u \in \mathcal{R}(\Omega)$, the ridge of Ω. If $u \neq x_0$, the straight line through u and x_0 meets the boundary of Ω in a unique point by Lemma 5.1.3. This point must be y and hence it follows that $u = p(x_0)$. The same conclusion holds if $u = x_0$. □

Theorem 5.1.10. *The functions p and r are continuous on Ω if and only if $\mathcal{R}(\Omega)$ is closed relative to Ω.*

Proof. If $\mathcal{R}(\Omega)$ is closed, the proof of the continuity of p follows that of Theorem 5.1.9. Conversely, if p is continuous, so is the map $x \mapsto p(x) - x$ and hence $\mathcal{R}(\Omega) = \{x \in \Omega : p(x) - x = 0\}$ is closed. □

A corollary of Theorem 5.1.10 is that if Ω is connected and $\mathcal{R}(\Omega)$ is closed then $\mathcal{R}(\Omega)$ is connected. However, as noted earlier, the connectedness of $\mathcal{R}(\Omega)$ is established in [95] for any connected Ω. It is shown in [80] that the ridge of a domain need not be closed, and hence p need not be continuous. The condition in Theorem 5.1.9 is not necessary for the continuity of p as is easily seen in the case of an ellipse Ω.

5.1.2 Simple ridges in $\mathbb{R}^2$

The following notion is prompted by the criterion in Theorem 5.1.9.

Definition 5.1.11. *A ridge point x_0 of $\Omega \subset \mathbb{R}^2$ is said to be* **simple** *if there exist a neighbourhood $U(x_0)$ of x_0 and $\delta > 0$ such that, for every $x \in U(x_0) \cap \mathcal{R}(\Omega)$, $\operatorname{card} N(x) = 2$ and $w(x) \geq \delta$.*

If x_0 is a simple ridge point and $N(x_0) = \{y_1, y_2\}$, it follows from Lemma 5.1.1(iii) that if V_1, V_2 are closed disjoint neighbourhoods of y_1, y_2 on $\partial\Omega$, there is a neighbourhood U of x_0 such that $U \subset U(x_0)$ and $\mathcal{R}(\Omega) \cap U = \{x : d_{V_1}(x) = d_{V_2}(x)\}$, where $d_{V_i}(x)$ denotes the distance from x to V_i. Since the near sets on V_1 and V_2 of each $x \in \mathcal{R}(\Omega) \cap U$ consist of only one element, it follows from Theorem 5.1.5 that d_{V_1} and d_{V_2} are continuously differentiable on $\mathcal{R}(\Omega) \cap U$ although d is not; also

$$n_i := \nabla d_{V_i}(x_0) = (x_0 - y_i)/|x_0 - y_i| \quad (i = 1, 2)$$

is the inward normal to $\partial\Omega$ at y_i. By the Implicit Function Theorem, there exists a neighbourhood V of x_0 such that $\mathcal{R}(\Omega) \cap V$ is the set of points on a C^1 arc with unit tangent t at x_0 perpendicular to $n_1 - n_2$. By the local compactness of $\mathbb{R}^2$, there is a maximal C^1 curve in $\mathcal{R}(\Omega)$ containing x_0; it can be expressed in parametric form $x = u(s), s \in J$, where s is arc length measured from some chosen point of $\mathcal{R}(\Omega)$ and J is an open interval (possibly infinite). The curve is regular in the sense that, for all $s \in J, u'(s) \neq 0$.

Let

$$\Omega_1 := \{x \in \Omega : p(x) = u(s) \ \text{ for some } \ s \in J\}$$

and

$$\mu(s) := \text{meas}\{x \in \Omega_1 : \tau(x) \le s\} \tag{5.1.1}$$

where for $x \in \Omega_1, \tau(x)$ denotes the unique value of s such that $u(s) = p(x)$. Then, by Theorem 5.1.9, $\tau : \Omega_1 \to J$ is continuous. If $\rho = d \circ u$ and $s_0 = \tau(x_0)$, then $\rho = d_{V_1} \circ u$ in a neighbourhood of s_0 and hence, by Theorem 5.1.5, $\rho \in C^1(J)$ with $\rho'(s) = (\nabla d_{V_1} \circ u)(s) \cdot u'(s)$ and $|\rho'| \le 1$. The function μ in (5.1.1) is positive, non-decreasing, continuous on the right and bounded above by $|\Omega|$. With this notation we have

Lemma 5.1.12. *In (5.1.1), the inverse function $s = s(\mu)$ is locally absolutely continuous on J with respect to the measure $d\mu$.*

Proof. Let $u(s), u(\sigma)$ be neighbouring simple ridge points of Ω_1 and, with $x_0 = u(s)$, let the points of V_1 which are nearest $u(s), u(\sigma)$ be denoted by y_1, z_1 respectively. Suppose that the circle centre $u(s)$ which passes through y_1 meets the line from $u(\sigma)$ to z_1 at a point v in Ω. If x lies in the set bounded by the line from $u(s)$ to y_1, the line from $u(\sigma)$ to v, the arc of the ridge from $u(s)$ to $u(\sigma)$, and the circular arc from y_1 to v, then $\tau(x)$ must lie on the ridge between $u(s)$ and $u(\sigma)$. For, by Lemma 5.1.3, the line from x to the nearest point of $\partial\Omega$ cannot intersect the lines from $u(s)$ to y_1 and $u(\sigma)$ to z_1. Thus the measure of the set bounded by the lines from $u(s)$ to y_1 and $u(\sigma)$ to z_1 and the ridge between $u(s)$ and $u(\sigma)$, together with the similar set on the other side of the ridge, does not exceed $|\mu(s) - \mu(\sigma)|$. Thus to first order in $|s - \sigma|$, we have

$$\frac{1}{2}|s - \sigma|(w \circ u)(s) \le |\mu(s) - \mu(\sigma)|$$

and, as $\sigma \to s$,

$$|s - \sigma| \le 2\{(w \circ u)(s)\}^{-1}|\mu(s) - \mu(\sigma)|\{1 + o(1)\},$$

whence the lemma. □

The sets defined for $\varepsilon > 0$ by

$$A_\varepsilon = \{x \in \Omega : (\rho \circ \tau)(x) \le \varepsilon\}$$

have a crucial role in estimating the measure of non-compactness of embedding maps later in this chapter. If $\mathcal{R}(\Omega)$ is closed, the sets A_ε are relatively closed subsets of Ω. If $|\Omega| < \infty$, each $\Omega \backslash A_\varepsilon$ is bounded and satisfies the cone condition, and the set

$$\mathcal{A} := \{A_\varepsilon : \varepsilon > 0\}$$

satisfies the following conditions:

1. for each $\varepsilon > 0$, the embedding

$$W_p^1(\Omega) \hookrightarrow L_p(\Omega \backslash A_\varepsilon) \tag{5.1.2}$$

is compact;

2. $\mathcal{A}$ is finer than the *filter base*

$$\mathcal{A}_0 := \{A : A = \Omega \backslash \Omega', \Omega' \subset\subset \Omega\}. \qquad (5.1.3)$$

The significance of these properties will become apparent in Section 5.2. Recall that a *filter base* $\mathcal{B}$ in Ω is a family of non-empty subsets of Ω which is such that if $\Omega_1, \Omega_2 \in \mathcal{B}$, then there is some $\Omega_3 \in \mathcal{B}$ such that $\Omega_3 \subset \Omega_1 \cap \Omega_2$. We shall later use the notion of convergence along a filter base in our analysis of $\alpha(E)$.

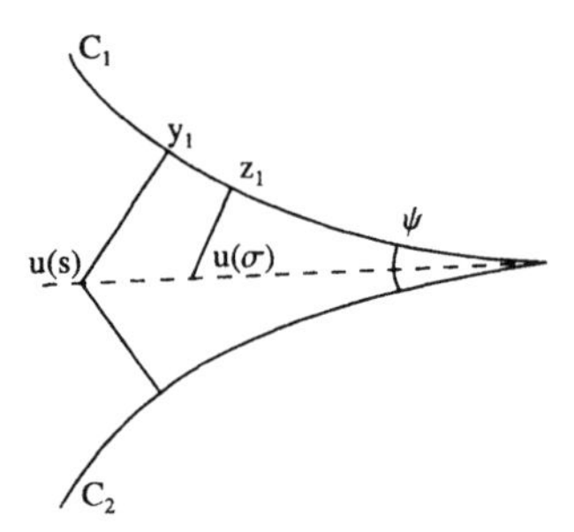

Fig. 5.3. Simple ridge Ω_2

To proceed with our objective of motivating the definition of a GRD, suppose that A_ε is the union of a finite number of sets like the set Ω_2 say in Figure 5.3, in each of which the ridge points are simple and lie on a C^1 curve which meets $\partial\Omega$ at the join of two C^2 arcs C_1, C_2 say, intersecting at an interior angle $\psi \in (0, \pi)$. This can be seen to be the case if $\partial\Omega$ is a bounded simple curve in $\mathbb{R}^2$ which is piecewise C^2, and ε is small enough. Let $\tau : \Omega \to J$ be the map defined above by $u(s) = p(x), \tau(x) = s, s \in J$.

Lemma 5.1.13. *The function τ is Lipschitz continuous if ε is small enough.*

Proof. Let the C^2 arcs bounding Ω_2 be parameterised by arc lengths s_1, s_2 and for $s \in J$ let $y_i = y_i(s_i) \in C_i (i = 1, 2)$ be such that $\{y_1, y_2\} = N(u(s))$. Furthermore, let ε be less than the minimum radius of curvature of C_1 and C_2 and take $\rho \circ \tau(x) \le \varepsilon$ in Ω_2, $\rho = d \circ u$. Then

$$u(s) = y_i(s_i) + \rho(s) n_i(s_i) \quad (i = 1, 2) \qquad (5.1.4)$$

and

$$\frac{dy_i}{ds_i} = t_i, \quad \frac{dt_i}{ds_i} = \kappa_i n_i, \quad \frac{dn_i}{ds_i} = -\kappa_i t_i, \qquad (5.1.5)$$

where $t_i = t_i(s_i)$ is the tangent vector, $n_i = n_i(s_i)$ is the inward normal vector and $\kappa_i = \kappa_i(s_i)$ is the curvature of C_i at $y_i = y_i(s_i)$. We prove firstly that s is a differentiable function of s_i. Define

$$f(s, s_1) = e \cdot \{u(s) - y_1(s_1) - \rho(s) n_1(s_1)\}$$

where e is a constant vector such that $t_1 \cdot e \neq 0$. Then, with $t = t(s) = u'(s)$, the tangent vector to the ridge at $u(s)$, we obtain, on using Theorem 5.1.5,

$$\begin{aligned}\frac{\partial f}{\partial s} &= e \cdot \{t - \rho' n_1\} \\ &= e \cdot \{t - [(\nabla d_{C_1} \circ u) \cdot u'] n_1\} \\ &= e \cdot \{t - (n_1 \cdot t) n_1\} \\ &= e \cdot (t \cdot t_1) t_1 \\ &= (t \cdot t_1)(e \cdot t_1) \\ &= (\sin \phi)(e \cdot t_1) \neq 0,\end{aligned}$$

where 2ϕ is the angle between n_1 and n_2; note that $2\rho \sin \phi = w \circ u$. By the Implicit Function Theorem, we therefore infer that s is a differentiable function of s_1 and, from (5.1.4) and (5.1.5),

$$(u' - \rho' n_1) \frac{ds}{ds_1} = (1 - \rho \kappa_1) t_1,$$

whence

$$\frac{ds}{ds_1} = (1 - \rho \kappa_1)/(t \cdot t_1) = (1 - \rho \kappa_1)/ \sin \phi. \tag{5.1.6}$$

A similar result holds for ds/ds_2.

Let $x, z \in \Omega_1$ be such that $\tau(x) = s, \tau(z) = s + \varepsilon$ and let $y_1(s_1 + \delta)$ be the nearest point to $p(z)$ on C_1. By Lemma 5.1.1(iii) and Theorem 5.1.9, it follows that $\delta \to 0$ and $\varepsilon \to 0$ as $|x - z| \to 0$. Clearly

$$|x - z| \geq \min\{|[u(s + \varepsilon) - u(s)] \cdot t_1|, |[y_1(s_1 + \delta) - y_1(s_1)] \cdot t_1|\}.$$

By (5.1.4) and (5.1.5),

$$\begin{aligned}y_1(s_1 + \delta) - y_1(s_1) &= \delta t_1 + \frac{1}{2}\delta^2 \kappa_1 n_1 + o(\delta^2), \\ u(s + \varepsilon) &= y_1(s_1 + \delta) + \{\rho(s) + O(\varepsilon)\}\{n_1 - \delta \kappa_1 t_1 + O(\delta^2)\},\end{aligned}$$

whence

$$\begin{aligned}\{y_1(s_1 + \delta) - y_1(s_1)\} \cdot t_1 &= \delta + o(\delta^2), \\ \{u(s + \varepsilon) - u(s)\} \cdot t_1 &= (1 - \rho \kappa_1)\delta + O(\delta^2) + O(\varepsilon \delta).\end{aligned}$$

Since $1 - \rho \kappa_1$ is strictly positive, by our initial choice of ε, it follows from these estimates and (5.1.6) that, for z in a neighbourhood of x,

$$|z - x| \geq c\varepsilon |\sin \phi|$$

for some $c > 0$. Hence

$$|\tau(x) - \tau(z)|/|x - z| = O(1/|\sin \phi|) = O(1)$$

since $\phi \to \frac{1}{2}(\pi - \psi) \in (0, \frac{1}{2}\pi)$ as $u(s)$ tends to $\partial \Omega$. The same argument applies to the C_2 side of the ridge of Ω_2 and therefore we have proved that τ is Lipschitz continuous on Ω_2. □

5.2 Generalised ridged domains

The following definition is motivated by the properties of the ridge of an open set established in the last section, especially those in Lemma 5.1.13 and the preceding discussion concerning the sets Ω_1 and Ω_2 . The tree Γ referred to in the definition below is allowed to have a finite or infinite number of edges as long as each vertex is of finite degree, and, in particular, Γ may be an interval $[a,b), \infty < a < b \leq \infty$.

Before proceeding, we recall that, from Rademacher's Theorem (Theorem 1.2.1), a function g which is Lipschitz continuous on an interval is differentiable almost everywhere, and we can define its 'derivative' everywhere by setting

$$g'(t) := \limsup_{n\to\infty}\{n[g(t+n^{-1}) - g(t)\}.$$

Definition 5.2.1. *A domain Ω in $\mathbb{R}^n (n \geq 1)$ with $|\Omega| < \infty$ will be called a* **generalised ridged domain** *(***GRD** *for short) if there exist real-valued functions u, ρ, τ, a tree Γ and positive constants $\alpha, \beta, \gamma, \delta$ such that the following conditions are satisfied :*

(i) $u : \Gamma \to \Omega, \rho : \Gamma \to \mathbb{R}^+ \equiv (0,\infty)$ are Lipschitz;

(ii) $\tau : \Omega \to \Gamma$ is surjective and for each $x \in \Omega$ there exists a neighbourhood $V(x)$ such that for all $y \in V(x), |\tau(x) - \tau(y)|_\Gamma \leq \gamma|x-y|$, where $|\cdot|_\Gamma$ denotes the metric on Γ : thus τ is uniformly locally Lipschitz;

(iii) $|x - u\circ\tau(x)| \leq \alpha(\rho\circ\tau(x))$ for all $x \in \Omega$;

(iv) $|u'(t)| + |\rho'(t)| \leq \beta$ for all $t \in \Gamma$;

(v) with $B_t := B(u(t), \rho(t))$ and $C(x) := \{y : sy + (1-s)x \in \Omega$ for all $s \in [0,1]\}$, we have that for all $x \in \Omega, C(x) \cap B_{\tau(x)}$ contains a ball $B(x)$ such that $|B(x)|/|B_{\tau(x)}| \geq \delta_1 > 0$.

The curve $t \mapsto u(t) : \Gamma \to \Omega$ will be called a **generalised ridge** *of Ω.*

From (v) it follows that for each $\varepsilon > 0$ the set

$$\Omega(\varepsilon) := \{x : x \in \Omega, \rho\circ\tau(x) > \varepsilon\} \tag{5.2.1}$$

lies in a bounded open subset Ω_ε of Ω which satisfies a cone condition, that is, there is a cone $C(\varepsilon)$ such that each $x \in \overline{\Omega_\varepsilon}$ is the vertex of a cone congruent to $C(\varepsilon)$ which lies in $\overline{\Omega_\varepsilon}$. Consequently, the embedding

$$W_p^1(\Omega) \hookrightarrow L_p(\Omega(\varepsilon)) \tag{5.2.2}$$

is compact (see [2], Remark 6.3(4)), and hence if the embedding $W_p^1(\Omega) \hookrightarrow L_p(\Omega)$ fails to be compact, it is due to what happens at the set of points where the ridge meets the boundary of Ω. This gives extra precision to the implication of Remark 4.2.14(*ii*) that the measure of non-compactness of $E(\Omega)$ depends only on the boundary of Ω. For all bounded domains Ω the embedding $W_p^1(\Omega) \hookrightarrow L_p(\Omega')$ is compact for any open set $\Omega' \subset\subset \Omega$. If Ω is a GRD we are therefore able to isolate the singular points on $\partial\Omega$ which give rise to the non-compactness of the embedding $E(\Omega)$.

In Section 5.4 we consider embeddings $E = E(\Omega) : W(X,Y)(\Omega) \to X(\Omega)$, when X, Y are rearrangement-invariant Banach function spaces, and determine the values of $\alpha(E)$ as limits along filter bases. The relevant filter bases are composed of the following sets, their choice being governed by (5.1.2), (5.1.3) and the remarks of the last paragraph. In what follows, we shall adopt the convention of referring to a connected subset of Γ as a subtree : this can be achieved by adding its boundary points, and hence creating new edges from existing ones. Let

$$\begin{aligned}\mathcal{A}(\Gamma) := \{\Lambda : \Lambda \subset \Gamma \ \text{ non-empty and relatively closed}, \overline{\Gamma \setminus \Lambda} \\ \subset \Gamma \ \text{ a compact subtree } \}\end{aligned} \tag{5.2.3}$$

and

$$\mathcal{A}(\Omega) := \{\tau^{-1}(\Lambda) : \Lambda \in \mathcal{A}(\Gamma)\}. \tag{5.2.4}$$

Note that if Γ has an infinite number of edges, then the boundary of $\Gamma \setminus \Lambda$ is finite for any $\Lambda \in \mathcal{A}$, by Lemma 2.6.1, and hence Λ is a finite union of closed, disjoint subtrees of Γ which are rooted at the boundary points of $\Gamma \setminus \Lambda$. If Γ is an interval $[a, b)$ then the sets Λ are of the form

$$\Lambda = \Lambda(\varepsilon) = \begin{cases} [b-\varepsilon, b) & \text{if } b < \infty, \\ [\varepsilon^{-1}, \infty) & \text{if } b = \infty, \end{cases} \tag{5.2.5}$$

for suitable $\varepsilon > 0$.

Lemma 5.2.2. *The set $\mathcal{A}(\Omega)$ in (5.2.4) is a filter base of relatively closed subsets of Ω which satisfy the following conditions ;*

(i) for each $A \in \mathcal{A}(\Omega)$ the embedding $W_p^1(\Omega) \hookrightarrow L_p(\Omega \setminus A)$ is compact;

(ii) $\mathcal{A}(\Omega)$ is finer than the filter base

$$\mathcal{A}_0(\Omega) := \{A : A = \Omega \setminus \Omega', \Omega' \subset\subset \Omega\}.$$

Proof. If $\emptyset \in \mathcal{A}(\Omega)$, then $\tau^{-1}(\Lambda) = \emptyset$ for some $\Lambda \in \Gamma$, which contradicts the assumed surjectivity of τ. Let $A_i = \tau^{-1}(\Lambda_i), i = 1, 2$, belong to $\mathcal{A}(\Omega)$. Define $\overline{\Gamma \setminus \Lambda}$ to be the union of $\overline{\Gamma \setminus \Lambda_1}$ and $\overline{\Gamma \setminus \Lambda_2}$ and a path connecting them. Then $\Lambda \in \mathcal{A}(\Gamma)$ and

$$\tau^{-1}(\Lambda) \subset \tau^{-1}(\Lambda_1) \cap \tau^{-1}(\Lambda_2) = A_1 \cap A_2.$$

Consequently $\mathcal{A}(\Omega)$ is a filter base.

Let $A = \tau^{-1}(\Lambda) \in \mathcal{A}(\Omega)$. Then $\Omega \setminus A = \tau^{-1}(\Gamma \setminus \Lambda)$ and hence $\tau(\Omega \setminus A) \subset \overline{(\Gamma \setminus \Lambda)}$. It follows from Definition 5.2.1 that $\Omega \setminus A$ is bounded and on it $\rho \circ \tau$ is bounded away from zero. Thus by Definition 5.2.1(*v*), $\Omega \setminus A$ lies in a bounded open subset of Ω which satisfies the cone condition and consequently $W_p^1(\Omega \setminus A) \hookrightarrow L_p(\Omega \setminus A)$ is compact. Thus (i) is proved.

Finally, let $\Omega' \subset\subset \Omega$. Then $\tau(\Omega')$ is relatively compact and is contained in a compact subtree of Γ, and hence $\tau(\overline{\Omega'}) \subset (\Gamma')^0$, the interior of some compact subtree Γ' of Γ. Let $\Lambda = \Gamma \setminus (\Gamma')^0 = \overline{\Gamma \setminus (\Gamma')}$. Then $\Lambda \in \mathcal{A}(\Gamma)$ and

$$\tau^{-1}(\Lambda) = \Omega \setminus \tau^{-1}[(\Gamma')^0] \subset \Omega \setminus \overline{\Omega'} \subset \Omega \setminus \Omega'.$$

The lemma is therefore proved. □

Since $\mathcal{A}(\Omega)$ is a filter base it is directed by reverse inclusion, that is, by the order relation $\succ$ where $A_1 \succ A_2$ if $A_1 \subset A_2$. Therefore, if $\{\psi_A\}$ is a family in $\mathbb{R}$ indexed by $\mathcal{A} = \mathcal{A}(\Omega)$, the pair $(\{\psi_A\}, \succ)$ is a *net* in $\mathbb{R}$. It converges to a limit ψ in $\mathbb{R}$, written $\lim_{\mathcal{A}} \psi_A = \psi$, if for each neighbourhood U of ψ in $\mathbb{R}$, there is an $A_0 \in \mathcal{A}$ such that $\psi \in U$ for all $A \succ A_0$ in $\mathcal{A}$. Similarly, $\lim_{\mathcal{A}(\Gamma)} \psi_\Lambda$ can be defined.

The map τ in Definition 5.2.1 enables us to define a positive Borel measure μ on Γ. Since

$$\int_\Omega F \circ \tau(x) dx, \quad F \in C_0(\Gamma)$$

is a positive linear functional on $C_0(\Gamma)$, it follows by the Riesz Representation Theorem for $C_0(a, b)$ that there exists a positive finite measure μ on Γ such that

$$\int_\Gamma F(t) d\mu(t) := \int_\Omega F \circ \tau(x) dx, \quad F \in C_0(\Gamma). \tag{5.2.6}$$

For any open subset Γ_0 of Γ we have

$$\mu(\Gamma_0) = |\tau^{-1}(\Gamma_0)|. \tag{5.2.7}$$

The map $F \mapsto F \circ \tau : C_0(\Gamma) \to L_p(\Omega)$ extends by continuity to a map

$$T : L_p(\Gamma, \mu) \to L_p(\Omega) \tag{5.2.8}$$

which satisfies

$$TF(x) = F \circ \tau(x) \quad \text{for a.e. } x \in \Omega. \tag{5.2.9}$$

Also, if $L_p(\Gamma, \mu)$ is endowed with the standard norm given by

$$\|F\|_{p,\Gamma,\mu} := \left\{ \int_\Gamma |F(t)^p d\mu(t) \right\}^{1/p},$$

then T is an isometry:

$$\|TF\|_{p,\Omega} = \|F\|_{p,\Gamma,\mu}. \tag{5.2.10}$$

In fact μ is given explicitly by the co-area formula of Theorem 1.2.4. For if $\nabla\tau(x)$ does not vanish on a set of positive measure,

$$\int_\Omega (F\circ\tau)(x)dx = \int_\Gamma F(t)\int_{\tau^{-1}(t)} |\nabla\tau(x)|^{-1}dH^{n-1}(x)dt, \qquad (5.2.11)$$

where H^{n-1} denotes $(n-1)$-dimensional Hausdorff measure. Hence μ is locally absolutely continuous with respect to Lebesgue measure and since $|\nabla\tau(x)|\le\gamma$ in Definition 5.2.1(ii),

$$\frac{d\mu}{dt} = \int_{\tau^{-1}(t)} |\nabla\tau(x)|^{-1}dH^{n-1}(x) \ge \frac{1}{\gamma}H^{n-1}(\tau^{-1}(t)). \qquad (5.2.12)$$

Therefore, if $|\nabla\tau(x)|\ne 0$ a.e., dt is locally absolutely continuous with respect to $d\mu$; we shall assume hereafter that this is always satisfied. If $n=2$ and $\tau^{-1}(t)$ is a rectifiable curve in Ω for a.e. $t\in\Gamma$ then its 1-dimensional Hausdorff measure is equal to its length, $l(t)$ say, (see [90], Lemma 3.2) and hence we have

$$\frac{d\mu}{dt} \ge \gamma^{-1}l(t). \qquad (5.2.13)$$

Consequently, dt is absolutely continuous with respect to $d\mu$ on any compact subset of Γ on which $l(\cdot)$ is positive.

Example 5.2.3. (Horn-shaped domain)

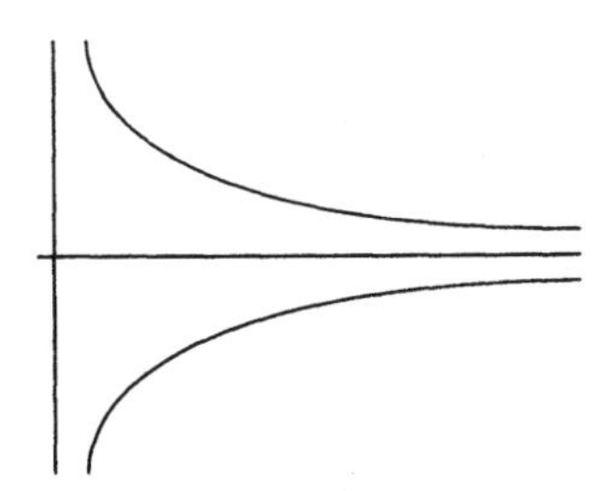

Fig. 5.4. Horn-shaped domain

Let $x=(x_1,x_2,\cdots,x_n)\in\mathbb{R}^n$ be written as (x_1,x'), where $x'=(x_2,\cdots,x_n)\in\mathbb{R}^{n-1}$, and define

$$\Omega := \{x=(x_1,x')\in\mathbb{R}^n : 0<x_1<\infty, |x'|<\Phi(x_1)\}, \qquad (5.2.14)$$

where Φ is smooth and bounded. This is a GRD with the positive half-axis as generalized ridge, so that in the notation of Definition 5.2.1, $u(t)=(t,0,0,\cdots)$, and $\tau(x_1,x')=x_1$. In fact the half-axis is clearly the ridge and skeleton of Ω. We have

$$\int_0^s F(t)d\mu(t) = \int_0^s F(x_1)dx_1 \int_{|x'|<\Phi(x_1)} dx'$$
$$= \omega_{n-1} \int_0^s F(x_1)\Phi(x_1)^{n-1}dx_1,$$

where ω_{n-1} is the measure of the unit ball in $\mathbb{R}^{n-1}$. Consequently,

$$\mu'(t) = \omega_{n-1}\Phi(t)^{n-1}. \tag{5.2.15}$$

Example 5.2.4. (Rooms and passages)

Let $\{h_k\}$ and $\{\delta_k\}$, for $k \in \mathbb{N}$, be infinite sequences of positive numbers such that for some positive constant a,

$$\sum_{k-1}^{\infty} h_k = b < \infty, \quad 0 < a \leq h_{k+1}/h_k \leq 1, \quad 0 < \delta_{2k} \leq h_{2k+1}, \tag{5.2.16}$$

and let

$$H_k := \sum_{j=1}^{k} h_j, \quad k \in \mathbb{N}. \tag{5.2.17}$$

Then the rooms and passages domain Ω is defined to be the union of the rooms R_k and passages P_{k+1} given by

$$\begin{aligned} R_k &= (H_k - h_k, H_k) \times (-\frac{1}{2}h_k, \frac{1}{2}h_k), \\ P_{k+1} &= [H_k, H_k + h_{k+1}] \times (-\frac{1}{2}\delta_{k+1}, \frac{1}{2}\delta_{k+1}), \end{aligned} \tag{5.2.18}$$

for $k = 1, 3, 5, \cdots$; see Figure 5.5.

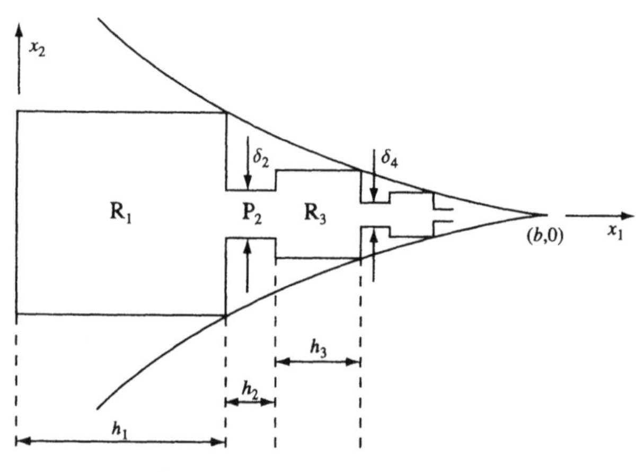

Fig. 5.5. Rooms and passages

We shall see that Ω is a GRD with the interval $[0,b)$ a generalised ridge, $u(t)=(t,0)$ and the following choices for the functions ρ and τ. In a passage P_δ of width δ and length h_δ we set

$$\rho(t)=\delta/2, \quad \tau(x_1,x_2)=x_1, \quad (x_1,x_2)\in P_\delta. \tag{5.2.19}$$

Then $u'(t)=(1,0), \rho'(t)=0$ and $\nabla\tau(x_1,x_2)=(1,0)$. In Definition 5.2.1 we therefore have $\alpha=\gamma=1, \beta=1$ and $\delta_1=1$. Also

$$\mu'(t)=\delta. \tag{5.2.20}$$

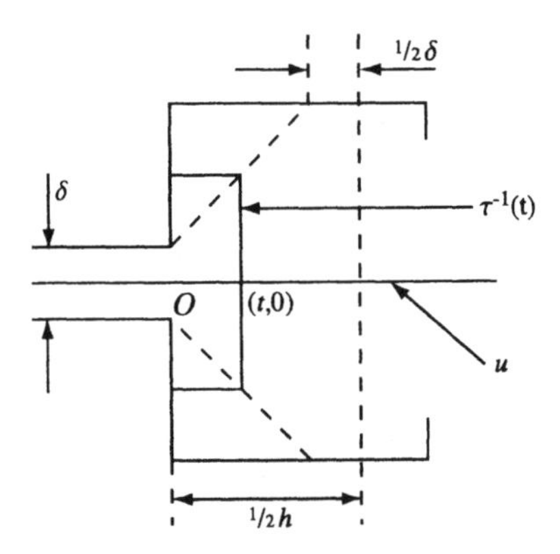

Fig. 5.6. The maps τ and u

In a room R of side h immediately succeeding P_δ the choice is more delicate. First we let (taking O to be the origin in Figure 5.6)

$$\rho(t)=\begin{cases}\delta & \text{if } 0\le t\le \frac{1}{2}\delta,\\ 2t & \text{if } \frac{1}{2}\delta\le t\le \frac{1}{2}h,\end{cases} \tag{5.2.21}$$

and

$$\tau(x_1,x_2)=\max(x_1,|x_2|-\frac{1}{2}\delta) \tag{5.2.22}$$

for $0\le x_1\le \frac{1}{2}h$. Then τ,ρ are Lipschitz and

$$1\le |u'(t)|+|\rho(t)|\le 3, \quad |\nabla\tau(x_1,x_2)|=1.$$

Therefore in Definition 5.2.1, $\gamma=1$ and $\beta=3$. For every $x\in R$, the smallest cone with vertex x and containing $B_{\tau(x)}\cap R$ lies in R and hence we can take $B(x)$ in Definition 5.2.1(v) to have half the radius of $B_{\tau(x)}$ so that $\delta_1=1/4$. If $x_1\le |x_2|-\frac{1}{2}\delta$, then $\tau(x)=|x_2|-\frac{1}{2}\delta$ and

$$|x-u\circ\tau(x)|\le 2\max\left\{\left|x_1-|x_2|+\frac{1}{2}\delta\right|,|x_2|\right\}$$

$$= 2|x_2| \leq 2\rho \circ \tau(x),$$

while if $x_1 > |x_2| - \frac{1}{2}\delta$, then $\tau(x) = x_1$ and

$$|x - u \circ \tau(x)| = 2|x_2| \leq 2\rho \circ \tau(x).$$

Hence $\alpha = 2$.

If $0 < t < \frac{1}{2}(h - \delta)$, then from (5.2.7),

$$\mu(t) - \mu(0) = t(\delta + 2t),$$

whence

$$\mu'(t) = 4t + \delta. \tag{5.2.23}$$

For $\frac{1}{2}(h - \delta) < t < \frac{1}{2}h$, we have $\mu'(t) = h$. If R is immediately followed by a passage P_ε of width ε and length h_ε we proceed in an analogous way and obtain

$$\mu'(t) = \begin{cases} h & \text{if } \frac{1}{2}h < t < \frac{1}{2}(h + \varepsilon), \\ 4(h - t) + \varepsilon & \text{if } \frac{1}{2}(h + \varepsilon) < t < h. \end{cases} \tag{5.2.24}$$

Thus in $P_\delta \cup R \cup P_\varepsilon$ we have

$$\mu'(t) = \begin{cases} \delta & \text{if } -h_\delta < t < 0, \\ 4t + \delta & \text{if } 0 < t < \frac{1}{2}(h - \delta), \\ h & \text{if } \frac{1}{2}(h - \delta) < t < \frac{1}{2}(h + \varepsilon), \\ 4(h - t) + \varepsilon & \text{if } \frac{1}{2}(h + \varepsilon) < t < h, \\ \varepsilon & \text{if } h < t < h + h_\varepsilon. \end{cases} \tag{5.2.25}$$

The ridge (and skeleton) of Ω is clearly made up of line segments and parabolic curves.

Other examples of GRD are given in [80] and [82]. These include spirals, interlocking combs and snowflake-type domains.

5.3 Measure of non-compactness

Let $X = X(\Omega), Y = Y(\Omega)$ be Banach function spaces defined on a domain Ω, which is arbitrary for now, and let $\alpha(E)$ be the measure of non-compactness of the embedding

$$E = E(\Omega) : W(X, Y)(\Omega) \to X(\Omega),$$

i.e.,

$$\alpha(E) = \inf \{\|E - P\| : P \in \mathcal{F}(W(X, Y), X)\}$$

where $\mathcal{F}$ denotes the set of bounded linear maps of finite rank between the indicated spaces and $\|\cdot\|$ is the operator norm. We obviously have

$$0 \le \alpha(E) \le 1$$

and if $\alpha(E) = 0$ then E is compact. The converse is also true if $X(\Omega)$ has the approximation property and then

$$\alpha(E) = \inf \{ \|E - P\| : P \in \mathcal{K}(W(X,Y), X)\}$$

where $\mathcal{K}$ denotes the compact linear operators.

The present section has some overlap with Section 4.2. Indeed the ultimate abstract result in Section 5.3 is Theorem 5.3.9, which proves the equivalence of the Poincaré inequality in the form involving the integral mean and $\alpha(E) < 1$. This is reminiscent of Theorem 4.2.6 and Corollary 4.2.7; see also Remark 4.2.14. Even the proofs have a similar flavour. However, the results here are specially adapted for use with GRDs. The essential difference is that for GRDs, the $\alpha(E)$ is determined in terms of a limit along the filter base (5.2.4), whereas the coarser filter base $\mathcal{A}_0$ of Lemma 5.2.2 has to be used in general, as was done in Section 4.2.

We shall use the notation

$$\|f|X(\Omega')\| := \|\chi_{\Omega'} f|X(\Omega)\|$$

for $\Omega' \subset \Omega$, and for $R \in (0, \infty), \tau_R$ will stand for the dilation operator

$$\tau_R f(x) := f(Rx), \quad x \in \mathbb{R}^n \tag{5.3.1}$$

and $I_{1,R}$ the Riesz potential operator of Remark 3.5.7(i), namely

$$(I_{1,R} f)(x) := \int_{B(x,R)} \frac{|f(y)|}{|y-x|^{n-1}} dy, \tag{5.3.2}$$

where $B(x,R) = \{y \in \mathbb{R}^n : |y - x| < R\}$. If X and Y are rearrangement-invariant we have from [15] that

$$\tau_R : X(\Omega) \to X(\Omega), \quad \tau_R : Y(\Omega) \to Y(\Omega). \tag{5.3.3}$$

We shall always assume that

$$I_{1,1} : Y(\Omega) \to X(\Omega). \tag{5.3.4}$$

Example 5.3.1. Let $0 < p < \infty, 0 < q \le \infty$ and $X(\Omega) = L_{p,q}(\Omega)$. Then it is readily shown that

$$(\tau_R f)^*(t) = f^*(R^n t)$$

and this gives, for $q < \infty$,

$$\|\tau_R f|L_{p,q}(\Omega)\| = \|t^{\frac{1}{p}-\frac{1}{q}} f^*(R^n t)|L_q(0,\infty)\| = R^{-n/p}\|f|L_{p,q}(\Omega)\|.$$

Thus

$$\|\tau_R|X \to X\| = R^{-n/p}$$

when $q < \infty$, and the result continues to hold for $q = \infty$.

Remark 5.3.2. (i) Note that

$$I_{1,R} = R(\tau_{1/R} \circ I_{1,1} \circ \tau_R) \tag{5.3.5}$$

so that (5.3.3) and (5.3.4) imply that $I_{1,R} : Y(\Omega) \to X(\Omega)$.

(ii) Examples of pairs of spaces X, Y for which (5.3.4) is satisfied are: for $p, q \in (1, \infty]$,

$$I_{1,1} : L_p(\Omega) \to L_q(\Omega) \quad \text{if} \quad 0 \le \frac{1}{p} - \frac{1}{q} < \frac{1}{n};$$

$$I_{1,1} : L_{p,r}(\Omega) \to L_{q,r}(\Omega) \quad \text{if} \quad 0 \le \frac{1}{p} - \frac{1}{q} < \frac{1}{n}, \quad 1 \le r \le \infty.$$

These and many other examples of such pairs may be found in [88], Section 12.1.

Lemma 5.3.3. *Let $X = X(\Omega), Y = Y(\Omega)$ be Banach function spaces satisfying (5.3.3) and (5.3.4). If a sequence $\{f_m\} \subset W(X,Y)$ is such that $\|f_m|X\| = 1$ and $\|\nabla f_m|Y\| \to 0$ as $m \to \infty$, then there exists a subsequence $\{f_{m(j)}\}$ and a constant c which satisfy $\|f_{m(j)} - c|X(\Omega')\| \to 0$ for every $\Omega' \subset\subset \Omega$. If $\chi_\Omega \notin X$ then $c = 0$.*

Proof. Let $\Omega = \bigcup_{n=1}^{\infty} Q_n^o$, where the Q_n^o are the interiors of closed cubes Q_n, and, for any cube Q and $f \in W(X,Y)$, set $Pf = \chi_Q f_Q$ where

$$f_Q = |Q|^{-1} \int_Q f(x)dx.$$

Then, by Definition 3.1.1 (*P5*),

$$\begin{aligned} \|Pf|X(\Omega)\| &= |Q|^{-1} \left| \int_Q f(x)dx \right| \|\chi_Q|X(\Omega)\| \\ &\le \left(|Q|^{-1} C_Q \|\chi_Q|X(\Omega)\|\right) \|f|X(\Omega)\| \\ &\le C'_Q \|f|W(X,Y)\|, \end{aligned} \tag{5.3.6}$$

where C_Q, C'_Q are positive constants depending on Q. Hence P is a bounded operator from $W(X,Y)$ to X of finite rank, and so $P \in \mathcal{K}(W(X,Y), X)$. Since Q is a John domain, it follows from Theorem 4.3.29 that for all $x \in Q$,

$$\begin{aligned} |f(x) - f_Q| &\lesssim \int_Q \frac{|\nabla f(y)|}{|x-y|^{n-1}} dy \\ &\lesssim \int_{B(x,R)} \frac{|\nabla f(y)|}{|x-y|^{n-1}} dy = (I_{1,R}|\nabla f|)(x), \end{aligned}$$

where $R = \text{diam} Q \approx |Q|^{1/n}$: here and hereafter we set $|\nabla f| = 0$ outside Ω. It follows that

$$\begin{aligned}\|f - f_Q|X(Q)\| &= \|\chi_Q(f - f_Q)|X(\Omega)\| \\ &\lesssim \|\chi_Q(I_{1,R}|\nabla f|)|X(\Omega)\| \\ &\lesssim \|I_{1,R}(|\nabla f|)|X(\Omega)\| \\ &\lesssim \|I_{1,R}|Y(\Omega) \to X(\Omega)\|\|\nabla f|Y(\Omega\|.\end{aligned}$$

Thus, by hypothesis and (5.3.5), $f_m - Pf_m \to 0$ in $X(Q)$. Since $P \in \mathcal{K}(W(X,Y),X)$ and $\{f_m\}$ is bounded in $W(X,Y)$, there exists a subsequence $\{f_{m(j)}\}$ such that $\{Pf_{m(j)}\}$ converges to a limit ϕ, say, in $X(\Omega)$, and hence in $X(Q)$. This in turn implies that $f_{m(j)} \to \phi$ in $X(Q)$ and since P is continuous on $X(Q)$ (see the inequality preceding (5.3.6)), $Pf_{m(j)} \to P\phi$ in $X(Q)$. Consequently $\phi = P\phi$, a constant, in $X(Q)$. We now proceed by the usual diagonalization procedure. There exists a $\phi_1 : \mathbb{N} \to \mathbb{N}$ with ϕ strictly increasing, such that

$$\|f_{\phi(j)} - c_1|X(Q_1)\| \to 0,$$

where c_1 is a constant. Define by induction sequences $\{\phi_m\}, \{c_m\}$ such that

$$\lim_{j\to\infty} \|f_{\phi_m \circ \phi_{m-1} \circ \cdots \circ \phi_1(j)} - c_m|X(Q_m)\| = 0.$$

Let $\psi_m = \phi_m \circ \phi_{m-1} \circ \cdots \circ \phi_1(m)$. Then, for all j, $\lim_{m\to\infty} \|f_{\psi_m} - c_m|X(Q_j)\| = 0$. If $Q_{j_1} \cap Q_{j_2} \neq \emptyset$, then $c_{j_1} = c_{j_2}$ and so, by the connectedness of Ω, the c_{j_i} are all equal. Since every Ω' is contained in a finite union of the Q_n^o the first part of the result follows.

Let $\Omega_l, l = 1, 2, \cdots$, be bounded domains such that $\Omega_l \subset\subset \Omega_{l+1} \subset\subset \Omega$ and $\Omega = \bigcup_{i=1}^\infty \Omega_l$. Then $\{\chi_{\Omega_l}\}$ is increasing and converges to χ_Ω on Ω. Therefore by property *(P3)* in Definition 3.1.1, $\|\chi_{\Omega_l}|X(\Omega)\| \to \|\chi_\Omega|X(\Omega)\|$. If $\chi_\Omega \notin X$ we must therefore have $c = 0$ since

$$\begin{aligned}c\|\chi_{\Omega_l}|X(\Omega)\| &\leq \|f_{m(j)} - c|X(\Omega_l)\| + \|f_{m(j)}|X(\Omega_l)\| \\ &\leq \text{constant}.\end{aligned}$$

Note that we have used the notation $\|f|X(\Omega')\|$ for $\|\chi_{\Omega'} f|X(\Omega)\|$. □

We shall have need of the following general version of a result used in the proof of Theorem 2.3.1.

Lemma 5.3.4. *Let $X = X(\Omega), Y = Y(\Omega)$ be Banach function spaces and suppose that $C_0^\infty(\Omega)$ is dense in X. Given $\varepsilon > 0$ and $P \in \mathcal{F}(W(X,Y),X)$, the set of all bounded linear operators of finite rank, there exists $R \in \mathcal{F}(W(X,Y),X)$ and $\Omega' \subset\subset \Omega$ such that $\|P - R|W(X,Y) \to X\| < \varepsilon$ and the range of R lies in $C_0^\infty(\Omega')$. If $\Omega_0 \subset \Omega, P \in \mathcal{F}(W(X,Y),X)$ and $\varepsilon > 0$ are given, there exists $R \in \mathcal{F}(W(X,Y),X)$ such that $\|(P - R)|X(\Omega_0)\| < \varepsilon\|f|W(X,Y)\|$ and the range of R is a subset of $C_0^\infty(\Omega_0)$.*

Proof. As P is of finite rank, there exist linearly independent functions $u_i \in X(\Omega), i = 1, 2, \cdots, N$ such that for all $f \in W(X, Y)$,

$$Pf = \sum_{i=1}^{N} c_i(f) u_i.$$

Since all norms are equivalent on the finite-dimensional range of P, we have

$$\sum_{i=1}^{N} |c_i(f)| \leq K\|Pf|X\| \leq K\|P\|\|f|W(X,Y)\|.$$

Choose $\phi_i \in C_0^\infty(\Omega)$ such that $\|u_i - \phi_i|X\| < \varepsilon/K\|P\|, i = 1, 2, \cdots, N$, and set $Rf = \sum_{i=1}^{N} c_i(f)\phi_i$. Then, supp $Rf \subset \bigcup_{i=1}^{N}$ supp $\phi_i \subset\subset \Omega$,

$$\begin{aligned} \|Rf|X\| &\leq \sum_{i=1}^{N} |c_i(f)|\|\phi_i|X\| \\ &\leq K\|P\|\big(\max_{1\leq i\leq N} \|\phi_i|X\|\big)\|f|W(X,Y)\| \end{aligned}$$

and

$$\begin{aligned} \|Pf - Rf|X\| &\leq \sum_{i=1}^{N} |c_i(f)|\|u_i - \phi_i|X\| \\ &< \varepsilon\|f|W(X,Y)\|. \end{aligned}$$

The rest of the lemma is proved in a similar way. □

Theorem 5.3.5. *Let $X(\Omega), Y(\Omega)$ be Banach function spaces which satisfy (5.3.3) and (5.3.4), and suppose that $C_0^\infty(\Omega)$ is dense in $X(\Omega)$. Suppose further that $\chi_\Omega \notin X(\Omega)$. Then if $\alpha(E) < 1$, the inequality (4.2.13) holds, i.e., there exists a positive constant K such that for all $f \in W(X,Y)(\Omega)$,*

$$\|f|X(\Omega)\| \leq K\|\nabla f|Y(\Omega)\|. \tag{5.3.7}$$

Proof. Suppose that $\alpha(E) < 1$. Then there exists $P \in \mathcal{F}(W(X,Y), X)$ and $k < 1$ such that

$$\|f - Pf|X(\Omega)\| \leq k\big(\|\nabla f|Y(\Omega)\| + \|f|X(\Omega)\|\big)$$

for all $f \in W(X,Y)(\Omega)$. In view of Lemma 5.3.4 we may suppose that $Pf(x) = 0$ outside a set $\Omega' \subset\subset \Omega$. Hence

$$\|f|X(\Omega \setminus \Omega')\| \leq k\big(\|\nabla f|Y(\Omega)\| + \|f|X(\Omega \setminus \Omega')\| + \|f|\Omega'\|\big)$$

and so

$$\|f|X(\Omega \setminus \Omega')\| \leq k(1-k)^{-1}\big(\|\nabla f|Y(\Omega)\| + \|f|X(\Omega')\|\big). \tag{5.3.8}$$

This implies the inequality (5.3.7). For otherwise there exists a sequence $\{f_m\}$ satisfying the hypothesis of Lemma 5.3.3 and so we have that a subsequence $\{f_{m(j)}\}$ exists which converges to 0 in $X(\Omega')$. But then (5.3.8) implies that $f_{m(j)} \to 0$ in $X(\Omega)$ contrary to $\|f_{m(j)}|X(\Omega)\| = 1$. □

Remark 5.3.6. Theorem 5.3.5 is similar to Lemma 4.2.4, with $\alpha(E)$ in place of the quantity A of (4.2.6) which is shown in Remark 4.2.14 to equal $\beta(E)$ when $X(\Omega)$ has absolutely continuous norm; see [46], Theorem V.5.7 for cases in which $\alpha(E) = \beta(E)$. In the case of $X(\Omega) = Y(\Omega) = L_p(\Omega)$ the inequality (5.3.7) is impossible when $|\Omega| = \infty$ and hence $|\Omega| = \infty$ implies $\alpha(E) = 1$. To see this we argue as follows.

For each $k \in \mathbb{N}$ put $B_k := \{x \in \mathbb{R}^n : |x| \leq k\}, \Omega_k = \Omega \cap B_k$, and let $u_k \in C_0^\infty(\mathbb{R}^n)$ be such that

$$u_k(x) = \begin{cases} 1, & x \in B_k, \\ 0, & x \notin B_{k+1} \end{cases}$$

with $0 \leq u_k(x) \leq 1$ and $|\nabla u_k(x)| \leq 2$ for all $x \in \mathbb{R}^n$. Then the restriction of u_k to Ω lies in $W_p^1(\Omega)$, so that if (5.3.7) holds,

$$|\Omega_k| \leq \int_\Omega |u_k(x)|^p \leq K^p \int_\Omega |\nabla u_k(x)|^p dx \leq (2K)^p |\Omega_{k+1} \setminus \Omega_k|.$$

Thus, with $\Omega_0 = \emptyset$, we have

$$\sum_{j=0}^{k-1} |\Omega_{j+1} \setminus \Omega_j| \leq (2K)^p |\Omega_{k+1} \setminus \Omega_k|$$

and hence

$$\left(1 + (2K)^{-p}\right) \sum_{j=0}^{k-1} |\Omega_{j+1} \setminus \Omega_j| \leq \sum_{j=0}^{k} |\Omega_{j+1} \setminus \Omega_j|.$$

With $A_k = \sum_{j=0}^k |\Omega_{j+1} \setminus \Omega_j|$ this gives

$$A_k \geq \left(1 + (2K)^{-p}\right) A_{k-1} \geq \left(1 + (2K)^{-p}\right)^{k-1} A_1.$$

But $A_k \leq |B_{k+1}| = \omega_n (k+1)^n$, and so

$$\omega_n (k+1)^n \geq \left(1 + (2K)^{-p}\right)^{k-1} A_1$$

for all $k \in \mathbb{N}$. Since this is clearly impossible, our assertion follows.

Theorem 5.3.7. *Suppose that $X(\Omega)$ satisfies the bounded approximation property and that $C_0^\infty(\Omega)$ is dense in $X(\Omega)$. Let Ξ be a set of measurable subsets of Ω which satisfy the following conditions:*

1. *each compact subset of Ω is contained in some $X \in \Xi$;*

2. *for every $X \in \Xi$, the map $E_X : f \mapsto \chi_X f$ is in $\mathcal{K}(W(X,Y)(\Omega), X(\Omega))$.*

Then

$$\alpha(E) = \inf_{X \in \Xi} \|E - E_X | W(X,Y) \to X\|. \tag{5.3.9}$$

Proof. Clearly

$$\alpha(E) \leq \inf_{X \in \Xi} \|E - E_X\|.$$

If $\alpha > \alpha(E)$ there are, by Lemma 5.3.4, a $P \in \mathcal{F}(W(X,Y), X)$ and an $\Omega' \subset\subset \Omega$ such that for all $f \in W(X,Y)(\Omega)$,

$$\|Ef - Pf | X(\Omega)\| \leq \alpha \|f | W(X,Y)(\Omega)\|$$

and

$$\operatorname{supp} Pf \subset \Omega'.$$

Then $\|\chi_{\Omega \setminus \Omega'} f | X(\Omega)\| \leq \alpha \|f | W(X,Y)(\Omega\|$ and by hypothesis 2, there is an $X_1 \in \Xi$ such that

$$\begin{aligned} \|(E - E_{X_1}) f | X(\Omega)\| &= \|\chi_{\Omega \setminus X_1} f | X(\Omega)\| \\ &\leq \|\chi_{\Omega \setminus \Omega'} f | X(\Omega)\| \\ &\leq \alpha \|f | W(X,Y)(\Omega)\|. \end{aligned}$$

Thus $\|E - E_{X_1}\| \leq \alpha$ and the theorem follows. □

In our applications of Theorem 5.3.7 we need to have at our disposal the following family $\mathcal{A}$ of subsets of Ω which constitute a filter base if, and only if, E is not compact. It is the analogue of the filter base in Lemma 5.2.2 for Banach function spaces. The family $\mathcal{A}$ consists of all those relatively closed subsets of Ω which satisfy the following conditions:

1. for each $A \in \mathcal{A}$, the embedding

$$W(X,Y)(\Omega) \to X(\Omega \setminus A) \tag{5.3.10}$$

is compact;

2. $\mathcal{A}$ is finer than the filter base

$$\mathcal{A}_0 := \{A : A = \Omega \setminus \Omega', \Omega' \subset\subset \Omega\}. \tag{5.3.11}$$

Note that $\emptyset \notin \mathcal{A}$ if, and only if, E is not compact. The following result is the core of our method for estimating $\alpha(E)$.

Corollary 5.3.8. *Let the hypothesis of Theorem 5.3.7 be satisfied and let $\mathcal{A}$ be a filter base. For $A \in \mathcal{A}$ define*

$$\psi_A := \sup\left\{\frac{\|f + Hf | X(A)\|}{\|f | W(X,Y(\Omega)\|}; f \in W(X,Y)(\Omega), f \neq 0\right\}, \tag{5.3.12}$$

where $H \in \mathcal{K}(W(X,Y), X)$ is fixed. Then ψ_A decreases with A and $\alpha(E) = \lim_{\mathcal{A}} \psi_A$.

Proof. It is clear that ψ_A is decreasing, i.e. $\psi_{A_1} \leq \psi_{A_2}$ if $A_1 \subset A_2$. The result then follows from the theorem and its proof on using the fact that $\alpha(E) = \alpha(E + H)$. □

Theorem 5.3.9. *Suppose that $|\Omega| < \infty$, (5.3.3) and (5.3.4) are satisfied, and $C_0^\infty(\Omega)$ is dense in $X(\Omega)$. Then $\alpha(E) < 1$ if, and only if, there exists a positive constant K such that, for all $f \in W(X,Y)(\Omega)$,*

$$\|f - f_\Omega | X(\Omega)\| \leq K \|\nabla f | Y(\Omega)\|, \tag{5.3.13}$$

where $f_\Omega = |\Omega|^{-1} \int_\Omega f(x)dx$.

Proof. In the proof of Theorem 5.3.5 we saw that $\alpha(E) < 1$ implies that (5.3.8) is satisfied for some $k < 1, \Omega' \subset\subset \Omega$ and for all $f \in W(X,Y)(\Omega)$. Suppose that (5.3.13) is not satisfied. Then there exists a sequence $\{f_m\}$ in $W(X,Y)(\Omega)$ such that $g_m = f_m - (f_m)_\Omega$ satisfies

$$\|\nabla g_m | Y(\Omega)\| \to 0, \quad \|g_m | X(\Omega)\| = 1, \quad (g_m)_\Omega = 0. \tag{5.3.14}$$

By Lemma 5.3.3, a subsequence $\{g_{m(k)}\}$ converges to a constant c in $X(\Omega')$ and hence on using (5.3.8) with $f = g_{m(k)} - c$, we deduce that $g_{m(k)} \to c$ in $X(\Omega)$. By (5.3.14),

$$c = \lim_{k\to\infty} (g_{m(k)})_\Omega = 0,$$

which contradicts $\|g_{m(k)} | X(\Omega)\| = 1$. Therefore we conclude that (5.3.13) is valid.

Suppose now that (5.3.13) is satisfied. The map $f \mapsto f_\Omega$ is in the space of bounded linear finite rank operators $\mathcal{F}(W(X,Y)(\Omega), X(\Omega))$ and hence by Lemma 5.3.4, given $\varepsilon > 0$, there exists an $R \in \mathcal{F}(W(X,Y)(\Omega), X(\Omega))$ and $A \in \mathcal{A}$ (since $\mathcal{A}$ is finer than the family $\mathcal{A}_0$ in (5.3.11)) such that for all $f \in W(X,Y)(\Omega)$,

$$\|f_\Omega - Rf | X(\Omega)\| < \varepsilon \|f | W(X,Y)(\Omega)\|, \quad \text{supp } Rf \subset \Omega \setminus A.$$

Hence by (5.3.13),

$$\begin{aligned} \|f - Rf | X(\Omega)\| &\leq K \|\nabla f | Y(\Omega)\| + \varepsilon \|f | W(X,Y)(\Omega)\| \\ &\leq (K + \varepsilon) \|\nabla f | Y(\Omega)\| + \varepsilon \|f | X(\Omega)\|. \end{aligned}$$

Therefore

$$\|f | X(A)\| \leq (K + \varepsilon) \|\nabla f | Y(\Omega)\| + \varepsilon \|f | X(\Omega)\|$$

and so

$$\|f | X(A)\| \leq [\frac{(K + \varepsilon)}{(1 + K)}] \|f | W(X,Y)(\Omega)\|.$$

It now follows from Corollary 5.3.8 that

$$\alpha(E) \leq [\frac{(K + \varepsilon)}{(1 + K)}] < 1.$$

□

Suppose that the hypotheses of Theorem 5.3.9 hold and define the space

$$W_M(X,Y)(\Omega) := \{f \in W(X,Y)(\Omega) : f_\Omega = 0\} \tag{5.3.15}$$

with norm

$$\|f|W_M(X,Y)(\Omega)\| := \|\nabla f|Y((\Omega)\|. \tag{5.3.16}$$

By (5.3.13), (5.3.16) defines a norm on $W_M(X,Y)(\Omega)$ which is equivalent to the norm $\|\cdot|W(X,Y)(\Omega)\|$ inherited from $W(X,Y)(\Omega)$, and we have the topological isomorphism

$$W(X,Y)(\Omega) = \mathbb{C}\dot{+}W_M(X,Y)(\Omega).$$

Define the natural embedding

$$E_M = E_M(\Omega) : W_M(X,Y)(\Omega) \hookrightarrow X(\Omega), \tag{5.3.17}$$

and set

$$\alpha(E_M) := \inf\{\|E_M - P\| : P \in \mathcal{F}(W_M(X,Y)(\Omega), X_M(\Omega))\}. \tag{5.3.18}$$

Note that E_M actually maps $W(X,Y)(\Omega)$ into the space

$$X_M(\Omega) := \{f \in X(\Omega) : f_\Omega = 0\}. \tag{5.3.19}$$

The space $X_M(\Omega)$ with the $X(\Omega)$ norm is topologically isomorphic to the quotient space $X(\Omega)/\mathbb{C}$ with norm $\|[f]|X(\Omega)/\mathbb{C}\| = \inf_{c\in\mathbb{C}} \|f - c|X(\Omega)\|$; recall that the quotient space consists of equivalence classes $[\cdot]$, two elements belonging to the same equivalence class if, and only if, they differ by a constant. We have for all $f \in X(\Omega)$,

$$\inf_{c\in\mathbb{C}} \|f - c|X(\Omega)\| \le \|f - f_\Omega|X(\Omega)\| \le K(\Omega) \inf_{c\in\mathbb{C}} \|f - c|X(\Omega)\|, \tag{5.3.20}$$

where

$$K(\Omega) = 1 + |\Omega|^{-1}\|\chi_\Omega|X'(\Omega)\|\|\chi_\Omega|X(\Omega)\| \tag{5.3.21}$$

and X' denotes the associate space of X. The first inequality in (5.3.20) is obvious, while the second follows from

$$\|f - f_\Omega|X(\Omega)\| \le \|f - c|X(\Omega)\| + |c - f_\Omega|\|\chi_\Omega|X(\Omega)\|$$

since, by Hölder's inequality (Theorem 3.1.8),

$$\begin{aligned}|c - f_\Omega| &= |\Omega|^{-1}\Big|\int_\Omega [c - f(x)]\,dx\Big| \\ &\le |\Omega|^{-1}\|\chi_\Omega|X'(\Omega)\|\|f - c|X(\Omega)\|.\end{aligned}$$

Note that $\chi_\Omega \in X(\Omega) \cap X'(\Omega)$ by properties (*P4*) and (*P5*) in Definition 3.1.1, since $|\Omega| < \infty$.

In analogy with Theorem 5.3.7 we have

Theorem 5.3.10. *Suppose that $X(\Omega)$ has the bounded approximation property and that $C_0^\infty(\Omega)$ is dense in $X(\Omega)$. Let Ξ be a set of measurable subsets of Ω which satisfy the following conditions:*

1. *each compact subset of Ω is contained in some $X \in \Xi$;*
2. *for every $X \in \Xi$, the map $E_X : f \mapsto \chi_X f$ is in $\mathcal{K}(W_M(X,Y)(\Omega), X(\Omega))$.*

Then

$$\alpha(E_M) = \inf_{X \in \Xi} \|E_M - E_X|W(X,Y) \to X\|. \tag{5.3.22}$$

In the following corollary, $\mathcal{A}$ is the filter base in Corollary 5.3.8.

Corollary 5.3.11. *Let the hypothesis of Theorem 5.3.7 be satisfied and let $\mathcal{A}$ be a filter base. For $A \in \mathcal{A}$ define*

$$\phi_A := \sup\left\{ \frac{\|f + Hf|X(A)\|}{\|f|W_M(X,Y)(\Omega)\|} : f \in W_M(X,Y)(\Omega), f \neq 0 \right\}, \tag{5.3.23}$$

where $H \in \mathcal{K}(W_M(X,Y), X)$ is fixed. Then ϕ_A decreases with A and $\alpha(E_M) = \lim_{\mathcal{A}} \phi_A$.

If E_M is bounded, then $\alpha(E_M) < \infty$. The next theorem shows that in this case $\alpha(E) < 1$. We shall exploit this fact in Section 5.7 to determine whether or not $\alpha(E) < 1$.

Theorem 5.3.12. *Let $|\Omega| < \infty$ and suppose that $E_M(\Omega)$ is bounded. Then*

$$\frac{\alpha(E_M)}{1 + \|E_M\|} \leq \alpha(E) \leq \frac{\alpha(E_M)}{1 + \alpha(E_M)}. \tag{5.3.24}$$

In particular $\alpha(E)$ and $\alpha(E_M)$ are zero together. Thus if $X(\Omega)$ has the bounded approximation property, E is compact if, and only if, E_M is compact.

Proof. Given $\delta > 0$ there exists a $P \in \mathcal{F}(W(X,Y)(\Omega), X(\Omega))$ such that for all $f \in W(X,Y)(\Omega)$,

$$\|f - Pf|X(\Omega)\| \leq (\alpha(E) + \delta)\|f|W(X,Y)(\Omega)\|.$$

Since $g = f - f_\Omega \in W_M(X,Y)(\Omega)$ and $g \mapsto Pg \in \mathcal{F}(W_M(X,Y), X(\Omega))$, it follows from

$$\|g - Pg|X(\Omega)\| \leq (\alpha(E) + \delta)\{\|\nabla g|Y(\Omega)\| + \|E_M\|\|\nabla g|Y(\Omega)\|\}$$

that $\alpha(E_M) \leq \alpha(E)\{1 + \|E_M\|\}$.

Next choose $P \in \mathcal{F}(W_M(X,Y)(\Omega), X(\Omega))$ such that for all $g \in W_M(X,Y)(\Omega)$,

$$\|g - Pg|X(\Omega)\| < (\alpha(E_M) + \delta)\|\nabla g|Y(\Omega)\|.$$

Let $g = f - f_\Omega, f \in W(X,Y)(\Omega)$, and let $P_1 f = P(f - f_\Omega) + f_\Omega$. Then

$$P_1 \in \mathcal{F}(W(X,Y)(\Omega), X(\Omega))$$

and for all $f \in W(X,Y)(\Omega)$,

$$\|f - P_1 f|X(\Omega)\| \leq (\alpha(E_M) + \delta)\|\nabla f|Y(\Omega)\|.$$

Hence, $N(f) := \inf\{\|f - P_1 f|X(\Omega)\| : P_1 \in \mathcal{F}(W(X,Y)(\Omega), X(\Omega))\|$ satisfies

$$\begin{aligned} N(f) &\leq \frac{(\alpha(E_M) + \delta)}{(1 + [\alpha(E_M) + \delta])}\{\|\nabla f|Y(\Omega)\| + N(f)\} \\ &\leq \frac{(\alpha(E_M) + \delta)}{(1 + [\alpha(E_M) + \delta])}\{\|\nabla f|Y(\Omega)\| + \|f|X(\Omega)\|\} \end{aligned}$$

as $N(f) \leq \|f|X(\Omega)\|$. Since δ is arbitrary the proof is complete. □

5.4 Analysis on GRD

In this section Ω is assumed to be a GRD and $X(\Omega), Y(\Omega)$ rearrangement-invariant BFSs. For every measurable subset Γ_0 of the generalised ridge Γ of Ω,

$$\mu(\Gamma_0) = |\tau^{-1}(\Gamma_0)| \tag{5.4.1}$$

where μ is the measure defined by the Lipschitz map $\tau : \Omega \to \Gamma$ in Section 5.2. It follows that if T is the map defined in (5.2.9), namely, $TF = F \circ \tau$, then

$$\mu\{t \in \Gamma : |F(t)| > \lambda\} = |\{x \in \Omega : |TF(x)| > \lambda\} \tag{5.4.2}$$

and

$$F_\mu^*(t) = (TF)^*(t). \tag{5.4.3}$$

This motivates the following.

Definition 5.4.1. *Let $X = X(\Omega)$ be a rearrangement-invariant BFS on the GRD Ω. The* **representation space** *$\tilde{X} = \tilde{X}(\Gamma, \mu)$ is the rearrangement-invariant BFS defined on Γ by*

$$\tilde{X} \equiv \tilde{X}(\Gamma, \mu) := \{F : TF = F \circ \tau \in X(\Omega)\} \tag{5.4.4}$$

with norm

$$\|F|\tilde{X}(\Gamma, \mu)\| = \|TF|X(\Omega)\| = \|(TF)^*|X(0, |\Omega|)\|. \tag{5.4.5}$$

The last norm *represents* that of $X(\Omega)$ in the sense described in [15], Theorem II.4.10 (see also Theorem 3.3.4).

If $X(\Omega)$ is the Lorentz space $L_{p,q}(\Omega)$, then $\tilde{X}(\Gamma, \mu) = L_{p,q}(\Gamma, \mu)$ and

$$\|F|\tilde{X}(\Gamma, \mu)\| = \|F|L_{p,q}(\Gamma, \mu)\|. \tag{5.4.6}$$

5.4.1 The map T and its approximate inverse M

The map T is an isometry from $L_p(\Gamma,\mu)$ to $L_p(\Omega)$ and, indeed, in view of (5.4.5), from $\tilde{X}(\Gamma,\mu)$ to $X(\Omega)$ for a general rearrangement-invariant BFS $X(\Omega)$. In fact we can say much more. To this end, let us denote by $\tilde{W}(\tilde{X},\tilde{Y})(\Gamma,\mu) \equiv \tilde{W}(\tilde{X}(\Gamma,\mu),\tilde{Y}(\Gamma,\mu))$ the completion of the set

$$\{F : F \in \mathrm{Lip}_{loc}(\Gamma) \cap \tilde{X}(\Gamma,\mu), F' \in \tilde{Y}(\Gamma,\mu)\} \tag{5.4.7}$$

with respect to the norm

$$\|F|\tilde{W}(X,Y)(\Gamma,\mu)\| := \|F|\tilde{X}(\Gamma,\mu)\| + \|F'|\tilde{Y}(\Gamma,\mu)\|, \tag{5.4.8}$$

where $\mathrm{Lip}_{loc}(\Gamma)$ denotes the set of functions which are locally Lipschitz on Γ. If $\tilde{X}(\Gamma,\mu) = \tilde{Y}(\Gamma,\mu) = L_p(\Gamma,\mu)$, we write $W_p^1(\Gamma,\mu)$ for $\tilde{W}(\tilde{X},\tilde{Y})(\Gamma,\mu)$. We then have

Lemma 5.4.2. *Let $F \in \mathrm{Lip}_{loc}(\Gamma)$ and $F' \in \tilde{Y}(\Gamma,\mu)$. Then $\nabla(TF) \in Y(\Omega)$ and*

$$\|\nabla(TF)|Y(\Omega)\| \le \gamma\|F'|\tilde{Y}(\Gamma,\mu)\|, \tag{5.4.9}$$

where γ is the constant in Definition 5.2.1(ii). Hence, T is a bounded linear map from $\tilde{W}(\tilde{X},\tilde{Y})(\Gamma,\mu)$ to $W(X,Y)(\Omega)$.

Proof. We have from Definition 5.2.1(ii) that

$$|\nabla(TF)| = |F' \circ \tau||\nabla\tau| \le \gamma|F' \circ \tau|,$$

and hence

$$\begin{aligned}\|\nabla(TF)|Y(\Omega)\| &\le \gamma\|F' \circ \tau|Y(\Omega)\| \\ &= \gamma\|F'|\tilde{Y}(\Gamma,\mu)\|.\end{aligned}$$

□

The approximate inverse M of T is defined as follows : if $B_t := B(u(t),\rho(t))$ is the ball in Ω occurring in Definition 5.2.1 (v) and $f \in L_{1,loc}(\Omega)$,

$$(Mf)(t) := \frac{1}{|B_t|}\int_{B_t} f(x)dx \quad ,t \in \Gamma. \tag{5.4.10}$$

To analyse its properties we need the Hardy-Littlewood *maximal function* (cf. Remark 3.5.2(ii)) :

$$\mathbb{M}f(x) := \sup_{R>0}\{|B(x,R)|^{-1}\int_{B(x,R)} |f(y)|dy\}, \tag{5.4.11}$$

and the Riesz potential operator defined in (5.3.2), namely

$$I_{1,R}f(x) := \int_{B(x,R)} \frac{|f(y)|}{|y-x|^{n-1}} dy. \tag{5.4.12}$$

It is proved in [88], Theorem 12.8 (ii), that, in particular, $\mathbb{M}$ maps the Lorentz space $L_{p,r}(\Omega)$ boundedly into itself when $1 < p \leq \infty$ and $1 \leq r \leq \infty$, and $I_{1,R}$ does likewise for all values of p, r in $[1, \infty]$.

The map $\mathbb{M}$ is not linear, but we adopt the notation

$$\|\mathbb{M}|X \to X\| := \sup\left\{\frac{\|\mathbb{M}f|X\|}{\|f|X\|} : f \in X, f \neq 0\right\}.$$

Lemma 5.4.3. *Let $X = X(\Omega)$ be such that $\mathbb{M} : X \to X$. Then, for every $f \in X$,*

$$|Mf(\tau(x))| \leq (\alpha+1)^n \mathbb{M}f(x) \quad , x \in \Omega \tag{5.4.13}$$

and

$$\|Mf|\tilde{X}(\Gamma,\mu)\| \leq (\alpha+1)^n \|\mathbb{M}|X \to X\| \|f|X(\Omega)\|, \tag{5.4.14}$$

where α is the constant in Definition 5.2.1(iii).

Proof. We set $f(x) = 0$ for $x \notin \Omega$. Then with $k = \alpha + 1$,

$$\begin{aligned}
|(Mf \circ \tau)(x)| &\leq |B_{\tau(x)}|^{-1} \int_{B(\tau(x)} |f(z)| dz \\
&\leq |B(0, k\rho \circ \tau(x))| |B(0, \rho \circ \tau(x))|^{-1} \\
&\times \left\{ |B(0, k\rho \circ \tau(x))|^{-1} \int_{B(0, k\rho\circ\tau(x))} |f| \right\} \\
&\leq (\alpha+1)^n (\mathbb{M}f)(x).
\end{aligned}$$

Hence

$$\begin{aligned}
\|Mf|\tilde{X}(\Gamma,\mu)\| &= \|TMf|X(\Omega)\| \\
&\leq (\alpha+1)^n \|\mathbb{M}f|X(\Omega)\| \\
&\leq (\alpha+1)^n \|\mathbb{M}|X \to X\| \|f|X(\Omega)\|.
\end{aligned}$$

□

Lemma 5.4.4. *Let Ω_1 be a measurable subset of Ω and assume that*

$$k(\Omega_1) := \sup_{\Omega_1} \rho[\tau(x)] < \infty. \tag{5.4.15}$$

Suppose that, with $R = \max(\alpha+1, 2)k(\Omega_1)$,

$$\mathbb{M} : X(\Omega_1) \to X(\Omega_1), \quad I_{1,R} : Y(\Omega_1) \to Y(\Omega_1). \tag{5.4.16}$$

Then, for all $f \in C^1(\Omega)$ with $|\nabla f| \in Y(\Omega)$,

$$\|f - TMf|X(\Omega_1)\| \le \frac{(\alpha+2)^n}{n\delta_1}\{(1 + 2^n\|\mathbb{M}|X(\Omega_1) \to X(\Omega_1)\|\}$$
$$\times \|I_{1,R}|Y(\Omega_1) \to X(\Omega_1)\|\|\nabla f|Y(\Omega)\|, \qquad (5.4.17)$$

where δ_1 is the constant in Definition 5.2.1(v).

Proof. Let $C(x)$ be the cone in Definition 5.2.1(v) and $C'(x) = C(x) \cap B_{\tau(x)}$. Then, since $C'(x) \subset B(x, r(x))$, where $r(x) = (\alpha + 1)\rho[\tau(x)]$,

$$\begin{aligned}
|f(x) - f_{C'(x)}| &\le \frac{1}{|C'(x)|}\int_0^1\int_{C'(x)} |\nabla f(x + t(y-x))||y-x|dydt \\
&\le \frac{1}{|C'(x)|}\int_0^1\int_{B(x,r(x))} |\nabla f(x + t(y-x))||y-x|dydt \\
&\le \frac{1}{|C'(x)|}\int_0^1\int_{B(x,tr(x))} |\nabla f(z)||z-x|t^{-n-1}dzdt \\
&\le \frac{1}{|C'(x)|}\int_{B(x,r(x))} |\nabla f(z)||z-x|\left(\int_{|z-x|/r(x)}^\infty t^{-n-1}dt\right)dz \\
&\le \frac{r(x)^n}{n|C'(x)|}\int_{B(x,r(x))}\frac{|\nabla f(z)|}{|z-x|^{n-1}}dz \qquad (5.4.18)
\end{aligned}$$

where we have set $|\nabla f| = 0$ outside Ω.

Similarly, for $y \in C'(x)$,

$$\begin{aligned}
|f(y) - TMf(x)| &\le \frac{1}{|B_{\tau(x)}|}\int_0^1\int_{B_{\tau(x)}} |\nabla f(y + t(z-y))||z-y|dzdt \\
&\le \frac{2^n}{n}\int_{B(y,2\rho[\tau(x)])}\frac{|\nabla f(z)|}{|z-y|^{n-1}}dz. \qquad (5.4.19)
\end{aligned}$$

Therefore,

$$|f_{C'(x)} - TMf(x)| \le \frac{2^n}{n|C'(x)|}\int_{C'(x)}\int_{B(y,2\rho[\tau(x)])}\frac{|\nabla f(z)|}{|z-y|^{n-1}}dzdy, \qquad (5.4.20)$$

and hence, from (5.4.18) and (5.4.20),

$$\begin{aligned}
|f(x) - TMf(x)| &\le \frac{r(x)^n}{n|C'(x)|}\int_{B(x,r(x))}\frac{\nabla f(z)|}{|z-x|^{n-1}}dz \\
&+ \frac{2^n}{n|C'(x)|}\int_{C'(x)}\int_{B(y,2\rho[\tau(x)])}\frac{|\nabla f(z)|}{|z-y|^{n-1}}dzdy. \qquad (5.4.21)
\end{aligned}$$

Thus, for $x \in \Omega_1$, since $C'(x) \subset B(x, r(x)) \subset B(x, R)$,

$$|f(x) - TMf(x)| \le \frac{(\alpha+1)^n}{n\delta_1}I_{1,R}(|\nabla f|)(x)$$

$$+ 2^n \frac{|B(x,R)|}{n|\mathcal{C}'(x)|}\left(\frac{1}{|B(x,R)|}\int_{B(x,R)} I_{1,R}(|\nabla f|(y)dy\right)$$

$$\leq \frac{(\alpha+2)^n}{n\delta_1}\{I_{1,R}(|\nabla f|(x) + 2^n \mathbb{M}(I_{1,R}(|\nabla f|)(x)\}$$

and this gives (5.4.17). □

In (5.3.5) we noted that $I_{1,R} = R\tau_{1/R}I_{1,1}\tau_R$ where τ_R is the dilation operator in (5.3.1), that is, $(\tau_R f)(x) = f(Rx)$. Hence, when $X(\Omega) = Y(\Omega) = L_{p,q}(\Omega)$, we get from [81], Example 3.1 that the constant on the right-hand side of (5.4.17) is bounded by $k(\Omega_1)$ and hence tends to zero as Ω_1 approaches any point on the boundary which lies on the generalised ridge. We shall see in Theorem 5.5.1 below that for $\Omega_a = \{x \in \Omega : \rho[\tau(x)] \geq a > 0\}$, the embedding $W(X,Y)(\Omega) \hookrightarrow Z(\Omega_a)$ is usually compact. Hence, in such a case, the embedding $W(X,Y)(\Omega) \hookrightarrow Z(\Omega)$ differs from TM by a compact operator. This is why we refer to the operator M as an approximate inverse of T.

Lemma 5.4.5. *If $f \in C^1(\Omega), |\nabla f| \in Y(\Omega)$ and $\mathbb{M} : Y(\Omega) \to Y(\Omega)$, then $Mf \in \mathrm{Lip}_{loc}(\Gamma)$ and*

$$\|(Mf)'|\tilde{Y}(\Gamma,\mu)\| \leq \beta(\alpha+1)^n \|\mathbb{M}|Y(\Omega) \to Y(\Omega)\| \||\nabla f|Y(\Omega)\|, \tag{5.4.22}$$

where β is the constant in Definition 5.2.1 (iv).

Proof. We first prove that for all $t \in \Gamma, Mf \in \mathrm{Lip}_{loc}(\Gamma)$ and

$$|(Mf)'(t)| \leq \beta \mathbb{M}(|\nabla f|)(t). \tag{5.4.23}$$

We have

$$\begin{aligned} F_\rho(u) &:= |B(u,\rho)|^{-1}\int_{B(u,\rho)} f(y)dy \\ &= \rho^n |B(0,\rho)|^{-1}\int_{B(0,1)} f(u+\rho w)dw \\ &= |B(0,1)|^{-1}\int_{B(0,1)} f(u+\rho w)dw. \end{aligned}$$

Hence

$$\begin{aligned} (\partial/\partial\rho)F_\rho(u) &= |B(0,1)|^{-1}\int_{B(0,1)} \nabla f(u+\rho w)\cdot w dw \\ &= |B(0,\rho)|^{-1}\int_{B(u,\rho)} \nabla f(y)\cdot(y-u)\rho^{-1}dy \end{aligned}$$

and

$$\nabla_u F_\rho(u) = |B(0,1)|^{-1} \int_{B(0,1)} \nabla f(u+\rho w)dw$$
$$= |B(0,\rho)|^{-1} \int_{B(u,\rho)} \nabla f(y)dy.$$

Thus, if $F(t) = (Mf)(t)$, we have $F \in \mathrm{Lip}_{loc}(\Gamma)$ and, for all $t \in \Gamma$,

$$\begin{aligned}|F'(t)| &\leq |(\partial F/\partial \rho)\rho'(t) + \nabla_u F \cdot u'(t)| \\ &\leq |B(u,\rho)|^{-1} \int_{B(u,\rho)} |\nabla f(y)|\{|\rho'(t)| + |u'(t)|\}dy \\ &\leq \mathbb{M}(|\nabla f|)(t)\{|\rho'(t)| + |u'(t)|\} \\ &\leq \beta \mathbb{M}(|\nabla f|)(t),\end{aligned}$$

which proves (5.4.23). It follows that

$$\begin{aligned}\|(Mf)'|\tilde{Y}(\Gamma,\mu)\| &= \|(Mf)' \circ \tau|Y(\Omega)\| \\ &\leq \beta\|T\mathbb{M}(|\nabla f|)|Y(\Omega)\| \\ &\leq \beta(\alpha+1)^n\|\mathbb{M}(|\nabla f|)|Y(\Omega)\|\end{aligned}$$

by (5.4.13). The lemma is therefore proved. □

5.4.2 Equivalent embeddings

In what follows we shall assume that

$$C^1(\Omega) \cap W(X,Y)(\Omega) \quad \text{is dense in } W(X,Y)(\Omega). \tag{5.4.24}$$

If $X = Y = L_p, 1 \leq p < \infty$, (5.4.24) is a consequence of the well-known result of Meyers and Serrin in [174] concerning the denseness of $C^\infty(\Omega) \cap W_p^k(\Omega)$ in $W_p^k(\Omega)$ for any $k \in \mathbb{N}$.

Recall that $\tilde{W}(\tilde{X},\tilde{Y})(\Gamma,\mu) = \tilde{W}(\tilde{X}(\Gamma,\mu),\tilde{Y}(\Gamma,\mu))$ denotes the completion of

$$\{F : F \in \mathrm{Lip}_{loc}(\Gamma) \cap \tilde{X}(\Gamma,\mu), F' \in \tilde{Y}(\Gamma,\mu)\} \tag{5.4.25}$$

with respect to the norm

$$\|F|\tilde{W}(\Gamma,\mu)\| := \|F|\tilde{X}(\Gamma,\mu)\| + \|F'|\tilde{Y}(\Gamma,\mu)\|. \tag{5.4.26}$$

Our main tool for characterising the embeddings $W(X,Y)(\Omega) \hookrightarrow Z(\Omega)$ is the next theorem which equates the problem with an analogous one on Γ.

Theorem 5.4.6. *Suppose that (5.4.24) is satisfied,* $\mathbb{M}$ *maps each of the spaces* $X(\Omega), Y(\Omega), Z(\Omega)$ *continuously into itself,* $I_{1,R} : Y(\Omega) \to Z(\Omega)$*, and*

$$k(\Omega) := \sup_\Omega \rho \circ \tau < \infty. \tag{5.4.27}$$

Then the embeddings

1. $W(X,Y)(\Omega) \hookrightarrow Z(\Omega)$,
2. $\tilde{W}(\tilde{X},\tilde{Y})(\Gamma,\mu) \hookrightarrow \tilde{Z}(\Gamma,\mu)$,

are equivalent. Moreover the constants

$$\begin{aligned} C_\Omega := \inf\{A : & \|f|Z\| \le A\|\nabla f|Y\| + B\|f|X\| \\ & \textit{for some } B > 0 \textit{ and all } f \in W(X,Y)\}, \\ c_\Gamma := \inf\{a : & \|F|\tilde{Z}\| \le a\|F'|\tilde{Y}\| + b\|F|\tilde{X}\| \\ & \textit{for some } b > 0 \textit{ and all } F \in \tilde{W}(\tilde{X},\tilde{Y})\} \end{aligned}$$

satisfy

$$\begin{aligned} \gamma^{-1}c_\Gamma &\le C_\Omega \\ &\le (\alpha+2)^n\{\frac{1}{n\delta_1}(1+2^n\|\mathbb{M}|Z \to Z\|)\|I_{1,R}|Y \to Z\| \\ &+ \beta\|\mathbb{M}|Y \to Y\|\ c_\Gamma\} \end{aligned} \tag{5.4.28}$$

with

$$R = \max(\alpha+1, 2)k(\Omega).$$

Proof. To simplify the notation we shall write $\|f\|_X$, $\|F\|_{\tilde{X}}$ for $\|f|X(\Omega\|$, $\|F|\tilde{X}(\Gamma,\mu)\|$ respectively, and similarly for the other spaces involved.

Assume (1) and let $A > C_\Omega$. Then, there exists B such that for all $F \in \tilde{W}(\tilde{X},\tilde{Y})$,

$$\begin{aligned} \|F\|_{\tilde{Z}} \equiv \|TF\|_Z &\le A\|\nabla(TF)\|_Y + B\|TF\|_X \\ &\le \gamma A\|F'\|_{\tilde{Y}} + B\|F\|_{\tilde{X}} \end{aligned}$$

by Lemma 5.4.2, yielding (2) and the first inequality in (5.4.28).

Conversely, suppose (2) is satisfied and let $a > c_\Gamma$. Then, for some b and all $f \in W(X,Y)$,

$$\begin{aligned} \|f\|_Z &\le \|f - TMf\|_Z + \|TMf\|_Z \\ &= \|f - TMf\|_Z + \|Mf\|_{\tilde{Z}} \\ &\le \frac{(\alpha+2)^n}{n\delta_1}\{(1+2^n\|\mathbb{M}|Z \to Z\|)\|I_{1,R}|Y \to Z\|\}\,\|\nabla f\|_Y \\ &+ a\|(Mf)'\|_{\tilde{Y}} + b\|Mf\|_{\tilde{X}} \end{aligned}$$

(by Lemma 5.4.4)

$$\begin{aligned} &\le \frac{(\alpha+2)^n}{n\delta_1}\{(1+2^n\|\mathbb{M}|Z \to Z\|)\|I_{1,R}|Y \to Z\|\}\,\|\nabla f\|_Y \\ &+ a\beta(\alpha+1)^n\|\mathbb{M}|Y \to Y\|\|\nabla f\|_Y + b(\alpha+1)^n\|\mathbb{M}|X \to X\|\|f\|_X \end{aligned}$$

by Lemmas 5.4.3 and 5.4.5, whence (1) and the second inequality in (5.4.28). □

Remark 5.4.7. In [7], Ariño and Muckenhoupt prove that $\mathbb{M}$ is bounded on a Lorentz space $L_{p,q}$ for $p \in (1,\infty)$ and $q \in [1,\infty)$. Therefore, in view of Remark 5.3.2, with $X = L_{p,q}, Y = L_{p,r}$ and $Z = L_{s,t}$, the conditions on $\mathbb{M}$ and $I_{1,R}$ in Theorem 5.4.6 are satisfied when $|\Omega| < \infty$ if

$$p, s \in (1,\infty), \quad q, r, t \in [1,\infty), \quad 0 \leq \frac{1}{p} - \frac{1}{s} < \frac{1}{n}.$$

5.4.3 Equivalent Poincaré inequalities

The next theorem is the core of our strategy for analysing the Poincaré inequality in $W(X,Y)(\Omega)$. When used in conjunction with Corollary 5.3.11 and Theorem 5.3.12 it yields precise information on the constant $\alpha(E)$ and the compactness, or otherwise, of the embedding E.

Theorem 5.4.8. *Let $\Gamma_1 \subset \Gamma$ be such that $\mu(\Gamma_1) < \infty$ and with $\Omega_1 = \tau^{-1}(\Gamma_1)$ suppose that $I_{1,R} : Y(\Omega_1) \to X(\Omega_1)$ and $\mathbb{M}$ maps $X(\Omega_1), Y(\Omega_1)$ boundedly into themselves. Then, the following two statements are equivalent :*

$$(1): \quad \|f - f_{\Omega_1}|X(\Omega_1)\| \leq D(\Omega_1)\|\nabla f|Y(\Omega)\| \quad \textit{for all} \quad f \in W(X,Y)(\Omega),$$

$$(2): \quad \|F - F_{\Gamma_1}|\tilde{X}(\Gamma_1,\mu)\| \leq d(\Gamma_1)\|F'|\tilde{Y}(\Gamma,\mu)\| \quad \textit{for all} \quad F \in \tilde{W}(\tilde{X},\tilde{Y})(\Gamma,\mu),$$

where $F_{\Gamma_1} := \frac{1}{\mu(\Gamma_1)}\int_{\Gamma_1} F(t)d\mu(t)$ and $\|F|\tilde{X}(\Gamma_1,\mu)\| := \|TF|X(\Omega_1)\|$. The least constants $D(\Omega_1), d(\Gamma_1)$ in (1) and (2) satisfy

$$\begin{aligned}\gamma^{-1}d(\Gamma_1) &\leq D(\Omega_1)\\ &\leq K(\Omega_1)(\alpha+2)^n\{\frac{1}{n\delta_1}(1+2^n\|\mathbb{M}|X(\Omega_1)\to X(\Omega_1)\|)\\ &\times \|I_{1,R}|Y(\Omega_1)\to X(\Omega_1)\|\\ &+ \beta\|\mathbb{M}|Y(\Omega_1)\to Y(\Omega_1)\|\ d(\Gamma_1)\},\end{aligned} \tag{5.4.29}$$

where $R = \max(\alpha+1,2)k(\Omega_1)$, and $k(\Omega_1), K(\Omega_1)$ are defined in (5.4.15) and (5.3.21).

Proof. Let (1) be satisfied, and let F be a member of (5.4.25). Then, by Lemma 5.4.2, $TF \in W(X,Y)(\Omega)$ and $F_{\Gamma_1} = (TF)_{\Omega_1}$ since $\mu(\Gamma_1) = |\Omega_1|$. Hence,

$$\begin{aligned}\|F - F_{\Gamma_1}|\tilde{X}(\Gamma_1,\mu)\| &= \|TF - (TF)_{\Omega_1}|X(\Omega_1)\|\\ &\leq D(\Omega_1)\|\nabla(TF)|Y(\Omega)\|\\ &\leq \gamma D(\Omega_1)\|F'|\tilde{Y}(\Gamma_1,\mu)\|,\end{aligned}$$

whence (2) and the first inequality in (5.4.29).

Conversely, assume (2) and let $f \in C^1(\Omega) \cap W(X,Y)(\Omega)$. Then, by (5.3.20),

$$
\begin{aligned}
\|f - f_{\Omega_1}|X(\Omega_1)\| &\le K(\Omega_1)\|f - (Mf)_{\Gamma_1}|X(\Omega_1)\| \\
&\le K(\Omega_1)\{\|f - TMf|X(\Omega_1)\| + \|TMf - (Mf)_{\Gamma_1}|X(\Omega_1)\|\} \\
&\le K(\Omega_1)\{\|f - TMf|X(\Omega_1)\| + \|Mf - (Mf)_{\Gamma_1}|\tilde{X}(\Gamma_1,\mu)\|\} \\
&\le K(\Omega_1)\{\frac{(\alpha+2)^n}{n\delta_1}(1 + 2^n\|\mathbb{M}|X(\Omega_1) \to X(\Omega_1)\|) \\
&\times \|I_{1,R}|Y(\Omega_1) \to X(\Omega_1)\|\|\nabla f|Y(\Omega)\| \\
&+ d(\Gamma_1)\|(Mf)'|\tilde{Y}(\Gamma,\mu)\|\}
\end{aligned}
$$

by Lemma 5.4.4. On using Lemma 5.4.5, (1) and the second inequality in (5.4.29) are established. □

5.5 Compactness of E

5.5.1 Local compactness

We saw in Section 5.2 (see (5.2.1)) that as long as we keep away from the points on the boundary which meet the ridge of a GRD Ω, the corresponding subset Ω' satisfies a local cone condition and the embedding $E(\Omega')$ is compact. We begin this section by investigating the corresponding property for embeddings $W(X,Y)(\Omega) \hookrightarrow Z(\Omega)$ for general rearrangement-invariant BFSs $X(\Omega), Y(\Omega)$ and $Z(\Omega)$.

Theorem 5.5.1. *Let $\Omega_a := \{x \in \Omega : \rho[\tau(x)] \ge a > 0\}$, and suppose that $I_{1,1} : Y(\Omega_a) \to Z(\Omega_a)$ and*

$$\lim_{R\to 0+} \|I_{1,R}|Y(\Omega_a) \to Z(\Omega_a)\| = 0. \tag{5.5.1}$$

Then the embedding $W(X,Y)(\Omega_a) \hookrightarrow Z(\Omega_a)$ is compact.

Proof. Let $\mathcal{T}_h$ denote a tesselation of $\mathbb{R}^n$ by cubes of side h. For sufficiently small h, we have that for every $x \in \Omega_a$ there is a $Q_x \in \mathcal{T}_h$ such that $Q_x \subset \Omega_a$ and $\operatorname{dist}(x, Q_x) \le kh$ for some fixed $k \ge 1$. Therefore, as in (4.3.5)(cf. (5.4.18)), for $f \in C^1(\Omega)$,

$$
\begin{aligned}
|f(x) - f_{Q_x}| &\le \frac{1}{|Q_x|}\int_{Q_x} |f(x) - f(y)|dy \\
&\le k^n\left(\frac{1}{|kQ_x|}\int_{kQ_x} |f(x) - f(y)|dy\right) \\
&\lesssim k^n \int_{kQ_x} \frac{|\nabla f(y)|}{|x-y|^{n-1}}dy,
\end{aligned} \tag{5.5.2}
$$

where kQ_x denotes the cube of side kh which is concentric to Q_x. For $Q \in \mathcal{T}_h$ and $Q \subset \Omega_a$, define

$$\Omega_Q := \{x \in \Omega_a : Q_x = Q\}$$

and, for $f \in W(X,Y)(\Omega_a)$,

$$P_h f(x) := \sum_{Q \subset \Omega_a, Q \in \mathcal{T}_h} \chi_{\Omega_Q}(x) f_Q. \tag{5.5.3}$$

Since there are only finitely many cubes in Ω_a, P_h is of finite rank. Also

$$\begin{aligned}\|P_h f|Z(\Omega_a)\| &\leq \sum_{Q \subset \Omega_a, Q \in \mathcal{T}_h} \|\chi_{\Omega_Q}(x)|Z(\Omega_a)\||f_Q| \\ &\leq K(a,h)\|f|X(\Omega_a)\| \\ &\leq K(a,h)\|f|W(X,Y)(\Omega_a)\|.\end{aligned}$$

On using (5.5.2), we have for $f \in W(X,Y)(\Omega_a)$,

$$\begin{aligned}\|f - P_h f|Z(\Omega_a)\| &\leq \|\sum_{Q \subset \Omega_a, Q \in \mathcal{T}_h} \chi_{\Omega_Q}|f - f_Q||Z(\Omega_a)\| \\ &\lesssim k^n \|\sum_Q \chi_{\Omega_Q} \int_{kQ_x} \frac{|\nabla f(y)|}{|\cdot - y|^{n-1}} dy|Z(\Omega_a\|.\end{aligned}$$

Since $kQ_x \subset B(x, jkh)$ for some fixed $j > 0$, this yields

$$\begin{aligned}\|f - P_h f|Z(\Omega_a)\| &\lesssim k^n \|\sum_Q \chi_{\Omega_Q} I_{jkh}(|\nabla f|)|Z(\Omega_a)\| \\ &\lesssim k^n \|I_{jkh}(|\nabla f|)|Z(\Omega_a)\|.\end{aligned}$$

The theorem follows from this. □

In some instances we can prove that (5.5.1) is also necessary for compactness in Theorem 5.5.1.

Example 5.5.2. Let Ω_a be the domain in Theorem 5.5.1, $Y(\Omega_a) = L_{p,q}(\Omega_a)$, $Z(\Omega_a) = L_{r,s}(\Omega_a)$, where $1 \leq p < n$ and $1 \leq q, r, s \leq \infty$. Then

1. for arbitrary $X(\Omega_a)$, the embedding $W(X,Y)(\Omega_a) \hookrightarrow Z(\Omega_a)$ is compact if $r < p^* := np/(n-p)$;
2. if $X(\Omega_a) = L_{u,v}(\Omega_a)$ with $u \leq p^*$ and the embedding $W(X,Y)(\Omega_a) \hookrightarrow Z(\Omega_a)$ is compact, then $r < p^*$.

Proof. 1. This follows from Theorem 5.5.1 since, by (5.3.5) and Example 5.3.1,

$$\begin{aligned}\|I_{1,R}|Y(\Omega_a) \to Z(\Omega_a)\| &\leq R\|\tau_R|Y(\Omega_a) \to Y(\Omega_a)\| \\ &\quad \times \|\tau_{1/R}|Z(\Omega_a) \to Z(\Omega_a)\| \\ &= R^{n(\frac{1}{r} - \frac{1}{p*})} \\ &\to 0\end{aligned}$$

as $R \to 0$.

2. Choose $\eta \in C_0^\infty(\mathbb{R}^n)$ such that, for $k \in \mathbb{N}$ large enough, supp $\eta \subset k\Omega_a$ and set $f_k(x) = k^{n/p^*}(\tau_k\eta)(x)$. Then, by Example 5.3.1,

$$\|f_k|X(\Omega_a)\| = k^{n/p^*-n/u}\|\eta|L_{u,v}(\mathbb{R}^n)\|, \tag{5.5.4}$$

$$\begin{aligned}\|\nabla f_k|Y(\Omega_a)\| &= k^{n/p^*+1}\|\tau_k(|\nabla\eta|)|L_{p,q}(\mathbb{R}^n)\| \\ &= k^{n/p^*+1-n/p}\|\nabla\eta|L_{p,q}(\mathbb{R}^n)\| \\ &= \|\nabla\eta|L_{p,q}(\mathbb{R}^n)\|. \end{aligned}\tag{5.5.5}$$

Hence $\{f_k\}$ is bounded in $W(X,Y)(\Omega_a)$, and consequently is precompact in $Z(\Omega_a)$. Suppose that $r \geq p^*$ in (2) and let $f_k \to f$ in $L_{p^*,s}(\Omega_a)$. If $\Omega' \subset \Omega_a$ is of finite volume, then $f_k \to f$ in $L_{l,s}(\Omega')$, for any $l < p^*$, since $L_{p^*,s}(\Omega') \hookrightarrow L_{l,s}(\Omega')$. But

$$\|f_k|L_{l,s}(\Omega')\| \leq k^{n/p^*-n/l}\|\eta|L_{l,s}(\mathbb{R}^n)\| \to 0$$

whence $\|f|L_{l,s}(\Omega')\| = 0$ and $f = 0$ a.e. in Ω_a. This contradicts

$$\|f_k|L_{p^*,s}(\Omega_a)\| = \|\eta|L_{p^*,s}(\mathbb{R}^n)\|,$$

and hence $r < p^*$ as asserted.

□

5.5.2 Measure of non-compactness

We recall from Theorem 5.3.12 that the measure of non-compactness $\alpha(E)$ of E is estimated by the analogous quantity $\alpha(E_M)$ for E_M : $W_M(X,Y)(\Omega) \hookrightarrow X(\Omega)$. In particular, E and E_M are compact simultaneously. We assume throughout this section that $|\Omega| < \infty$ and that *E_M is bounded.*

We shall proceed to obtain two-sided estimates for $\alpha(E_M)$ in terms of quantities given by Poincaré-type inequalities on members of the filter base $\mathcal{A}(\Omega)$ defined in (5.2.4). Let $A \in \mathcal{A}(\Omega)$ and so $A = \tau^{-1}(\Lambda)$ for some Λ in the set $\mathcal{A}(\Gamma)$ defined in (5.2.3). We have already observed in Section 5.2 that Λ is a finite union of relatively closed disjoint subtrees of $\Gamma, \Lambda_i (i \in \mathcal{N}_\Lambda)$, say, which are rooted at the boundary points of $\Gamma \setminus \Lambda$. If Γ is an interval $[a,b)$ then Λ is a subinterval $[c,b)$; thus in this case the index set $\mathcal{N}_\Lambda$ is a singleton. Set $A_i = \tau^{-1}(\Lambda_i), i \in \mathcal{N}_\Lambda$, these being disjoint and closed in Ω and $A = \tau^{-1}(\Lambda) = \cup_{i \in \mathcal{N}_\Lambda} A_i$. Let

$$H_\Lambda F := \sum_{i \in \mathcal{N}_\Lambda} \chi_{\Lambda_i} F_{\Lambda_i}, \quad F_{\Lambda_i} = \frac{1}{\mu(\Lambda_i)} \int_{\Lambda_i} F(t) d\mu(t), \tag{5.5.6}$$

$$h_A f := \sum_{i \in \mathcal{N}_\Lambda} \chi_{A_i} f_{A_i}, \quad f_{A_i} = \frac{1}{|A_i|} \int_{A_i} f(x) dx. \tag{5.5.7}$$

We shall require that for all $A \in \mathcal{A}(\Omega)$,

$$K_X := \sup\{\|h_A f|X(A)\| : \|f|X(A)\| = 1\} < \infty. \tag{5.5.8}$$

As we are assuming that $|\Omega| < \infty$ this is satisfied since

$$\begin{aligned} \|h_A f|X(A)\| &\leq \sum_{\mathcal{N}_A} \|\chi_{A_i} f_{A_i}|X(A)\| \\ &= \sum_{\mathcal{N}_A} \Big(\frac{1}{|A_i|}\Big|\int_{A_i} f(x)dx\Big| \|\chi_{A_i}|X(A)\|\Big) \qquad (5.5.9) \\ &\leq \sum_{\mathcal{N}_A} \left(\frac{\|\chi_{A_i}|X'(\Omega)\|\|\chi_{A_i}|X(\Omega)\|}{|A_i|}\right) \|f|X(A)\|, \qquad (5.5.10) \end{aligned}$$

by Hölder's inequality (Theorem 3.1.8). If $X(\Omega) = L_p(\Omega)$, we have

$$\|h_A f|L_p(\Omega)\| \leq \|f|L_p(\Omega)\|. \tag{5.5.11}$$

Lemma 5.5.3. *Let $A \in \mathcal{A}(\Omega)$ and define*

$$\theta_A = \sup\{\|f - h_A f|X(A)\| : f \in W_M(X,Y)(\Omega), \|\nabla f|Y(\Omega)\| = 1\}. \tag{5.5.12}$$

Then, if $C_0^\infty(\Omega)$ is dense in $X(\Omega)$, we have

$$\alpha(E_M) \leq \inf_{\mathcal{A}(\Omega)} \theta_A \leq \limsup_{\mathcal{A}(\Omega)} \theta_A \leq (1 + K_X)\alpha(E_M). \tag{5.5.13}$$

Proof. For a fixed $A_0 \in \mathcal{A} = \mathcal{A}(\Omega)$, let

$$\theta_A^0 = \sup\{\|f - h_{A_0} f|X(A)\| : f \in W_M(X,Y)(\Omega), \|\nabla f|Y(A)\| = 1\}.$$

Then, as in Corollaries 5.3.8 and 5.3.11, since h_{A_0} is of finite rank,

$$\alpha(E_M) = \inf_{\mathcal{A}} \theta_A^0 = \lim_{\mathcal{A}} \theta_A^0.$$

Note that, since we are assuming that E_M is continuous, the norm (5.3.16) of $W_M(X,Y)(\Omega)$ is equivalent to the $W(X,Y)(\Omega)$ norm; also $h_A : W_M(X,Y)(\Omega) \to X(\Omega)$ is bounded by (5.5.10). Hence $\alpha(E_M) \leq \theta_{A_0}^0$, and so

$$\alpha(E_M) \leq \inf_{\mathcal{A}} \theta_A \leq \limsup_{\mathcal{A}} \theta_A.$$

Furthermore

$$\|f - h_A f|X(A)\| \leq \|f - h_{A_0} f|X(A)\| + \|h_{A_0} f - h_A f|X(A)\|,$$

and, by (5.5.8),

$$\|h_{A_0} f - h_A f|X(A)\| = \|\sum \chi_{A_i}[h_{A_0} f - f_{A_i}]|X(A)\|$$

$$\begin{aligned} &= \| \sum \chi_{A_i}(h_{A_0}f - f)_{A_i} | X(A) \| \\ &\leq K_X \| h_{A_0}f - f | X(A) \|. \end{aligned}$$

Hence

$$\| f - h_A f | X(A) \| \leq (1 + K_X) \| f - h_{A_0} f | X(A) \|,$$

whence

$$\begin{aligned} \limsup_{\mathcal{A}} \theta_A &\leq (1 + K_X) \lim_{\mathcal{A}} \theta_A^0 \\ &= (1 + K_X)\alpha(E_M) \end{aligned}$$

and the lemma is proved. □

Lemma 5.5.4. *Suppose that the maximal operator* $\mathbb{M}$ *maps* $X(\Omega)$ *and* $Y(\Omega)$ *into themselves. For* $\Lambda \in \mathcal{A}(\Gamma)$ *define*

$$\phi_\Lambda := \sup \{ \| F - H_\Lambda F | \tilde{X}(\Lambda, \mu) \| : \tag{5.5.14}$$

$$F \in \tilde{W}(\tilde{X}, \tilde{Y})(\Gamma, \mu), \| F' | \tilde{Y}(\Gamma, \mu) \| = 1 \}. \tag{5.5.15}$$

Then there exists $C > 0$ *such that, with* $A = \tau^{-1}(\Lambda)$,

$$\gamma^{-1} \phi_\Lambda \leq \theta_A \leq C[(1 + K_X)\mathcal{C}(A) + \phi_\Lambda], \tag{5.5.16}$$

where

$$\mathcal{C}(A) = \| I_{1,R} | Y(A) \to X(A) \| \tag{5.5.17}$$

and

$$R = \max(\alpha + 1, 2) k(A), \quad k(A) = \sup_A \rho[\tau(x)]. \tag{5.5.18}$$

Proof. If $F \in \tilde{W}(\tilde{X}, \tilde{Y})(\Gamma, \mu)$, then $TF \in W(X,Y)(\Omega)$ and $TH_\Lambda F = h_A(TF)$. Also, $f \in W(X,Y)(\Omega)$ implies $f - f_\Omega \in W_M(X,Y)(\Omega)$ and

$$\begin{aligned} \| f - h_A f | X(A) \| &= \| (f - f_\Omega) - h_A(f - f_\Omega) | X(A) \| \\ &\leq \theta_A \| \nabla f | Y(\Omega) \|. \end{aligned}$$

Hence

$$\begin{aligned} \| F - H_\Lambda F | \tilde{X}(\Lambda, \mu) \| &= \| \chi_\Lambda [F - H_\Lambda F] | \tilde{X}(\Lambda, \mu) \| \\ &= \| T(\chi_\Lambda F) - T(\chi_\Lambda H_\Lambda F) | X(\Omega) \| \\ &= \| \chi_A (TF - TH_\Lambda F) | X(\Omega) \| \\ &= \| TF - h_A(TF) | X(A) \| \\ &\leq \theta_A \| \nabla (TF) | Y(\Omega) \| \\ &\leq \gamma \theta_A \| F' | \tilde{Y}(\Gamma, \mu) \| \end{aligned}$$

by Lemma 5.4.2. Hence $\phi_\Lambda \leq \gamma \theta_A$.

Let $f \in W(X,Y)(\Omega)$. Then

$$\begin{aligned}
\|f - h_A f|X(A)\| &\leq \|f - TMf|X(A)\| + \|TMf - h_A(TMf)|X(A)\| \\
&\quad + \|h_A f - h_A(TMf)|X(A)\| \\
&\leq \|f - TMf|X(A)\| + \|TMf - TH_\Lambda Mf|X(A)\| \\
&\quad + K_X\|f - TMf|X(A)\| \\
&\leq (1 + K_X)\|f - TMf|X(A)\| + \|Mf - H_\Lambda Mf|\tilde{X}(\Lambda,\mu)\| \\
&\leq (1 + K_X)C\mathcal{C}(A)\|\nabla f|Y(\Omega)\| + \phi_\Lambda\|(Mf)'|\tilde{Y}(\Gamma,\mu)\|
\end{aligned}$$

(by Lemma 5.4.4)

$$\leq C\{(1 + K_X)\mathcal{C}(A) + \phi_\Lambda\}\|\nabla f|Y(\Omega)\|$$

by Lemma 5.4.5. □

Corollary 5.5.5. *Let the hypotheses of Lemmas 5.5.3 and 5.5.4 be satisfied and*

$$\liminf_{\mathcal{A}(\Omega)} \mathcal{C}(A) = 0, \tag{5.5.19}$$

where $\mathcal{C}(A)$ is defined in (5.5.17). Then

$$[(1 + K_X)\gamma]^{-1}\phi_+ \leq \alpha(E_M) \leq C\phi_-, \tag{5.5.20}$$

where

$$\phi_+ := \limsup_{\mathcal{A}(\Gamma)} \phi_\Lambda, \tag{5.5.21}$$

$$\phi_- := \liminf_{\mathcal{A}(\Gamma)} \phi_\Lambda. \tag{5.5.22}$$

Remark 5.5.6. 1. If $\lim_{\mathcal{A}(\Omega)} k(A) = 0$ then $\mathcal{C}(A) \to 0$ for the Lorentz spaces X, Y in Remark 5.3.2 (*ii*), in view of what is noted in Example 5.3.1.

2. If $\liminf_{\mathcal{A}(\Omega)} \rho \circ \tau \neq 0$, then $\rho \circ \tau$ is bounded away from zero on Ω and hence $\Omega = \Omega_a$, for some $a > 0$ in the notation of Theorem 5.5.1. It follows that E and thus E_M are compact. Consequently $\alpha(E_M) = 0$ and (5.5.20) is still satisfied since $\phi_+ = \phi_- = 0$ by (5.5.13) and (5.5.16).
3. It is proved in [80], Corollary 4.8, that when Γ is an interval, (5.5.20) continues to hold even if $\lim_{\mathcal{A}(\Omega)} k(A) \neq 0$.

The case $\Gamma_1 = \Gamma$ and hence $\Omega_1 = \Omega$ of Theorem 5.4.8 gives the equivalence of the two Poincaré inequalities

$$\|f - f_\Omega|X(\Omega)\| \leq D(\Omega)\|\nabla f|Y(\Omega)\| \quad (f \in W(X,Y)(\Omega)), \tag{5.5.23}$$

$$\|F - F_\Gamma|\tilde{X}(\Gamma,\mu)\| \leq d(\Gamma)\|F'|\tilde{Y}(\Gamma,\mu)\| \quad (F \in \tilde{W}(\tilde{X},\tilde{Y})(\Gamma,\mu)) \tag{5.5.24}$$

and (5.4.29) relates the best constants $D(\Omega), d(\Gamma)$. We now show that (5.5.24) can be described in terms of the boundedness of the Hardy operator $H : \tilde{Y}(\Gamma,\mu) \to \tilde{X}(\Gamma,\mu)$ defined for $f \in \tilde{Y}(\Gamma,\mu)$ by

$$Hf(x) := \int_a^x f(t)dt, \quad a \in \Gamma.$$

Suppose that

$$\|\chi_{(a,c)}\frac{dt}{d\mu}|\tilde{X}'(\Gamma,\mu)\| < \infty \tag{5.5.25}$$

for all $c \in \Gamma$, where $\tilde{X}'(\Gamma,\mu)$ is the associate space of $\tilde{X}(\Gamma,\mu)$. Then $H : \tilde{Y}(\Gamma,\mu) \to \tilde{X}(\Gamma,\mu)$ is bounded if, and only if, the same is true for

$$H_c f(x) := \int_c^x f(t)dt,$$

For the difference satisfies

$$\begin{aligned}\|\int_a^c f(t)dt|\tilde{X}(\Gamma,\mu)\| &= \|\int_\Gamma (\chi_{(a,c)}(t)\frac{dt}{d\mu})f(t)d\mu(t)|\tilde{X}(\Gamma,\mu)\| \\ &\le |\int_\Gamma \chi_{(a,c)}(t)\frac{dt}{d\mu}f(t)d\mu(t)|\|\chi_\Gamma|\tilde{X}(\Gamma,\mu)\| \\ &\le \left(\|\chi_{(a,c)}\frac{dt}{d\mu}|\tilde{X}'(\Gamma,\mu)\|\|\chi_\Gamma|\tilde{X}(\Gamma,\mu)\|\right)\|f|\tilde{X}(\Gamma,\mu)\|,\end{aligned}$$

by Hölder's inequality.

Let c be chosen such that there exist subtrees Γ_1, Γ_2 of Γ with $\Gamma_2 = \Gamma \setminus \Gamma_1, \overline{\Gamma_1} \cap \overline{\Gamma_2} = \{c\}$ and

$$K_c := \mu(\Gamma)^{-1} \max_{i=1,2}\left(\|\chi_{\Gamma_i}|\tilde{X}(\Gamma,\mu)\|\|\chi_{\Gamma_i}|\tilde{X}'(\Gamma,\mu)\|\right) < 1 \quad (i = 1,2). \tag{5.5.26}$$

Define

$$A_c(\Gamma_i) := \sup\left\{\|H_c f\chi_{\Gamma_i}|\tilde{X}(\Gamma,\mu)\| : \|f\chi_{\Gamma_i}|\tilde{Y}(\Gamma,\mu)\| = 1\right\}. \tag{5.5.27}$$

Theorem 5.5.7. *For every Γ_1, Γ_2 of the above form, $d(\Gamma)$ in (5.5.24) is finite if, and only if, $A_c(\Gamma_1)$ and $A_c(\Gamma_2)$ are finite. If $d(\Gamma)$ stands for the best constant in (5.5.24) and*

$$A_c(\Gamma) := \max\left(A_c(\Gamma_1), A_c(\Gamma_2)\right),$$

then

$$(1 - K_c)A_c(\Gamma) \le d(\Gamma) \le 2\left(1 + \mu(\Gamma)^{-1}\|\chi_\Gamma|\tilde{X}(\Gamma,\mu)\|\|\chi_\Gamma|\tilde{X}'(\Gamma,\mu)\|\right)A_c(\Gamma). \tag{5.5.28}$$

Proof. Let F lie in (5.4.25) and $F' = f \in \tilde{Y}(\Gamma,\mu)$, and set $F_c(x) = H_c f(x)$. Then

$$F(x) - F_\Gamma = F_c(x) - (F_c)_\Gamma$$

Suppose that $A_c(\Gamma) < \infty$. Then, with $\tilde{X} = \tilde{X}(\Gamma,\mu), \tilde{Y} = \tilde{Y}(\Gamma,\mu)$, we have

$$\begin{aligned}
\|F - F_\Gamma|\tilde{X}\| &= \|F_c - (F_c)_\Gamma|\tilde{X}\| \leq \|F_c|\tilde{X}\| + \big|(F_c)_\Gamma\big|\|\chi_\Gamma|\tilde{X}\| \\
&\leq \big(1 + \mu(\Gamma)^{-1}\|\chi_\Gamma|\tilde{X}'\|\|\chi_\Gamma|\tilde{X}\|\big)\|F_c|\tilde{X}\| \\
&\leq \big(1 + \mu(\Gamma)^{-1}\|\chi_\Gamma|\tilde{X}'\|\|\chi_\Gamma|\tilde{X}\|\big)\big(\|F_c\chi_{\Gamma_1}|\tilde{X}\| + \|F_c\chi_{\Gamma_2}|\tilde{X}\|\big) \\
&\leq 2\big(1 + \mu(\Gamma)^{-1}\|\chi_\Gamma|\tilde{X}'\|\|\chi_\Gamma|\tilde{X}\|\big)A_c(\Gamma)\|f|\tilde{Y}\|
\end{aligned}$$

on using Hölder's inequality and (5.5.27). Thus $d(\Gamma) < \infty$ and the second inequality in (5.5.28) is proved.

Conversely, suppose $d(\Gamma) < \infty$ and let $\operatorname{supp} f \subset \Gamma_1$. Then F_c lies in $\tilde{W}(\tilde{X},\tilde{Y})(\Gamma,\mu)$ and is also supported in Γ_1. By (5.5.24),

$$\begin{aligned}
d(\Gamma)\|f\chi_{\Gamma_1}|\tilde{Y}\| &\geq \|F_c - (F_c)_\Gamma|\tilde{X}\| \\
&\geq \|\big(F_c - (F_c)_\Gamma\big)\chi_{\Gamma_1}|\tilde{X}\| \\
&\geq \|F_c\chi_{\Gamma_1}|\tilde{X}\| - |(F_c)_\Gamma|\|\chi_{\Gamma_1}|\tilde{X}\| \\
&\geq \{1 - \mu^{-1}\|\chi_{\Gamma_1}|\tilde{X}\|\|\chi_{\Gamma_1}|\tilde{X}'\|\}\|F_c\chi_{\Gamma_1}|\tilde{X}\| \\
&\geq (1 - K_c)\|F_c\chi_{\Gamma_1}|\tilde{X}\|,
\end{aligned}$$

whence

$$A_c(\Gamma_1) \leq d(\Gamma)/(1 - K_c).$$

Similarly for $A_c(\Gamma_2)$ and the theorem is proved. □

Corollary 5.5.8. *Let $\tilde{X}(\Gamma,\mu) = \tilde{Y}(\Gamma,\mu) = L_p(\Gamma,\mu), 1 < p < \infty$, and $\Gamma = (a,b) \subset \mathbb{R}$. Suppose that dt is absolutely continuous with respect to $d\mu$ on Γ and that $dt/d\mu \in L_{p',loc}(\Gamma,\mu)$. For any $c \in [a,b)$ define*

$$N_c := \sup_{c \leq X < b} \mu(X,b)^{1/p}\Big(\int_c^X |\frac{dt}{d\mu}|^{p'} d\mu\Big)^{1/p'}. \tag{5.5.29}$$

Then, $d(\Gamma)$ in (5.5.24) satisfies

$$\lim_{c\to b-} N_c \lesssim d(\Gamma) \lesssim N_a. \tag{5.5.30}$$

Proof. In Theorem 5.5.7 we now have $\Gamma_1 = (a,c), \Gamma_2 = (c,b), \tilde{X}'(\Gamma,\mu) = L_{p'}(\Gamma,\mu)$ and so

$$K_c = \frac{1}{\mu(\Gamma)}\max\big\{\mu(\Gamma_1),\mu(\Gamma_2)\big\} < 1.$$

Also

$$A_c(\Gamma_i) = \|H_c | L_p(\Gamma_i, \mu) \to L_p(\Gamma_i, \mu)\|.$$

Since $f \in L_p(\Gamma, \mu)$ if, and only if, $f\big(\frac{d\mu}{dt}\big)^{1/p} \in L_p(\Gamma)$, it follows that

$$A_c(\Gamma_i) = \|T_c | L_p(\Gamma_i) \to L_p(\Gamma_i)\|,$$

where

$$T_c = \Big(\frac{d\mu}{dt}\Big)^{1/p} H_c \Big(\frac{dt}{d\mu}\Big)^{1/p}.$$

We now apply Theorem 2.2.1 with $q = p$ and

$$u(t) = \Big(\frac{dt}{d\mu}\Big)^{1/p}, \quad v(t) = \Big(\frac{d\mu}{dt}\Big)^{1/p}.$$

Thus from (2.2.3) and (2.2.4), we have

$$\begin{aligned} A_c(\Gamma_1)^p &\approx \sup_{a<X<c} \big\{ \|u\|^p_{p',(a,X)} \|v\|^p_{p,(X,c)} \big\} \\ &= \sup_{a<X<c} \Big\{ \Big(\int_a^X \Big|\frac{dt}{d\mu}\Big|^{1/(p-1)} \Big)^{p-1} \mu(X,c) \Big\}, \\ A_c(\Gamma_2)^p &\approx \sup_{c<X<b} \big\{ \|u\|^p_{p',(c,X)} \|v\|^p_{p,(X,b)} \big\} \\ &= \sup_{c<X<b} \Big\{ \Big(\int_c^X \Big|\frac{dt}{d\mu}\Big|^{1/(p-1)} \Big)^{p-1} \mu(X,b) \Big\}. \end{aligned}$$

Hence

$$\max\big\{A_c^p(\Gamma_1), A_c^p(\Gamma_2)\big\} \lesssim N_a^p.$$

Also, for any $c \in (a,b)$,

$$\max\big\{A_c^p(\Gamma_1), A_c^p(\Gamma_2)\big\} \gtrsim N_c^p,$$

whence (5.5.30). □

Since $\|I_{1,R} | L_p(\Omega) \to L_p(\Omega)\| \le R$ by Example 5.3.1 and (5.3.5) (see the proof of Example 5.5.2(i)), an immediate consequence of Theorem 5.4.8 is

Corollary 5.5.9. *Let $X(\Omega) = Y(\Omega) = L_p(\Omega), 1 < p < \infty$, where Ω is a GRD with generalised ridge $u(\Gamma)$, with $\Gamma = (a,b)$. Then, under the hypothesis of Corollary 5.5.8, the constant $D(\Omega)$ in Theorem 5.4.8 (with $\Omega_1 = \Omega$) satisfies*

$$\lim_{c\to b-} N_c \lesssim D(\Omega) \lesssim k(\Omega) + N_a, \tag{5.5.31}$$

where $k(\Omega) = \sup_\Omega \rho \circ \tau(x)$.

The case of Theorem 5.5.7 for Lorentz spaces on an interval Γ is covered within Example 5.7.1 below. Two-sided estimates for (5.5.27) when Γ is a tree are determined from Theorem 2.6.3 for L_p spaces and an analogue for Lorentz spaces given in [85], Theorem 7.2. Results for intervals Γ are known also for Orlicz spaces, and indeed for general BFSs; see [72].

5.6 Embedding Theorems

Theorem 5.4.6 provides a means of characterizing the possible range of target spaces $Z(\Omega)$ in the embedding $W(X,Y)(\Omega) \hookrightarrow Z(\Omega)$. To illustrate the technique we shall concentrate on the cases when $X(\Omega) = Y(\Omega) = L_p(\Omega), Z(\Omega) = L_q(\Omega, (\nu \circ \tau)dx), p, q \in (1,\infty)$ and $\Gamma = (a,b) \subset \mathbb{R}$. Hence

$$\begin{aligned} W(X,Y)(\Omega) &= W_p^1(\Omega), \\ \tilde{X}(\Gamma,\mu) = \tilde{Y}(\Gamma,\mu) &= L_p(\Gamma,\mu), \\ \tilde{Z}(\Gamma,\mu) &= L_q(\Gamma,\nu d\mu), \\ \tilde{W}(\tilde{X},\tilde{Y})(\Gamma,\mu) &= \tilde{W}(L_p(\Gamma,\mu), L_p(\Gamma,\mu)) =: \tilde{W}(\Gamma,\mu). \end{aligned} \tag{5.6.1}$$

No difficulties are introduced by the weight $\nu \circ \tau$ in $Z(\Omega)$, and we shall see in Example 5.6.2 that such a weight is required to deal with unbounded domains.

Lemma 5.6.1. *Let $p, q \in (1,\infty)$ and $a \leq c \leq h \leq d \leq b$. Then, for $F \in \tilde{W}(\Gamma,\mu) = \tilde{W}(\tilde{X},\tilde{Y})(\Gamma,\mu)$,*

$$\begin{aligned} &\|F|L_q((h,d),\nu d\mu)\| \\ &\leq \Big\|\frac{1}{\mu(c,\cdot)}\Big\{\int_c^{\cdot}\Big(\int_s^{\cdot}\psi^{p'}(x)d\mu(x)\Big)^{p/p'}d\mu(s)\Big\}\Big|L_{q/p}((h,d),\nu d\mu)\Big\|^{1/p} \\ &\times \|F'|L_p((c,d),\mu)\| \\ &+ \Big\|\frac{1}{\mu(c,\cdot)}\Big|L_{q/p}((h,d),\nu d\mu)\Big\|^{1/p}\|F|L_p((c,d),\mu)\| \end{aligned} \tag{5.6.2}$$

where $\psi(x) = dx/d\mu$, and

$$\begin{aligned} &\|F|L_q((c,h),\nu d\mu)\| \\ &\leq \Big\|\frac{1}{\mu(\cdot,d)}\Big\{\int_{\cdot}^{d}\Big(\int_{\cdot}^{s}\psi^{p'}(x)d\mu(x)\Big)^{p/p'}d\mu(s)\Big\}\Big|L_{q/p}((c,h),\nu d\mu)\Big\|^{1/p} \\ &\times \|F'|L_p((c,d),\mu)\| \\ &+ \Big\|\frac{1}{\mu(\cdot,d)}\Big|L_{q/p}((c,h),\nu d\mu)\Big\|^{1/p}\|F|L_p((c,d),\mu)\|. \end{aligned} \tag{5.6.3}$$

Also, for $a \leq s < t \leq b$,

$$\frac{|F(t)-F(s)|}{|t-s|^{\gamma}} \leq \Big(|t-s|^{-\gamma p'}\int_s^t \psi(x)^{p'}d\mu(x)\Big)^{1/p'}\|F'|L_p((a,b),\mu)\|. \tag{5.6.4}$$

Proof. From

$$F(t) = F(s) + \int_s^t F'(x)dx = F(s) + \int_s^t F'(x)\psi(x)d\mu(x) \tag{5.6.5}$$

and Hölder's inequality, we have for $t > s$,

$$|F(t)| \le |F(s)| + \Big(\int_s^t |F'(x)|^p d\mu(x)\Big)^{1/p}\Big(\int_s^t \psi(x)^{p'} d\mu(x)\Big)^{1/p'}.$$

On using the triangle inequality, this yields

$$\mu(c,t)^{1/p}|F(t)| \le \|F|L_p((c,t),\mu)\| + \Big\{\int_c^t \Big(\int_s^t |F'(x)|^p d\mu(x)\Big)\Big(\int_s^t \psi^{p'} d\mu(x)\Big)^{p/p'} d\mu(s)\Big\}$$

and

$$\|F|L_q((h,d),\nu d\mu)\| \le \Big\|\frac{1}{\mu(c,\cdot)^{1/p}}\|F|L_p((c,t),\mu)\|\Big|L_q((h,d),\nu d\mu)\Big\| + \Big\|\frac{1}{\mu(c,\cdot)^{1/p}}\Big\{\int_c^{\cdot}\Big(\int_s^{\cdot}|F'(x)|^p d\mu(x)\Big) \times \Big(\int_s^{\cdot}\psi^{p'}(x)d\mu(x)\Big)^{p/p'} d\mu(s)\Big\}^{1/p}\Big|L_q((h,d),\nu d\mu)\Big\|$$

whence (5.6.2). The proof of (5.6.3) is similar. The inequality (5.6.4) follows from (5.6.5). □

Example 5.6.2. (Horn-shaped domain) Let Ω be the GRD in Example 5.2.3, namely,

$$\Omega := \{x = (x_1, x') \in \mathbb{R}^n : 0 < x_1 < \infty, |x'| < \Phi(x_1)\}, \tag{5.6.6}$$

where Φ is smooth and bounded. On putting $c = 0, d = \infty, \phi(t) = \Phi(t)^{n-1}$ in (5.6.2) we derive

$$\|F|L_q((h,\infty),\nu d\mu)\| \le A_1(p,q,\nu,\mu)\|F'|L_p((0,\infty),\mu)\| + B_1(p,q,\nu,\mu)\|F|L_p((0,\infty),\mu)\|,$$

where

$$A_1(p,q,\nu,\mu)^q \approx \int_h^\infty \mu(0,t)^{-q/p} \times \Big[\int_0^t \Big(\int_s^t \phi^{-p'/p}(x)dx\Big)^{p/p'}\phi(s)ds\Big]^{q/p}\phi(t)\nu(t)dt \tag{5.6.7}$$

and

$$B_1(p,q,\nu,\mu)^q \approx \int_h^\infty \mu(0,t)^{-q/p}\phi(t)\nu(t)dt. \tag{5.6.8}$$

In this example we are forced to incorporate a weight ν in the space $\tilde{Z}$ in order to get a positive result. The requirement $I_{1,R} : L_p(\Omega) \to L_q(\Omega, \nu\circ\tau dx)$ is easily shown to be satisfied by standard methods (see [46], Lemma V.3.13), if $0 \le \frac{1}{p} - \frac{1}{q} < \frac{1}{n}$ and

$$\sup_{y\in\Omega}\int_{\Omega\cap B(y,R)} (\nu\circ\tau)(x)|x-y|^{r(n-1)}dx < \infty, \tag{5.6.9}$$

where $\frac{1}{r} = 1 + \frac{1}{q} - \frac{1}{p}$. Any bounded function ν satisfies (5.6.9).

- *Case 1* : $\phi(t) = e^{-\lambda t}, \lambda > 0$. Here $\mu(0,\infty) < \infty$, and so, with $h = 1$,

$$\begin{aligned} A_1(p,q,\nu,\mu)^q &\approx \int_1^\infty \Big[\int_0^t \big(1 - e^{-\lambda\frac{p'}{p}(t-s)}\big)^{p/p'} e^{-\lambda s} ds\Big]^{q/p} \\ &\quad \times e^{-\lambda t(1-\frac{q}{p})}\nu(t)dt \\ &\approx \int_1^\infty e^{-\lambda t(1-\frac{q}{p})}\nu(t)dt, \end{aligned} \tag{5.6.10}$$

$$B_1(p,q,\nu,\mu) \approx \int_0^\infty e^{-\lambda t}\nu(t)dt. \tag{5.6.11}$$

- *Case 2* : $\phi(t) = (t+1)^{-\theta}, \theta > 0$. On setting $s+1 = (t+1)z$ in (5.6.7),

$$\begin{aligned} &\int_0^t \Big(\int_s^t \phi^{-p'/p}(x)dx\Big)^{p/p'} \phi(s)ds \\ &= (t+1)^p \int_{1/(t+1)}^1 (1 - z^{\theta\frac{p'}{p}+1})^{p/p'} z^{-\theta} dz \\ &\approx \begin{cases} ((t+1)^p & \text{if} \quad 0 \le \theta < 1, \\ (t+1)^p \ \ln(t+1) & \text{if} \quad \theta = 1, \\ (t+1)^{p+\theta-1} & \text{if} \quad \theta > 1. \end{cases} \end{aligned}$$

Hence

$$A_1(p,q,\nu,\mu) \approx \begin{cases} \int_0^\infty (t+1)^{q(1-\theta)[\frac{1}{q}-\frac{1}{p}+\frac{1}{(1-\theta)}]-1}\nu(t)dt, & \theta < 1, \\ \int_0^\infty (t+1)^{q-1}\nu(t)dt, & \theta = 1, \\ \int_0^\infty (t+1)^{q(1-\theta)[\frac{1}{q}-\frac{1}{p}+\frac{1}{(1-\theta)}]-1}\nu(t)dt, & \theta > 1. \end{cases} \tag{5.6.12}$$

Also

$$B_1(p,q,\nu,\mu) \approx \begin{cases} \int_0^\infty (t+1)^{q(1-\theta)[\frac{1}{q}-\frac{1}{p}]-1}\nu(t)dt, & \theta < 1, \\ \int_0^\infty (t+1)^{-1}[\ln(t+1)]^{-q/p}\nu(t)dt, & \theta = 1, \\ \int_0^\infty (t+1)^{-\theta}\nu(t)dt, & \theta > 1. \end{cases} \tag{5.6.13}$$

Theorem 5.6.3. *Let Ω be the horn-shaped domain in (5.6.6) and set $\phi(t) = \Phi^{n-1}(t)$. Suppose that $p, q \in (1,\infty)$ and that (5.6.9) is satisfied. Then $W_p^1(\Omega) \hookrightarrow L_q(\Omega, \nu\circ\tau dx)$ if (5.6.10) and (5.6.11) are finite when $\phi(t) = e^{-\lambda t}(\lambda > 0)$, and if (5.6.12) and (5.6.13) are finite when $\phi(t) = (t+1)^{-\theta}(\theta \ge 0)$.*

Example 5.6.4. (Rooms and passages)

Let Ω be the rooms and passages domain analysed in Example 5.2.4, and to simplify the analysis take

$$h_{2j} = h^{-j}, h > 1; \quad \delta_{2j} = ch_{2j}^\alpha, \quad c > 0, \quad \alpha > 1. \tag{5.6.14}$$

From (5.2.25) with $\psi(t) := dt/d\mu$, we have when $p' \neq 2$,

$$\begin{aligned}&\int_{-\frac{1}{2}h_\delta}^{h+\frac{1}{2}h_\varepsilon} \psi(t)^{p'-1}dt \\ &= h_\delta/(2\delta^{p'-1}) + \{(2h-\delta)^{2-p'} - \delta^{2-p'}\}/\{4(2-p')\} \\ &+ (\varepsilon+\delta)/(2h^{p'-1}) + \{(2h-\varepsilon)^{2-p'} - \varepsilon^{2-p'}\}/\{4(2-p')\} + h_\varepsilon/(2\varepsilon^{p'-1}) \\ &\approx h_\delta/\delta^{p'-1} + h_\varepsilon/\varepsilon^{p'-1}\end{aligned}$$

by (5.2.16). If $p' = 2$,

$$\begin{aligned}&\int_{-\frac{1}{2}h_\delta}^{h+\frac{1}{2}h_\varepsilon} \psi(t)dt \\ &= h_\delta/(2\delta) + \frac{1}{4}\ln\{(2h-\delta)/\delta\} + (\varepsilon+\delta)/(2\delta) \\ &+ \frac{1}{4}\ln\{(2h-\varepsilon)/\varepsilon\} + h_\varepsilon/(2\varepsilon) \\ &\approx h_\delta/\delta + h_\varepsilon/\varepsilon,\end{aligned}$$

since $0 \leq \ln X \leq e^{-1}X$ for $1 \leq X < \infty$. Using the notation (5.2.17), let

$$c_m := H_{2m} - \frac{1}{2}h_{2m},$$

so that

$$c_{m+1} = H_{2m+1} + \frac{1}{2}h_{2m+2}.$$

It then follows that

$$\int_{c_m}^{c_{m+1}} \psi(x)^{p'} d\mu(x) \approx h^{-m[1-(p'-1)\alpha]}, \tag{5.6.15}$$

$$\mu(c_m, c_{m+1}) \approx h^{-m(1+\alpha)}. \tag{5.6.16}$$

On setting $c = c_1, h = c_2, d = b$ and $\nu(x) = 1$ in (5.6.2), we obtain

$$\begin{aligned}\|F|L_q((c_2,b),\mu)\| \leq A_2(p,q,\mu)\|F'|L_p((a,b),\mu)\| \\ + B_2(p,q,\mu)\|F|L_p((a,b),\mu)\|,\end{aligned}$$

where

$$\begin{aligned}A_2(p,q,\mu)^q = \sum_{m=2}^{\infty}\int_{c_m}^{c_{m+1}}\{\frac{1}{\mu(c_1,t)} \\ \times\Big[\int_{c_1}^{t}\Big(\int_s^t \psi^{p'/p}(x)dx\Big)^{p/p'}\mu'(s)ds\Big]\}^{q/p}\mu'(t)dt,\end{aligned} \tag{5.6.17}$$

and

$$B_2(p,q,\mu)^q = \sum_{m=2}^{\infty} \int_{c_m}^{c_{m+1}} \frac{1}{\mu(c_1,t)^{q/p}} \mu'(t)dt. \tag{5.6.18}$$

From (5.6.16), we have for $t \in (c_m, c_{m+1})$ and $s \in (c_j, c_{j+1}), j \leq m$, that

$$\begin{aligned}\int_s^t \psi^{p'/p}(x)dx &\approx \sum_{l=j}^{m} h^{-l[1-(p'-1)\alpha]} \\ &\approx \begin{cases} h^{-j\beta} & \text{if} \quad \beta := 1-(p'-1)\alpha > 0, \\ h^{-m\beta} & \text{if} \quad \beta < 0, \\ m^{-j} & \text{if} \quad \beta = 0. \end{cases}\end{aligned}$$

On substituting in (5.6.17) and using (5.6.16), it follows that

$$\begin{aligned}&\int_{c_1}^t \Big(\int_s^t \psi^{p'/p}(x)dx \Big)^{p/p'} \mu'(s)ds \\ &\approx \begin{cases} \sum_{j=1}^m h^{-j\beta p/p'} h^{-j(1+\alpha)} & \text{if} \quad \beta > 0, \\ \sum_{j=1}^m h^{-m\beta p/p'} h^{-j(1+\alpha)} & \text{if} \quad \beta < 0, \\ \sum_{j=1}^m (m-j) h^{-j(1+\alpha)} & \text{if} \quad \beta = 0, \end{cases} \\ &\approx \begin{cases} 1 & \text{if} \quad \beta > 0, \\ h^{-m\beta p/p'} & \text{if} \quad \beta > 0, \\ m & \text{if} \quad \beta = 0. \end{cases}\end{aligned}$$

Also,

$$\mu(c_1, t) \approx \sum_{l=1}^{m} \int_{c_l}^{c_{l+1}} \mu'(t)dt \approx 1. \tag{5.6.19}$$

Thus

$$A_2(p,q,\mu) \approx \begin{cases} \sum_{m=2}^\infty h^{-m(1+\alpha)} & \text{if} \quad \beta > 0, \\ \sum_{m=2}^\infty h^{-m[\beta q/p'+1+\alpha]} & \text{if} \quad \beta < 0, \\ \sum_{m=2}^\infty m h^{-m(1+\alpha)} & \text{if} \quad \beta = 0, \end{cases}$$

and $A_2(p,q,\mu) < \infty$ if, and only if, one of the following holds:

1. $\beta \geq 0$, i.e., $\alpha \leq p-1$;
2. $\beta < 0$ and $\frac{\beta q}{p'} + \alpha + 1 > 0$, i.e., $\alpha > p-1$ and $\frac{1}{q} > \frac{1}{p} - \frac{1}{\alpha+1}$.

Note that in view of Remark 5.3.2(ii), the hypothesis of Theorem 5.4.6 with $X = Y = L_p$ and $Z = L_q$ is satisfied if $0 \leq \frac{1}{p} - \frac{1}{q} < \frac{1}{n}$. We therefore have

Theorem 5.6.5. *Let Ω be the rooms and passages domain defined in Example 5.2.4 and let (5.6.14) be satisfied. Then, for $p \in (1,\infty)$, $W_p^1(\Omega)$ is continuously embedded in $L_q(\Omega)$ if one of the following holds:*

1. $1 < p < 1 + \alpha$ *and* $q \in [p, p_\alpha)$, *where* $\frac{1}{p_\alpha} = \frac{1}{p} - \frac{1}{1+\alpha}$;
2. $p \geq 1 + \alpha$ *and* $q \in [p, \infty)$.

It is proved in [86], Remark 6.4 that when $1 < p < 1 + \alpha$ and $q > p_\alpha$ the conclusion of the theorem is false.

5.7 The Poincaré inequality and $\alpha(E)$

Theorems 5.4.8 and 5.5.7 establish a method for analysing the validity of the Poincaré inequality (5.5.23), and Corollary 5.5.5 shows how upper and lower bounds can be obtained for $\alpha(E_M)$, and hence $\alpha(E)$ by Theorem 5.3.12. To illustrate the effectiveness of the method we consider examples in which Γ is an interval and X, Y are Lebesgue or Lorentz spaces. Examples of the general scheme for trees Γ and general BFS can be found in [86].

If Γ is an interval (a, b), the family $\mathcal{A}(\Gamma)$ in (5.2.3) consists of intervals $[R, b)$ for some $R \in (a, b)$. In Lemma 5.5.4, with $\Lambda = [R, b)$, we have

$$\phi_\Lambda = \sup\left\{ \|F - F_\Lambda|\tilde{X}(\Lambda, \mu)\| : F \in \tilde{W}(\Lambda, \mu), \|F'|\tilde{Y}(\Lambda, \mu)\| = 1\right\} := d_\Lambda \tag{5.7.1}$$

in the notation of (5.5.24) and (5.6.1). Let $c \in (R, b)$ and set $\Lambda_1 = (R, c), \Lambda_2 = (c, b)$. Then (5.5.25) is satisfied since $dt/d\mu \in L_{\infty,loc}(\Gamma)$ by assumption (see (5.2.12)). Also, we shall see that when $\tilde{X}(\Lambda_i, d\mu), i = 1, 2$, are Lebesgue or Lorentz spaces,

$$\eta_c := \mu(\Lambda)^{-1} \max_{i=1,2} \left\{ \|\chi_{\Lambda_i}|\tilde{X}'(\Lambda, \mu)\| \|\chi_\Lambda|\tilde{X}(\Lambda, \mu)\| \right\} < 1. \tag{5.7.2}$$

From (5.5.27), given $\varepsilon > 0$, there exists $f \in \tilde{Y}(\Lambda, \mu)$ with support in Λ_1 and $\|f|\tilde{Y}(\Gamma, \mu)\| = 1$, such that

$$\|H_c f|\tilde{X}(\Gamma, \mu)\| \geq A_c(\Lambda_1) - \varepsilon.$$

Moreover, since $F_c = H_c f$ is also supported in Λ_1,

$$\begin{aligned}\|H_c f|\tilde{X}(\Gamma, \mu)\| &\leq \|F_c - (F_c)_{\Lambda_1}|\tilde{X}(\Gamma, \mu)\| + \|(F_c)_{\Lambda_1}|\tilde{X}(\Gamma, \mu)\| \\ &\leq \phi_\Lambda + \eta_c \|F_c|\tilde{X}(\Gamma, \mu)\|,\end{aligned}$$

whence

$$\|H_c f|\tilde{X}(\Gamma, \mu)\| \leq (1 - \eta_c)^{-1} \phi_\Lambda.$$

When f is supported in Λ_2, we have a similar result and so

$$\max_{i=1,2} \left(A_c(\Lambda_i) - \varepsilon \right) \leq (1 - \eta_c)^{-1} \phi_\Lambda.$$

Therefore, from Theorem 5.5.7 and (5.7.1),

$$\phi_\Lambda \leq d_\Lambda \leq 2(1 - \eta_c)^{-1} \{1 + \mu^{-1} \|\chi_\Lambda|\tilde{X}(\Gamma, \mu)\| \|\chi_\Lambda|\tilde{X}'(\Gamma, \mu)\|\} \phi_\Lambda.$$

Example 5.7.1. Let $X(\Omega) = L_{r,s}(\Omega), Y(\Omega) = L_{p,q}(\Omega), p, q, r, s \in [1, \infty]$, so that $\tilde{X}(\Gamma, \mu) = L_{r,s}(\Gamma, \mu)$ and $\tilde{Y}(\Gamma, \mu) = L_{p,q}(\Gamma, \mu)$. To simplify the exposition we shall suppose that these are all BFSs. The associate space $\tilde{X}'(\Gamma, \mu)$ is $L_{r',s'}(\Gamma, \mu)$, where $r' = r/(r-1), s' = s/(s-1)$. If $f = \chi_{(\alpha,\beta)}, (\alpha, \beta) \subset \Gamma$, then

$$\begin{aligned} f_\mu^* &= \inf \left\{ \lambda > 0 : \mu\{x : |\chi_{(\alpha,\beta)}(x)| > \lambda\} \le t \right\} \\ &= \begin{cases} 1 & \text{if} \quad t < \mu(\alpha, \beta), \\ 0 & \text{if} \quad t \ge \mu(\alpha, \beta), \end{cases} \end{aligned}$$

and

$$\|\chi_{(\alpha,\beta)}|L_{r,s}(\Gamma, \mu)\| = \begin{cases} \mu(\alpha, \beta)^{1/s} & \text{if} \quad s \in [1, \infty), \\ \mu(\alpha, \beta)^{1/r} & \text{if} \quad s = \infty. \end{cases}$$

For $1 < s < \infty$, it follows that in (5.5.26) and (5.7.2), for the interval (R, b),

$$\begin{aligned} K_c &= \mu(R, b)^{-1} \max \left\{ \mu(R, c)^{1/s}, \mu(c, b)^{1/s'} \right\} < 1, \\ \eta_c &= \mu(R, b)^{-1} \max \left\{ \mu(R, c)^{1/s'}, \mu(R, b)^{1/s} \right\} < 1, \end{aligned}$$

and the same applies for $s = 1, \infty$.

In [205] and [55] it is proved that if

$$s \ge \max\{p, q\} \tag{5.7.3}$$

and Λ is any interval in $\Gamma, H_c : L_{p,q}(\Lambda, \mu) \to L_{r,s}(\Lambda, d\mu)$ if, and only if,

$$B_\Lambda := \sup_{x \in \Lambda} \left\{ \|\chi_\Lambda \chi_{(x,\infty)}|L_{r,s}(\Lambda, \mu)\| \|\chi_\Lambda \chi_{(0,x)} \frac{dt}{d\mu} |L_{p',q'}(\Lambda, \mu)\| < \infty, \right.$$

and in this case

$$\|H_c|L_{p,q}(\Lambda, \mu) \to L_{r,s}(\Lambda, \mu)\| \approx B_\Lambda.$$

Hence, with $\Lambda = (R, b), \Lambda_1 = (R, c), \Lambda_2 = (c, b)$, we have from Theorem 5.5.7 that

$$d_\Lambda \approx \max(B_{\Lambda_1}, B_{\Lambda_2}),$$

for all $R < c < b$. When $\Lambda = \Gamma$, this gives a criterion for the validity of the Poincaré inequality (5.5.24), and hence (5.5.23).

By Example 5.5.2 with $X(\Omega) = Z(\Omega) = L_{r,s}(\Omega)$ and $Y(\Omega) = L_{p,q}(\Omega)$, the conclusion of Theorem 5.5.1 holds if

$$1 \le p < n, \quad 1 \le r < p^* = np/(n-p). \tag{5.7.4}$$

We also have from Remark 5.3.2, and with $A = \tau^{-1}(\Lambda)$ and $R \approx k(A)$, that for all $f \in L_{p,q}(A)$ and $q, s < \infty$,

$$\begin{aligned} \|I_{1,R} f|L_{r,s}(A)\| &= R \|\tau_{1/R} I_{1,1} \tau_R f|L_{r,s}(A)\| \\ &= R^{1+n(\frac{1}{r} - \frac{1}{p})} \|I_{1,1}|L_{p,q}(\Omega) \to L_{r,s}(\Omega)\| \|f|L_{p,q}(A)\| \end{aligned}$$

and so, as $R \to 0$,

$$\|I_{1,R}|L_{p,q}(\Omega) \to L_{r,s}(\Omega)\| = R^{n(\frac{1}{r}-\frac{1}{p^*})}\|I_{1,1}|L_{p,q}(\Omega) \to L_{r,s}(\Omega)\| \to 0$$

if (5.7.4) is satisfied. Hence by Corollary 5.5.5, since $\mathbb{M}$ maps $X(\Omega)$ and $Y(\Omega)$ into themselves if $p, r \in (1,\infty)$ (see Remark 5.4.7) and $C_0^\infty(\Omega)$ is dense in $X(\Omega)$, we have if $p,q,r,s \in (1,\infty)$ and (5.7.4) is satisfied, that

$$\limsup_{R\to b} d_{(R,b)} \lesssim \alpha(E_M) \lesssim \liminf_{R\to b} d_{(R,b)}.$$

We therefore need to estimate

$$B_R(z) := \|\chi_{(z,b)}|L_{r,s}((R,b);d\mu)\|\,\Big\|\chi_{(R,z)}\frac{dt}{d\mu}|L_{p',q'}((R,b),d\mu)\Big\|; \tag{5.7.5}$$

the Poincaré inequality (5.5.24) (and (5.5.23)) holds if, and only if,

$$\sup_{a<s<b} B_a(s) < \infty$$

in which case $\alpha(E) \in [0,1)$, and E and E_M are compact if, and only if,

$$\lim_{R\to b}\big[\sup_{R<z<b} B_R(z)\big] = 0.$$

Example 5.7.2. (Horn-shaped domains) Let Ω be the domain in (5.6.6), i.e.

$$\Omega := \{x = (x_1, x') \in \mathbb{R}^n : 0 < x_1 < \infty, |x'| < \Phi(x_1)\},$$

and suppose that in addition to Φ being smooth and non-increasing we have

$$\int_0^\infty \Phi(x_1)^{n-1}dx_1 < \infty.$$

From (5.2.15),

$$\varphi(t) := \frac{d\mu}{dt} = \omega_{n-1}\Phi(t)^{n-1}. \tag{5.7.6}$$

In (5.7.5) we have if $q \in [1,\infty)$,

$$\begin{aligned}
&\|\chi_{(z,\infty)}|L_{r,s}((R,\infty),\mu)\| \\
&= \Big(\int_0^\infty s\lambda^{s-1}[\mu\{t \in (0,\infty) : |\chi_{(z,\infty)}(t)| > \lambda\}]^{s/r}d\lambda\Big)^{1/s} \\
&= \Big(\int_0^1 s\lambda^{s-1}\mu(z,\infty)^{s/r}d\lambda\Big)^{1/s} = \mu(z,\infty)^{1/r},
\end{aligned} \tag{5.7.7}$$

and, on putting $R_\lambda = \max\{R, \varphi^{-1}(1/\lambda)\}$, if $q \in (1,\infty]$,

$$\Big\|\chi_{(R,z)}\frac{dt}{d\mu}|L_{p',q'}((R,\infty),\mu)\Big\|^{q'}$$

$$= \int_0^\infty q'\lambda^{q'-1}\big[\mu\{t \in (0,\infty) : \frac{\chi_{(R,z)}(t)}{\varphi(t)} > \lambda\}\big]^{q'/p'} d\lambda$$

$$= \int_0^\infty q'\lambda^{q'-1}\mu(R_\lambda, z)^{q'/p'} d\lambda$$

$$= \int_0^{1/\varphi(R)} q'\lambda^{q'-1}\big(\int_R^z \varphi(t)dt\big)^{q'/p'} d\lambda$$

$$+ \int_{1/\varphi(R)}^{1/\varphi(z)} q'\lambda^{q'-1}\big(\int_{\varphi^{-1}(1/\lambda)}^z \varphi(t)dt\big)^{q'/p'} d\lambda$$

$$= \big(\int_R^z \varphi(t)dt\big)^{q'/p'} \varphi^{-q'}(R) + \int_{1/\varphi(R)}^{1/\varphi(z)} q'\lambda^{q'-1}\big(\int_{\varphi^{-1}(1/\lambda)}^z \varphi(t)dt\big)^{q'/p'} d\lambda.$$

Hence for $s \in [1,\infty)$ and $q \in (1,\infty]$,

$$B_R(z) = \big(\int_z^\infty \varphi(t)dt\big)^{1/r}\{\varphi^{-q'}(R)\big(\int_R^z \varphi(t)dt\big)^{q'/p'}$$
$$+ \int_{1/\varphi(R)}^{1/\varphi(z)} q'\lambda^{q'-1}\big(\int_{\varphi^{-1}(1/\lambda)}^z \varphi(t)dt\big)^{q'/p'} d\lambda\}^{1/q'}. \qquad (5.7.8)$$

Theorem 5.7.3. *Let Ω be defined by (5.6.6) with Φ smooth, non-increasing and $\int_0^\infty \Phi(x_1)^{n-1}dx_1 < \infty$. Let $X(\Omega) = L_{r,s}(\Omega), Y(\Omega) = L_{p,q}(\Omega), p, q, r, s \in [1,\infty]$, where (5.7.3) and (5.7.4) are also satisfied. Then with φ defined in (5.7.6), the Poincaré inequality (5.5.23) holds in the following cases if, and only if, the indicated inequalities hold:*

1. $\varphi(t) = (t+1)^{-\theta}, \theta > 1 : \frac{1}{r} \geq \frac{1}{p} + \frac{1}{(\theta-1)}$;
2. $\varphi(t) = e^{-\theta t}, \theta > 0 : r \leq p$;
3. $\varphi(t) = e^{-t^\theta}, \theta > 0 : r \leq p$.

In (1), (2) and case $0 < \theta \leq 1$ of (3), $\alpha(E) = 0$ if, and only if, the inequality is strict; if $\theta > 1$ in (3), $\alpha(E) = 0$ if, and only if, $r \leq p$.

Proof. (1) In (5.7.8) we have

$$\int_z^\infty \varphi(t)dt = (\theta-1)^{-1}(z+1)^{1-\theta},$$

$$\varphi^{-q'}(R)\big(\int_R^z \varphi(t)dt\big)^{q'/p'} \begin{cases} \lesssim (R+1)^{q'(\theta-1+p)/p}, \ R \leq z < \infty, \\ \gtrsim (R+1)^{q'(\theta-1+p)/p}, \ 2R \leq z < \infty, \end{cases}$$

and

$$\int_{1/\varphi(R)}^{1/\varphi(z)} q'\lambda^{q'-1}\big(\int_{\varphi^{-1}(1/\lambda)}^z \varphi(t)dt\big)^{q'/p'} d\lambda$$

$$= q'(\theta-1)^{-1}\int_{(R+1)^\theta}^{(z+1)^\theta} \lambda^{q'-1-(1-\frac{1}{\theta})\frac{q'}{p'}}\big(1 - [\frac{\lambda^{1/\theta}}{R+1}]^{\theta-1}\big)^{q'/p'} d\lambda$$

$$\begin{cases} \lesssim (z+1)^{q'(\theta-1+p)/p}, & R \le z < \infty, \\ \gtrsim (z+1)^{q'(\theta-1+p)/p}, & 4R \le z < \infty. \end{cases}$$

Hence

$$\sup_{R<z<\infty} B_R(z) \approx \sup_{R<z<\infty} \{z^{(\theta-1)(\frac{1}{p}-\frac{1}{r})+1}\},$$

whence the result.

(2) We now have in (5.7.8),

$$\int_z^\infty \varphi(t)dt = \theta^{-1}e^{-\theta z},$$

$$\varphi^{-q'}(R)\Big(\int_R^z \varphi(t)dt\Big)^{q'/p'} \begin{cases} \lesssim e^{\theta Rq'/p}, \ R \le z < \infty, \\ \gtrsim e^{\theta Rq'/p}, \ R+1 \le z < \infty, \end{cases}$$

and

$$\int_{1/\varphi(R)}^{1/\varphi(z)} q'\lambda^{q'-1}\Big(\int_{\varphi^{-1}(1/\lambda)}^z \varphi(t)dt\Big)^{q'/p'} d\lambda$$
$$= q'\theta^{-q'/p'} \int_{e^{\theta R}}^{e^{\theta z}} \lambda^{q'-1}\big[\lambda^{-1} - e^{-z\theta}\big]^{q'/p'} d\lambda$$
$$\lesssim e^{\theta \frac{q'}{p'} z}.$$

Hence

$$B_R(z) \begin{cases} \lesssim e^{\theta z(\frac{1}{p}-\frac{1}{r})}, & R < z < \infty, \\ \gtrsim e^{-\theta z/p + \theta R/p}, & R+1 < z < \infty, \end{cases}$$

and the result follows.

(3) On integration by parts

$$\int_z^\infty e^{-t^\theta} dt = \frac{1}{\theta} z^{1-\theta} e^{-z^\theta} - \Big(\frac{\theta-1}{\theta}\Big) \int_z^\infty e^{-t^\theta} t^{-\theta} dt$$
$$\le \frac{1}{\theta} z^{1-\theta} e^{-z^\theta},$$

and, if $z \ge 1$,

$$\int_z^\infty e^{-t^\theta} dt \ge \frac{1}{\theta} z^{1-\theta} e^{-z^\theta} - \Big(\frac{\theta-1}{\theta}\Big) \int_z^\infty e^{-t^\theta} dt$$

which yields

$$\int_z^\infty e^{-t^\theta} dt \begin{cases} \lesssim z^{1-\theta} e^{-z^\theta} \text{ if } & z > 0, \\ \gtrsim z^{1-\theta} e^{-z^\theta} \text{ if } & z \ge 1. \end{cases}$$

Now,

$$\int_R^z e^{-t^\theta}\,dt = -\frac{1}{\theta}\big[t^{1-\theta}e^{-t^\theta}\big]_R^z - \left(\frac{\theta-1}{\theta}\right)\int_R^z e^{-t^\theta}t^{-\theta}\,dt$$
$$\le \frac{1}{\theta}R^{1-\theta}e^{-R^\theta},$$

and if $R \ge 1$,

$$\int_R^z e^{-t^\theta}\,dt \ge \frac{1}{\theta}\big[R^{1-\theta}e^{-R^\theta} - z^{1-\theta}e^{-z^\theta}\big] - \left(\frac{\theta-1}{\theta}\right)\int_R^z e^{-t^\theta}\,dt,$$

whence

$$\int_R^z e^{-t^\theta}\,dt \begin{cases} \lesssim R^{1-\theta}e^{-R^\theta}, & \text{if} \quad z>0, \\ \gtrsim R^{1-\theta}e^{-R^\theta}, & \text{if} \quad z \ge 2R. \end{cases}$$

The final term in (5.7.8) is now

$$I_2 := q'\int_{e^{-z^\theta}}^{e^{-R^\theta}} \lambda^{-q'-1}\left(\int_{(\ln\frac{1}{\lambda})^{1/\theta}}^z e^{-t^\theta}\,dt\right)^{q'/p'} d\lambda,$$

where we have, on integration by parts,

$$\int_{(\ln\frac{1}{\lambda})^{1/\theta}}^z e^{-t^\theta}\,dt = -\frac{1}{\theta}\big[t^{1-\theta}e^{t^\theta}\big]_{(\ln\frac{1}{\lambda})^{1/\theta}}^z$$
$$- \left(\frac{\theta-1}{\theta}\right)\int_{(\ln\frac{1}{\lambda})^{1/\theta}}^z t^{-\theta}e^{-t^\theta}\,dt$$
$$\le \frac{1}{\theta}\lambda\Big(\ln\frac{1}{\lambda}\Big)^{-(1-\frac{1}{\theta})},$$

and so

$$I_2 \lesssim \int_{e^{-z^\theta}}^{e^{-R^\theta}} \lambda^{-q'-1+\frac{q'}{p'}}\Big[\ln\frac{1}{\lambda}\Big]^{-(1-\frac{1}{\theta})\frac{q'}{p'}}\,d\lambda$$
$$\lesssim (z^\theta)^{-(1-\frac{1}{\theta})\frac{q'}{p'}}\int_{e^{-z^\theta}}^{e^{-R^\theta}} \lambda^{-\frac{q'}{p}-1}\,d\lambda$$
$$\lesssim z^{-(\theta-1)\frac{q'}{p'}}\exp(z^\theta\tfrac{q'}{p}).$$

Hence, if $z \ge R$,

$$B_R(z) \lesssim (z^{1-\theta})e^{-z^\theta})^{1/r}\big\{e^{q'R^\theta}\big[R^{1-\theta}e^{-R^\theta}\big]^{q'/p'} + z^{(1-\theta)q'/p'}\exp(z^{\theta q'/p})\big\}^{1/q'}$$
$$\lesssim z^{(1-\theta)/r}e^{z^\theta(\frac{1}{p}-\frac{1}{r})}\big\{R^{(1-\theta)/p'} + z^{(1-\theta)/p'}\big\},$$

while if $z \ge 2R$,

$$B_R(z) \gtrsim (z^{1-\theta}e^{-z^\theta})^{1/r}\big\{e^{q'R^\theta}\big[R^{1-\theta}e^{-R^\theta}\big]^{q'/p'}\big\}^{1/q'}$$

$$\gtrsim z^{(1-\theta)/r} e^{-z^\theta/r} R^{(1-\theta/p'} e^{R^\theta/p}.$$

It follows that

$$\sup_{1 \le R < z < \infty} B_R(z) < \infty$$

if, and only if, $r \le p$, and

$$\lim_{R \to \infty} \sup_{R < z < \infty} B_R(z) = 0$$

if, and only if, $r < p$ for $0 < \theta \le 1$, and $r \le p$ for $\theta > 1$. The theorem is therefore proved. □

Example 5.7.4. (Rooms and passages) Let Ω be the domain in Examples 5.2.4 and 5.6.4 and suppose that (5.6.14) is satisfied. If in (5.5.29), $c \in [H_{2M} - \frac{1}{2}h_{2M}, H_{2M+1} + \frac{1}{2}h_{2M+2})$ and $X \in [H_{2k} - \frac{1}{2}h_{2k}, H_{2k+1} + \frac{1}{2}h_{2k+2})$ for $k > M$, it follows from (5.2.25), as for (5.6.19), that

$$N_c = \sup_{M \le k < \infty} N(M, k),$$

where

$$N(M,k) \approx \sum_{j=k}^{\infty} h^{-2j} \sum_{i=M}^{k} h^{-2i[1-\alpha(p'-1)]}.$$

From this we readily derive the following:

$$N(M,k) \approx \begin{cases} h^{-(p+1-\alpha)k} & \text{if} \quad \alpha > p-1, \\ h^{-2k-(p-1-\alpha)M} & \text{if} \quad \alpha < p-1, \\ h^{-2k}(k-M)^{p-1} & \text{if} \quad \alpha = p-1. \end{cases}$$

Hence, $N_0 < \infty$ if and only if $\alpha \le p+1$ and $\lim_{c \to b-} N_c = 0$ if and only if $\alpha < p+1$.

In view of Corollaries 5.5.5 and 5.5.9 we have therefore established the following theorem.

Theorem 5.7.5. *Let Ω be the rooms and passages domain in Example 5.7.4. Then the Poincaré inequality*

$$\|f - f_\Omega | L_p(\Omega)\| \le K \|\nabla f | L_p(\Omega)\|$$

holds for all $f \in W_p^1(\Omega)$ if, and only if, $\alpha \le p+1$. The embeddings $E : W_p^1(\Omega) \to L_p(\Omega), E_M : W_{p,M}^1(\Omega) \to L_{p,M}(\Omega)$ are compact if, and only if, $\alpha < p+1$.

5.8 Notes

5.1. The proof of Theorem 5.1.5 is taken from [80], but another proof may be found in [128], Lemma 2.1.29; see also Lemma 8.5.12 in [127]. We note the interesting result of Motzkin given in [128], Theorem 2.1.30, that if F is a closed set in $\mathbb{R}^n$, it is convex if, and only if, for every $x \notin F$, there is a unique point in F at minimal distance from x, or equivalently, in view of Theorem 5.1.5, that the distance function d_F is continuously differentiable on $\mathbb{R}^n \setminus F$.

Sets like the ridge and skeleton of a domain have been studied in relation to the theory of shape recognition. Some further remarks and references may be found in [80], Section 6.

5.2. The rooms and passages domain of Example 5.2.4 was analysed by Amick [6] and Fraenkel [94], and is similar to a domain considered by Courant and Hilbert [38] and by Maz'ya [171]. The class of GRDs is wide and includes many pathological examples. In [80], [82] examples other than the rooms and passages and horns are analysed, including "interlocking combs", "spirals" and a self-similar domain made up of rectangles. The Koch snowflake is also an example.

5.3. The treatment in this section is modeled on that in [86], which in turn is based on that for Lebesgue spaces in [80]. The case $X = Y = L_2$ was proved earlier in Amick [6].

General mapping properties of the Riesz potential $I_{1,1}$ and the maximal function $\mathbb{M}$ of (5.4.11) are given in [88], [89] and [70].

5.6. A different technique for investigating the target space of the embedding $E(\Omega)$ in the case of Lebesgue spaces X, Y is developed in Gol'dstein and Gurov [110]. Their method requires Ω to be locally homeomorphic to a domain with a smooth boundary via a bi-Lipschitz map satisfying specific conditions. Some rooms and passages domains are covered by the analysis and similar results to Theorem 5.6.5 are obtained.

The book [172] by Maz'ya and Poborchi gives a comprehensive treatment of the properties of Sobolev spaces on domains with cusps. For instance, for a domain with an outer peak and $1 < p \leq q < \infty$, they establish in Section 8.2 necessary and sufficient conditions for $W_p^1(\Omega)$ to be continuously embedded in specified target spaces depending on the values of q.

6

Approximation numbers of Sobolev embeddings

6.1 Introduction

Suppose $|\Omega| < \infty$ and let $W^1_{p,\mathbb{C}}(\Omega)$ denote the quotient space $W^1_p(\Omega)/\mathbb{C}$ with norm

$$\|[f]|W^1_{p,\mathbb{C}}(\Omega)\| := \|\nabla f\|_{p,\Omega}, \tag{6.1.1}$$

where $[\cdot]$ denotes an equivalence class in $W^1_{p,\mathbb{C}}(\Omega)$ of functions which differ a.e. by a constant. The quotient space $L_p(\Omega)/\mathbb{C}$ will be denoted by $L_{p,\mathbb{C}}(\Omega)$ and its norm by

$$\|[f]|L_{p,\mathbb{C}}(\Omega)\| \equiv \|[f]\|_{p,\Omega} := \inf_{c\in\mathbb{C}} \|f - c\|_{p,\Omega}. \tag{6.1.2}$$

From Theorem 5.3.9 and (5.3.20), with $E(\Omega) : W^1_p(\Omega) \to L_p(\Omega)$, we see that $\alpha(E(\Omega)) < 1$ if, and only if, for all $f \in W^1_p(\Omega)$,

$$\inf_{c\in\mathbb{C}} \|f - c\|_{p,\Omega} \lesssim \|\nabla f\|_{p,\Omega}.$$

This is equivalent to $W^1_{p,\mathbb{C}}(\Omega)$ being continuously embedded in $L_{p,\mathbb{C}}(\Omega)$; also $W^1_{p,\mathbb{C}}(\Omega)$ is a Banach space whose norm is equivalent to the standard quotient space norm $\inf_{c\in\mathbb{C}} \|f - c|W^1_p(\Omega)\|$. In this case, we denote the embedding $W^1_{p,\mathbb{C}}(\Omega) \hookrightarrow L_{p,\mathbb{C}}(\Omega)$ by $E_\mathbb{C}(\Omega)$. When $p = 2$, let $E^*_\mathbb{C}(\Omega) : L_{2,\mathbb{C}}(\Omega) \to W^1_{2,\mathbb{C}}(\Omega)$ be the dual map defined by

$$(E_\mathbb{C}(\Omega)f, g)_{L_{2,\mathbb{C}}(\Omega)} = (f, E^*_\mathbb{C}(\Omega)g)_{W^1_{2,\mathbb{C}}(\Omega)} \tag{6.1.3}$$

for all $f \in W^1_{2,\mathbb{C}}(\Omega), g \in L_{2,\mathbb{C}}(\Omega)$, where $(\cdot,\cdot)$ denotes the inner-product on the indicated space. In this case of $p = 2$, we have in $L_{2,\mathbb{C}}(\Omega)$,

$$E_\mathbb{C}(\Omega)E^*_\mathbb{C}(\Omega) = (-\Delta_{N,\Omega})^{-1}, \tag{6.1.4}$$

where $\Delta_{N,\Omega}$ denotes the Neumann Laplacian. To see this, recall that $-\Delta_{N,\Omega}$ in $L_{2,\mathbb{C}}(\Omega)$ is the unique self-adjoint operator with domain in $W^1_{2,\mathbb{C}}(\Omega)$ and satisfying

$$(u,\phi)_{W^1_{2,\mathbb{C}}(\Omega)} = (-\Delta_{N,\Omega}u, E_{\mathbb{C}}(\Omega)\phi)_{L_{2,\mathbb{C}}(\Omega)} \tag{6.1.5}$$

for all $u \in \mathcal{D}(-\Delta_{N,\Omega})$ and $\phi \in W^1_{2,\mathbb{C}}(\Omega)$. But

$$(-\Delta_{N,\Omega}u, E_{\mathbb{C}}(\Omega)\phi)_{L_{2,\mathbb{C}}(\Omega)} = (E^*_{\mathbb{C}}(\Omega)(-\Delta_{N,\Omega})u,\phi)_{W^1_{2,\mathbb{C}}(\Omega)} \tag{6.1.6}$$

and so $E^*_{\mathbb{C}}(\Omega)(-\Delta_{N,\Omega})u = u$, whence (6.1.4).

If $E_{\mathbb{C}}(\Omega)$ is compact and $p=2$, the approximation numbers $a_m(E_{\mathbb{C}}(\Omega))$ are given by

$$a_m(E_{\mathbb{C}}(\Omega)) = \kappa_m(E_{\mathbb{C}}(\Omega)) = \kappa_m(E^*_{\mathbb{C}}(\Omega)), \tag{6.1.7}$$

where $\kappa_m(\cdot)$ is the mth singular value, that is, $\kappa_m(E_{\mathbb{C}}(\Omega))^2$ is the mth eigenvalue of the compact self-adjoint operator $E_{\mathbb{C}}(\Omega)E^*_{\mathbb{C}}(\Omega)$ in $L_{2,\mathbb{C}}(\Omega)$, when the eigenvalues are arranged in decreasing order and repeated according to multiplicity; see [46], Theorem II.5.10. Hence, from (6.1.4), and since $E_{\mathbb{C}}(\Omega)$ and $E^*_{\mathbb{C}}(\Omega)$ have the same approximation numbers (see [46], Proposition II.2.5),

$$a^2_m(E_{\mathbb{C}}(\Omega)) = \lambda^{-1}_m(-\Delta_{N,\Omega}), \tag{6.1.8}$$

where $\lambda_m(-\Delta_{N,\Omega})$ is the mth eigenvalue of $-\Delta_{N,\Omega}$ in $L_{2,\mathbb{C}}(\Omega)$, that is, the mth positive eigenvalue of $-\Delta_{N,\Omega}$ as an operator in $L_2(\Omega)$ since 0 is then a simple eigenvalue. It follows from (6.1.6) that

$$\begin{aligned} \nu_{\mathbb{C}}(\varepsilon,\Omega) &:= \max\{m : a_m(E_{\mathbb{C}}(\Omega)) \geq \varepsilon\} && (6.1.9)\\ &= \sharp\{m : \lambda_m(-\Delta_{N,\Omega}) \leq \varepsilon^{-2}\} \\ &=: N(\varepsilon^{-2}, -\Delta_{N,\Omega}). && (6.1.10)\end{aligned}$$

Similarly, if $E_0(\Omega)$ is the embedding $\mathring{W}^1_p(\Omega) \hookrightarrow L_p(\Omega)$, where $\mathring{W}^1_p(\Omega)$ has norm

$$\|f\|_{\mathring{W}^1_p(\Omega)} = \|\nabla f\|_{p,\Omega} \tag{6.1.11}$$

then, in the case $p=2$,

$$\begin{aligned} \nu_0(\varepsilon,\Omega) &:= \max\{m : a_m(E_0(\Omega)) \geq \varepsilon\} && (6.1.12)\\ &= N(\varepsilon^{-2}, -\Delta_{D,\Omega}), && (6.1.13)\end{aligned}$$

where $\Delta_{D,\Omega}$ is the Dirichlet Laplacian in $L_2(\Omega)$.

The asymptotic behaviour of (6.1.9) as $\varepsilon \to 0$, for any $p \in (1,\infty)$, will be the main concern of this chapter. In the case $p=2$, the identities (6.1.8) and (6.1.10) enable the Courant-Weyl method of Dirichlet-Neumann bracketing to be used to give asymptotic results. The essence of this method is the tesselation of Ω by non-overlapping cubes Q_i and the inequalities

$$\begin{aligned} \sum_i N(\lambda, -\Delta_{D,Q_i}) &\leq N(\lambda, -\Delta_{D,\Omega}) \\ &\leq N(\lambda, -\Delta_{N,\Omega}) \end{aligned}$$

$$\leq \sum_i N(\lambda, -\Delta_{N,Q_i}) \tag{6.1.14}$$

which are consequences of the spectral theorem and the min-max principle. The known eigenvalues of $-\Delta_{D,Q_i}$ and $-\Delta_{N,Q_i}$ then provide lower and upper bounds for $N(\lambda, -\Delta_{D,\Omega})$ and $N(\lambda, -\Delta_{N,\Omega})$; see [46], Chapter XI, for details. This technique is not available for $p \neq 2$, but a partial analogue was established in [82], Section 5. A description of these results will be given in Section 6.3 below. Firstly, we need to examine the equivalence of various norms in $W^1_{p,\mathbb{C}}(\Omega)$ and $L_{p,\mathbb{C}}(\Omega)$ in order to be able to use the theory.

6.2 Some quotient space norms

The domain Ω is a connected open subset of $\mathbb{R}^n, n \geq 1$, of finite volume $|\Omega|$ and $1 < p < \infty$ throughout. The quotient space $L_{p,\mathbb{C}}(\Omega) \equiv L_p(\Omega)/\mathbb{C}$ has norm

$$\|[f]|L_{p,\mathbb{C}}(\Omega)\| \equiv \|[f]\|_{p,\Omega} := \inf_{c\in\mathbb{C}} \|f - c\|_{p,\Omega} \tag{6.2.1}$$

and functions are equivalent if and only if they differ a.e. by a constant. Another relevant space in subsequent analysis is

$$L_{p;g}(\Omega) := \{f \in L_p(\Omega) : \Omega_g(f) := \frac{\int_\Omega fg dx}{\int_\Omega g dx} = 0\} \tag{6.2.2}$$

where $g \in L_{p'}(\Omega)$, $\frac{1}{p} + \frac{1}{p'} = 1$, and $\int_\Omega g dx \neq 0$. Clearly $L_{p;g}(\Omega)$ is a closed subspace of $L_p(\Omega)$ and $f \mapsto f - \Omega_g(f)$ is a projection of $L_p(\Omega)$ onto $L_{p;g}(\Omega)$. We shall write

$$\|f|L_{p;g}(\Omega)\| = \|f - \Omega_g(f)\|_{p,\Omega}, \quad f \in L_p(\Omega) \tag{6.2.3}$$

and hence

$$\|f|L_{p;g}(\Omega)\| = \|f\|_{p,\Omega}, \quad f \in L_{p;g}(\Omega).$$

When $g = 1$ we denote $L_{p;1}(\Omega)$ by $L_{p,M}(\Omega)$ to be consistent with (5.3.19).

Theorem 6.2.1. *Let V_Ω denote the canonical map $V_\Omega : L_p(\Omega) \to L_{p,\mathbb{C}}(\Omega)$, given by $f \mapsto [f]$. Then, for all $f \in L_p(\Omega)$,*

$$\|V_\Omega f|L_{p,\mathbb{C}}(\Omega)\| \leq \|f|L_{p;g}(\Omega)\| \leq c(p,g)\|V_\Omega f|L_{p,\mathbb{C}}(\Omega)\|, \tag{6.2.4}$$

where

$$c(p,g) = 1 + \frac{|\Omega|^{1/p}\|g\|_{p',\Omega}}{|\int_\Omega g dx|}.$$

When $p = 2$, and $g = 1$,

$$\|V_\Omega f|L_{2,\mathbb{C}}(\Omega)\| = \|f|L_{2;1}(\Omega)\| = \|f - \Omega_1(f)\|_{2,\Omega}. \tag{6.2.5}$$

Proof. The first inequality in (6.2.4) is obvious, while the second is a consequence of

$$|\Omega_g(f)| \leq \frac{1}{|\int_\Omega g(x)dx|}\|g\|_{p',\Omega}\|f\|_{p,\Omega}$$

which follows from Hölder's inequality. The identity (6.2.5) follows from the fact that the map $f \mapsto f_\Omega$ is an orthogonal projection of $L_2(\Omega)$ onto $\mathbb{C}$. □

We now set

$$W^1_{p;g}(\Omega) := \{f : f \in W^1_p(\Omega), \Omega_g(f) = 0\} \tag{6.2.6}$$

with norm

$$\|f|W^1_{p;g}(\Omega)\| := \|\nabla f\|_{p,\Omega}$$

and define the embedding

$$E_g(\Omega) : W^1_{p;g}(\Omega) \to L_{p;g}(\Omega).$$

Note that $W^1_{p;g}(\Omega)$ is a closed subspace of $W^1_p(\Omega)$ if we assume (as we shall hereafter) that the Poincaré inequality

$$\|f - \Omega_g(f)\|_{p,\Omega} \lesssim \|\nabla f\|_{p,\Omega}, \quad f \in W^1_p(\Omega)$$

is satisfied, for then the norm on $W^1_{p;g}(\Omega)$ is equivalent to the $W^1_p(\Omega)$ norm and $f \mapsto \Omega_g(f)$ is bounded on $W^1_p(\Omega)$. We shall denote $W^1_{p;1}(\Omega)$ and $E_1(\Omega)$ by $W^1_{p,M}(\Omega)$ and $E_M(\Omega)$ respectively. We have the direct sum decomposition

$$W^1_p(\Omega) = W^1_{p;g}(\Omega) \dot{+} \mathbb{C} \tag{6.2.7}$$

determined by

$$f = [f - \Omega_g(f)] + \Omega_g(f)$$

since $\Omega_g(\Omega_g(f)) = \Omega_g(f)$, and the map $W : [f] \mapsto f - \Omega_g(f)$ is an isometry between $W^1_{p,\mathbb{C}}(\Omega)$ and $W^1_{p;g}(\Omega)$. The map W is also a topological isomorphism of $L_{p,\mathbb{C}}(\Omega)$ onto $L_{p;g}(\Omega)$, by Theorem 6.2.1, satisfying

$$\|[f]\|_{L_{p,\mathbb{C}}(\Omega)} \leq \|W[f]|L_{p;g}(\Omega)\| \tag{6.2.8}$$

$$= \|f - \Omega_g f\|_{p,\Omega} \leq c(p,g)\|[f]|L_{p,\mathbb{C}}(\Omega)\|. \tag{6.2.9}$$

Note that if $f \in W^1_{p;g}(\Omega)$ and $h \in [f]$, then $h = f + \Omega_g(h)$ with $W^{-1}f = [f]$; a similar relationship holds between $L_{p,\mathbb{C}}(\Omega)$ and $L_{p;g}(\Omega)$. In the analysis to follow, it is helpful to distinguish between the maps

$$W_1 : [f] \mapsto f - \Omega_g(f) : W^1_{p,\mathbb{C}}(\Omega) \to W^1_{p;g}(\Omega) \tag{6.2.10}$$

and

$$W_0 : [f] \mapsto f - \Omega_g(f) : L_{p,\mathbb{C}}(\Omega) \to L_{p;g}(\Omega). \tag{6.2.11}$$

From the preceding remarks, both maps are surjective, W_1 is an isometry and

$$1 \leq \|W_0\| \leq c(p,g). \tag{6.2.12}$$

It is readily seen that

$$E_g(\Omega) = W_0 E_{\mathbb{C}}(\Omega) W_1^{-1}, \tag{6.2.13}$$
$$E_{\mathbb{C}}(\Omega) = W_0^{-1} E_g(\Omega) W_1. \tag{6.2.14}$$

Theorem 6.2.2. *For all $f \in L_p(\Omega)$,*

$$\|V_\Omega f | L_{p,\mathbb{C}}(\Omega)\| = \min_{g\in\mathcal{S}} \|f - \Omega_g(f)\|_{p,\Omega}, \tag{6.2.15}$$

where

$$\mathcal{S} := \{g : g \in L_{p'}(\Omega), \int_\Omega g(x)dx \neq 0\}.$$

When $p = 2$ we have

$$\|V_\Omega f | L_{2,\mathbb{C}}(\Omega)\| = \|f - \Omega_1(f)\|_{2,\Omega}.$$

Proof. If $f_0 \in L_p(\Omega)$ and $V_\Omega f_0 = 0$, then $f_0 = c$, a constant, a.e. and, for any $g \in \mathcal{S}$,

$$\|V_\Omega f_0 | L_{p,\mathbb{C}}(\Omega)\| = \|c - \Omega_g(c)\|_{p,\Omega} = 0.$$

Hence, from (6.2.4),

$$\|V_\Omega f_0 | L_{p,\mathbb{C}}(\Omega)\| = \min_{g\in\mathcal{S}} \|f_0 - \Omega_g(f_0)\|_{p,\Omega}.$$

Suppose $\|V_\Omega f | L_{p,\mathbb{C}}(\Omega)\| \neq 0$. Then, by Lemma 2.4.2, there exists a unique constant $c = c(f)$ such that

$$\|V_\Omega f | L_{p,\mathbb{C}}(\Omega)\| = \|f - c\|_{p,\Omega}.$$

Since $f - c \neq 0$, the one-dimensional vector space $\{\lambda(f-c) : \lambda \in \mathbb{C}\}$ does not contain the constant function 1. Therefore, by the Hahn-Banach Theorem, there exists $l \in \left(L_p(\Omega)\right)^*$, the dual of $L_p(\Omega)$, such that

$$l(f-c) = 0,$$
$$l(1) = 1.$$

Hence, there exists $g_0 \in L_{p'}(\Omega)$ with

$$\int_\Omega [f(x) - c] g_0(x)dx = 0,$$

$$\int_{\Omega} g_0(x)dx = 1.$$

Therefore, $\Omega_{g_0}(f) = c$ and

$$\|V_{\Omega} f | L_{p,\mathbb{C}}(\Omega)\| = \|f - \Omega_{g_0}(f)\|_{p,\Omega}.$$

Since

$$\|V_{\Omega} f | L_{p,\mathbb{C}}(\Omega)\| \leq \|f - \Omega_g(f)\|_{p,\Omega}$$

for any $g \in \mathcal{S}$, (6.2.15) follows. The case $p = 2, g = 1$ was noted in (6.2.5). □

The approximation numbers of $E_g(\Omega)$ and $E_{\mathbb{C}}(\Omega)$ both feature in the analysis to follow. We assume throughout that the Poincaré inequality holds on Ω and hence that $E_g(\Omega)$ and $E_{\mathbb{C}}(\Omega)$ are continuous embeddings. They are related by

Lemma 6.2.3. *For any $g \in \mathcal{S}$,*

$$a_m(E_{\mathbb{C}}(\Omega)) \leq a_m(E_g(\Omega)) \leq c(p,g) a_m(E_{\mathbb{C}}(\Omega)) \tag{6.2.16}$$

where $c(p,g)$ is given in Theorem 6.2.1; when $g = 1$, $c(2,1) = 1$ and $c(p,1) = 2$ for $p \neq 2$. Therefore

$$\nu_{\mathbb{C}}(\varepsilon, \Omega) \leq \nu_g(\varepsilon, \Omega) \leq \nu_{\mathbb{C}}(\frac{\varepsilon}{c(p,g)}, \Omega), \tag{6.2.17}$$

where $\nu_{\mathbb{C}}(\varepsilon, \Omega) := \sharp\{m : a_m(E_{\mathbb{C}}(\Omega)) \geq \varepsilon\}$ and $\nu_g(\varepsilon, \Omega) := \sharp\{m : a_m(E_g(\Omega)) \geq \varepsilon\}$.

Proof. Let $P : W^1_{p,\mathbb{C}}(\Omega) \to L_{p,\mathbb{C}}(\Omega)$ be of rank $< m$. Then, from (6.2.13) and (6.2.14),

$$\begin{aligned}
&\|E_g(\Omega) - W_0 P W_1^{-1} | W^1_{p;g}(\Omega) \to L_{p;g}(\Omega)\| \\
&= \|W_0(E_{\mathbb{C}} - P) W_1^{-1} | W^1_{p;g}(\Omega) \to L_{p;g}(\Omega)\| \\
&\leq c(p,g) \|E_{\mathbb{C}} - P | W^1_{p,\mathbb{C}}(\Omega) \to L_{p,\mathbb{C}}(\Omega)\|.
\end{aligned}$$

Since $\text{rank}(W_0 P W_1^{-1}) \leq \text{rank} P < m$, the second inequality in (6.2.16) follows. The first inequality is proved similarly. The inequalities (6.2.17) are immediate consequences of (6.2.16). □

Results similar to the preceding ones in this section hold also on the tree Γ. The analogous spaces defined on Γ are the following:

$$L_{p,\mathbb{C}}(\Gamma, \mu) : \|[u] | L_{p,\mathbb{C}}(\Gamma, \mu)\| := \inf_{c \in \mathbb{C}} \|u - c | L_p(\Gamma, \mu)\|; \tag{6.2.18}$$

$$L_{p;h}(\Gamma,\mu) := \{u \in L_p(\Gamma,\mu) : \omega_h(u) := \frac{\int_\Gamma uh d\mu}{\int_\Gamma h d\mu} = 0\},$$

$$\|u|L_{p;h}(\Gamma,\mu)\| := \|u - \omega_h(u)\|_{L_p(\Gamma,\mu)}, \tag{6.2.19}$$

where $h \in L_{p'}(\Gamma,\mu)$ and $\int_\Gamma h d\mu \neq 0$;

$$L^1_{p,\mathbb{C}}(\Gamma,\mu) \ : \ \|[u]|L^1_{p,\mathbb{C}}(\Gamma,\mu)\| = \|u'|L_p(\Gamma,\mu)\|, \tag{6.2.20}$$

$$L^1_{p;h}(\Gamma,\mu) := \{u \in L^1_p(\Gamma,\mu) : \omega_h(u) = 0\},$$

$$\|u|L^1_{p;h}(\Gamma,\mu)\| := \|u'|L_p(\Gamma,\mu)\|. \tag{6.2.21}$$

We denote the associated embeddings by

$$I_\mathbb{C}(\Gamma) : L^1_{p,\mathbb{C}}(\Gamma,\mu) \to L_{p,\mathbb{C}}(\Gamma,\mu), \tag{6.2.22}$$

$$I_h(\Gamma) : L^1_{p;h}(\Gamma,\mu) \to L_{p;h}(\Gamma,\mu). \tag{6.2.23}$$

The maps

$$T_1 : [u] \mapsto u - \omega_h(u) : L^1_{p,\mathbb{C}}(\Gamma, d\mu) \to L^1_{p;h}(\Gamma,\mu), \tag{6.2.24}$$

$$T_0 : [u] \mapsto u - \omega_h(u) : L_{p,\mathbb{C}}(\Gamma,\mu) \to Lp; h(\Gamma,\mu) \tag{6.2.25}$$

are respectively an isometry and an isomorphism, with

$$1 \leq \|T_0\| \leq d(p,h) := 1 + \frac{\mu(\Gamma)^{1/p}\|h|L_{p'}(\Gamma,\mu)\|}{|\int_\Gamma h d\mu|}. \tag{6.2.26}$$

The analogue of Lemma 6.2.3 is

Lemma 6.2.4.

$$a_m(I_\mathbb{C}(\Gamma)) \leq a_m(I_h(\Gamma)) \leq d(p,h) a_m(I_\mathbb{C}(\Gamma)) \tag{6.2.27}$$

and

$$\nu_\mathbb{C}(\varepsilon,\Gamma) \leq \nu_h(\varepsilon,\Gamma) \leq \nu_\mathbb{C}(\frac{\varepsilon}{d(p,h)},\Gamma), \tag{6.2.28}$$

where $d(p,h)$ is given in (6.2.26); in particular $d(2,1) = 1$ and $d(p,1) = 2$ for $p \neq 2$. In (6.2.28),

$$\nu_\mathbb{C}(\varepsilon,\Gamma) := \max\{\mathrm{m} : \mathrm{a_m}(\mathrm{I_\mathbb{C}}(\Gamma)) \geq \varepsilon\},$$

and

$$\nu_h(\varepsilon,\Gamma) := \max\{\mathrm{m} : \mathrm{a_m}(\mathrm{I_h}(\Gamma)) \geq \varepsilon\}.$$

6.3 Dirichlet-Neumann bracketing in L_p

Our main concern in this section will be to obtain upper and lower bounds for the quantities

$$\nu_M(\varepsilon, \Omega) := \max\{m : a_m(E_M(\Omega)) \geq \varepsilon\}, \tag{6.3.1}$$

$$\nu_0(\varepsilon, \Omega) := \max\{m : a_m(E_0(\Omega)) \geq \varepsilon\} \tag{6.3.2}$$

where $E_M(\Omega) : W^1_{p,M}(\Omega) \to L_{p,M}(\Omega), E_0(\Omega) : \overset{\circ}{W}{}^1_p(\Omega) \to L_p(\Omega)$ are continuous embeddings and Ω is a domain of finite volume in $\mathbb{R}^n, n \geq 1$. In view of (6.2.17), estimates for $\nu_{\mathbb{C}}(\varepsilon, \Omega)$ will follow (since $\nu_M \equiv \nu_1$). Our approach is based on that in [82], Section 5, which attempts to mimic the Dirichlet-Neumann bracketing technique described in Section 6.1. To achieve this aim, the following additional quantities related to the approximation numbers of $E_M(\Omega)$ and $E_0(\Omega)$ are needed. Since $E_M(\Omega)$ is injective, it follows that, if S is a finite-dimensional linear subspace of $W^1_{p,M}(\Omega)$, the restriction of $E_M(\Omega)^{-1}$ to $E_M(\Omega)S$ is a bounded linear operator with bound

$$\alpha(S) := \sup_{u \in S} \left\{ \frac{\|\nabla u\|_{p,\Omega}}{\|E_M(\Omega)u\|_{p,\Omega}} \right\}. \tag{6.3.3}$$

Let $d(S)$ denote the dimension of S and define

$$\mu(\varepsilon, \Omega) := \max\{d(S) : \alpha(S) \leq 1/\varepsilon\}. \tag{6.3.4}$$

Similarly, define $\mu_0(\varepsilon, \Omega)$ with respect to $E_0(\Omega)$. We shall see in Lemma 6.3.2 below that, in the case $p = 2$, $\mu(\varepsilon, \Omega)$ and $\mu_0(\varepsilon, \Omega)$ coincide with $\nu(\varepsilon, \Omega)$ and $\nu_0(\varepsilon, \Omega)$ respectively.

Lemma 6.3.1. *We have for all* $\varepsilon > 0$,

$$\nu_0(\varepsilon, \Omega) \leq \nu_M(\varepsilon, \Omega) + 1, \tag{6.3.5}$$

$$\mu(\varepsilon, \Omega) \leq \nu_M(\varepsilon, \Omega), \tag{6.3.6}$$

$$\mu_0(\varepsilon, \Omega) \leq \nu_0(\varepsilon, \Omega). \tag{6.3.7}$$

Proof. Let $P : W^1_{p,M}(\Omega) \to L_{p,M}(\Omega)$ have rank $r(P)$, and define

$$P_0 f := f_\Omega + P(f - f_\Omega),$$

where f_Ω is the integral mean

$$f_\Omega := \frac{1}{|\Omega|} \int_\Omega f(x)dx,$$

that is, $f_\Omega = \Omega_1(f)$ in the notation of (6.2.2). The operator P_0 maps $\overset{\circ}{W}{}^1_p(\Omega)$ into $L_p(\Omega)$ and has rank $r(P_0) \leq r(P) + 1$. Suppose $r(P) + 2 \leq \nu_0(\varepsilon, \Omega)$. Then $r(P_0) \leq \nu_0(\varepsilon, \Omega) - 1$ and so

$$\|E_0(\Omega) - P_0|\mathring{W}_p^1(\Omega) \to L_p(\Omega)\| \geq a_{\nu_0(\varepsilon,\Omega)} \geq \varepsilon.$$

Hence, for any $\eta < \varepsilon$, there exists $g \in \mathring{W}_p^1(\Omega)$ such that

$$\|g - g_\Omega - P(g - g_\Omega)\|_{p,\Omega} \geq \eta \|\nabla g\|_{p,\Omega}.$$

Since $g - g_\Omega \in W_{p,M}^1(\Omega)$, it follows that $\|E_M(\Omega) - P|W_{p,M}^1(\Omega) \to L_p(\Omega)\| \geq \eta$ and so, since $\eta < \varepsilon$ is arbitrary, $\|E_M(\Omega) - P|W_{p,M}^1(\Omega) \to L_p(\Omega)\| \geq \varepsilon$. As this is true for any P of rank $r(P)$ satisfying $r(P) + 1 \leq \nu_0(\varepsilon, \Omega) - 1$, (6.3.5) is proved.

Next, let S be a subspace of $W_{p,M}^1(\Omega)$ of dimension $d(S)$ and $P : W_{p,M}^1(\Omega) \to L_{p,M}(\Omega)$ any bounded linear operator of rank $r(P) < d(S)$. Then if $\{e_1, \cdots, e_{d(S)}\}$ is a basis of S, there exist $\lambda_1, \cdots, \lambda_{d(S)}$, not all zero, such that

$$P(\sum_{i=1}^{d(S)} \lambda_i e_i) = 0.$$

Hence, with $\psi = \sum_{i=1}^{d(S)} \lambda_i e_i$ we have

$$\|(E_M(\Omega) - P)\psi\|_{p,\Omega} = \|E_M(\Omega)\psi\|_{p,\Omega} \geq \alpha(S)^{-1} \|\nabla\psi\|_{p,\Omega}$$

and consequently

$$\|E_M(\Omega) - P|W_{p,M}^1(\Omega) \to L_{p,M}(\Omega)\| \geq \alpha(S)^{-1}.$$

It follows that $a_{d(S)}(E_M(\Omega)) \geq \alpha(S)^{-1}$ and $a_{\mu(\varepsilon,\Omega)}(\Omega) \geq \varepsilon$. Therefore (6.3.6) is proved. A similar proof holds for (6.3.7). □

Lemma 6.3.2. *Let $p = 2$ and suppose that $E_M(\Omega)$ and $E_0(\Omega)$ are compact. Then for all $\varepsilon > 0$,*

$$\mu(\varepsilon, \Omega) = \nu_M(\varepsilon, \Omega) \tag{6.3.8}$$

$$= \nu_{\mathbb{C}}(\varepsilon, \Omega) = \sharp\{m : \lambda_m(-\Delta_{N,\Omega}) \leq \varepsilon^{-2}\}, \tag{6.3.9}$$

$$\mu_0(\varepsilon, \Omega) = \nu_0(\varepsilon, \Omega) \tag{6.3.10}$$

$$= \sharp\{m : \lambda_m(-\Delta_{D,\Omega}) \leq \varepsilon^{-2}\}, \tag{6.3.11}$$

where $\lambda_m(-\Delta_{N,\Omega}), \lambda_m(-\Delta_{D,\Omega})$ are respectively the mth positive eigenvalue of the Neumann Laplacian and the mth eigenvalue of the Dirichlet Laplacian in $L_2(\Omega)$, the eigenvalues being arranged in increasing order of magnitude and repeated according to multiplicity.

Proof. We showed in (6.1.8) that

$$a_m^2(E_{\mathbb{C}}(\Omega)) = \lambda_m(-\Delta_{N,\Omega}).$$

Since the maps W_0, W_1 in (6.2.10) and (6.2.11) are both isometric when $p = 2$ and $g = 1$, it follows that $a_m(E_{\mathbb{C}}(\Omega)) = a_m(E_M(\Omega))$ for all m and hence $\nu_{\mathbb{C}}(\varepsilon, \Omega) = \nu_M(\varepsilon, \Omega)$ for all $\varepsilon > 0$. Also,

$$E_{\mathbb{C}}^*(\Omega)E_{\mathbb{C}}(\Omega) = W_1^* E_M^*(\Omega)E_M(\Omega)W_1.$$

A consequence of this last identity and the Min-max Theorem (see [46], Theorem II.5.6) and (6.1.4) is that

$$\sharp\{m : \lambda_m(-\Delta_{N,\Omega}) \leq \varepsilon^{-2}\} = \max\{\dim S : S \in \mathcal{K}(\varepsilon^2)\},$$

where $\mathcal{K}(\varepsilon^2)$ is the set of closed linear subspaces S of $W^1_{p,M}(\Omega)$ such that, for all $f \in S$,

$$\begin{aligned}\varepsilon^2 \|\nabla f\|_{2,\Omega}^2 &\leq (E_{\mathbb{C}}^*(\Omega)E_{\mathbb{C}}(\Omega)f, f)_{W^1_{p,M}(\Omega)} \\ &= \|E_{\mathbb{C}}(\Omega)f\|_{2,\Omega}^2 \\ &= \|E_M(\Omega)f\|_{2,\Omega}^2.\end{aligned}$$

Therefore, in the notation (6.3.3), $\alpha(S) \leq \varepsilon^{-1}$ and $\dim S \leq \mu(\varepsilon, \Omega)$ for all $S \in \mathcal{K}(\varepsilon^2)$, which imply that $\nu_M(\varepsilon, \Omega) \leq \mu(\varepsilon, \Omega)$. The reverse inequality has already been established in (6.3.6) and so (6.3.8) and (6.3.9) are proved. The proofs of (6.3.10) and (6.3.11) are similar. □

Lemma 6.3.3. *Let $\Omega = \left(\bigcup_{i=1}^q \Omega_i\right) \cup N$, where the Ω_i are disjoint open subsets of $\mathbb{R}^n$ and N is a null set. Then*

$$\nu_M(\varepsilon, \Omega) \leq \Sigma_{i=1}^q [\nu_M(\varepsilon, \Omega_i) + 1].$$

Proof. Since $a_m(E_M(\Omega_i)) < \varepsilon$ when $m = \nu_M(\varepsilon, \Omega_i) + 1$, there exists a bounded linear operator $P_i : W^1_{p,M}(\Omega_i) \to L_{p,M}(\Omega_i)$ with $\operatorname{rank}(P_i) \leq \nu_M(\varepsilon, \Omega_i)$ and such that

$$\|E_M(\Omega_i) - P_i | W^1_{p,M}(\Omega_i) \to L_{p,M}(\Omega_i)\| =: \varepsilon_i < \varepsilon.$$

Let P be the operator defined on $W^1_{p,M}(\Omega)$ by

$$Pf := \Sigma_{i=1}^q \chi_{\Omega_i}\{f_{\Omega_i} + P_i(f - f_{\Omega_i})\}.$$

Then $Pf \in L_{p,M}(\Omega)$ and

$$\begin{aligned}\|f - Pf\|_{p,\Omega}^p &= \Sigma_{i=1}^q \|f - f_{\Omega_i} - P(f - f_{\Omega_i})\|_{p,\Omega_i}^p \\ &\leq \left(\max_{1\leq i\leq q} \varepsilon_i\right)^p \Sigma_{i=1}^q \|\nabla f\|_{p,\Omega_i}^p \\ &\leq \varepsilon^p \|\nabla f\|_{p,\Omega}^p.\end{aligned}$$

Furthermore,

$$\operatorname{rank}(P) \leq \Sigma_{i=1}^q \operatorname{rank}(P_i) + q \leq \Sigma_{i=1}^q \nu_M(\varepsilon, \Omega_i) + q.$$

Therefore

$$a_l(E_M(\Omega) < \varepsilon,$$

where $l = \Sigma_{i=1}^q \nu_M(\varepsilon, \Omega_i) + q + 1$, and so

$$\nu_M(\varepsilon, \Omega) \leq \Sigma_{i=1}^q \nu_M(\varepsilon, \Omega_i) + q.$$

□

Lemma 6.3.4. *Let the hypothesis of Lemma 6.3.3 be satisfied. Then*

$$\mu_0(\varepsilon, \Omega) \geq \Sigma_{i=1}^q \mu_0(\varepsilon, \Omega_i).$$

Proof. For each i there exists a subspace S_i of $\mathring{W}_p^1(\Omega_i)$ of dimension $\mu_0(\varepsilon, \Omega_i)$ and such that $\alpha(S_i) \leq \varepsilon^{-1}$. The direct sum S of these S_i is a subspace of $\mathring{W}_p^1(\Omega)$ of dimension $\Sigma_{i=1}^q \mu_0(\varepsilon, \Omega_i)$ and $\alpha(S) \leq \varepsilon^{-1}$, whence the result. □

Lemma 6.3.5. *Let $\Omega = \Omega_1 \cup \Omega_2 \cup N$, where Ω_1 and Ω_2 are disjoint open subsets of $\mathbb{R}^n$ and N is a null set. Suppose that for all $f \in \mathring{W}_p^1(\Omega)$, $\|f\|_{p,\Omega_2} \leq \varepsilon\|\nabla f\|_{p,\Omega_2}$. Then for all $\eta > \varepsilon$,*

$$\nu_0(\eta, \Omega) \leq \nu_M(\eta, \Omega_1) + 1.$$

Proof. Let $P : W_{p,M}^1(\Omega_1) \to L_{p,M}(\Omega_1)$ have finite rank and define

$$Qf := \{f_{\Omega_1} + P(f - f_{\Omega_1})\}\chi_{\Omega_1}$$

for $f \in \mathring{W}_p^1(\Omega)$. Then $\text{rank}(Q) \leq \text{rank}(P) + 1$ and

$$\begin{aligned}
\|(E_0(\Omega) - Q)f\|_{p,\Omega}^p &= \|(f - f_{\Omega_1}) - P(f - f_{\Omega_1})\|_{p,\Omega_1}^p + \|f\|_{p,\Omega_2}^p \\
&\leq \|E_M(\Omega_1) - P|W_{p,M}^1(\Omega_1) \to L_{p,M}(\Omega_1)\|^p \|\nabla f\|_{p,\Omega_1}^p \\
&\quad + \varepsilon^p \|\nabla f\|_{p,\Omega_2}^p \\
&\leq \big[\max\{\|E_M(\Omega_1) - P|W_{p,M}^1(\Omega_1) \to L_{p,M}(\Omega_1)\|, \varepsilon\} \\
&\quad \times \|\nabla f\|_{p,\Omega}\big]^p,
\end{aligned}$$

whence

$$\begin{aligned}
&\|E_0(\Omega) - Q|\mathring{W}_p^1(\Omega) \to L_p(\Omega)\| \leq \\
&\max\{\|E_M(\Omega_1) - P|W_{p,M}^1(\Omega_1) \to L_{p,M}(\Omega_1)\|, \varepsilon\}.
\end{aligned}$$

If $\text{rank}(P) + 1 \leq \nu_0(\eta, \Omega) - 1$, it follows that $\text{rank}(Q) < \nu_0(\eta, \Omega)$ and consequently $\|E_0(\Omega) - Q|\mathring{W}_p^1(\Omega) \to L_p(\Omega)\| \geq \eta$. Hence $\|E_M(\Omega_1) - P|W_{p,M}^1(\Omega_1) \to L_{p,M}(\Omega_1)\| \geq \eta$ and thus, since P is arbitrary, the lemma follows. □

Lemma 6.3.6. *Let Ω_1 be the image of Ω under an affine transformation which magnifies distances by a factor λ. Then $\nu_M(\varepsilon, \Omega_1) = \nu_M(\varepsilon\lambda^{-1}, \Omega)$ and similarly for ν_0, μ_0 and μ.*

Proof. Since Ω_1 is obtained from Ω by a similarity transformation $t \mapsto a + \lambda t$, the result is readily verified by a change of co-ordinates in the expressions for the approximation numbers. □

We are now in a position to give two theorems which will be important tools in our analysis of the asymptotic behaviour of $\nu_M(\varepsilon, \Omega)$ and $\nu_0(\varepsilon, \Omega)$ as $\varepsilon \to 0$

Theorem 6.3.7. *Let Q be an open cube in $\mathbb{R}^n$. Then*

$$\lim_{\varepsilon \to 0} \{\varepsilon^n \nu_M(\varepsilon, Q)\} = \inf_{\varepsilon > 0} \{\varepsilon^n \nu_M(\varepsilon, Q)\}. \tag{6.3.12}$$

Proof. For simplicity we shall prove the result for the case when Q is a square of side length 1 in $\mathbb{R}^2$. The general case is proved similarly.

Let R be a rectangle in $\mathbb{R}^2$. From the Poincaré inequality

$$\|f - f_R\|_{p,R} \lesssim \operatorname{diam} R \|\nabla f\|_{p,R}$$

on $W^1_{p,M}(R)$ it follows that $a_1(E_M(R)) \le c \operatorname{diam} R$ for some positive constant c, and hence $\nu_M(\varepsilon, R) = 0$ if $\operatorname{diam} R < \varepsilon / c$.

Let $\lambda \ge \lambda_0 \ge 1$. Then

$$\lambda = [\lambda/\lambda_0]\lambda_0 + \theta\lambda_0,$$

where $[\cdot]$ denotes the integer part and $0 \le \theta < 1$. We have that λQ can be expressed (modulo a null set) as a disjoint union of $[\lambda/\lambda_0]^2$ open squares congruent to $\lambda_0 Q$ together with an L-shaped region which can be cut up into $2[\lambda/\lambda_0] + 1$ rectangles R_j each of diameter less than $\sqrt{2}\lambda_0$. Each of these rectangles R_j is the union of $\{[c\lambda_0\sqrt{2}\delta^{-1}] + 1\}^2$ congruent rectangles of diameter less than δ/c and hence in view of the preceding paragraph and Lemma 6.3.3,

$$\nu_M(\delta, R_j) \le \{[c\lambda_0\sqrt{2}\delta^{-1}] + 1\}^2.$$

Therefore, by Lemma 6.3.3,

$$\nu_M(\delta, \lambda Q) \le [\lambda/\lambda_0]^2\{\nu_M(\delta, \lambda_0 Q) + 1\} + \{2[\lambda/\lambda_0] + 1\}\{[c\lambda_0\sqrt{2}\delta^{-1}] + 1\}^2$$

and Lemma 6.3.6 with $\delta = 1$ gives

$$\lambda^{-2}\{\nu_M(\lambda^{-1}, Q)\} \le \lambda_0^{-2}\{\nu_M(\lambda_0^{-1}, Q) + 1\} + O(\lambda_0/\lambda),$$

or, on setting $\varepsilon = \lambda^{-1}$ and $\varepsilon_0 = \lambda_0^{-1}$,

$$\varepsilon^2 \nu_M(\varepsilon, Q) \le \varepsilon_0^2 \nu_M(\varepsilon_0, Q) + O(\varepsilon/\varepsilon_0). \tag{6.3.13}$$

Given $\eta > 0$, choose ε_0 such that

$$\varepsilon_0^2\{\nu_M(\varepsilon_0, Q)\} < \inf_{\varepsilon>0} \{\varepsilon^2\nu_M(\varepsilon, Q)\} + \eta.$$

Then, since η is arbitrary, (6.3.13) yields

$$\limsup_{\varepsilon\to 0} \{\varepsilon^2\nu_M(\varepsilon, Q)\} \le \inf_{\varepsilon>0} \{\varepsilon^2\nu_M(\varepsilon, Q)\}.$$

Since

$$\liminf_{\varepsilon\to 0} \{\varepsilon^2\nu_M(\varepsilon, Q)\} \ge \inf_{\varepsilon>0} \{\varepsilon^2\nu_M(\varepsilon, Q)\}$$

the theorem is proved. □

Theorem 6.3.8. *Let Q be an open cube in $\mathbb{R}^n$. Then*

$$\lim_{\varepsilon\to 0}\{\varepsilon^n\mu_0(\varepsilon, Q)\} = \sup_{\varepsilon>0}\{\varepsilon^n\mu_0(\varepsilon, Q)\}. \tag{6.3.14}$$

Proof. As in the proof of the previous theorem (with $n = 2$ and $\lambda \ge \lambda_0 \ge 1$) we express λQ as the union of $[\lambda/\lambda_0]^2$ open squares congruent to $\lambda_0 Q$ together with an L-shaped region. Then by Lemmas 6.3.4 and 6.3.6,

$$\begin{aligned}\mu_0(\lambda^{-1}, Q) = \mu_0(1, \lambda Q) &\ge [\lambda/\lambda_0]^2\mu_0(1, \lambda_0 Q) \\ &\ge \{(\lambda/\lambda_0)^2 - 1\}\mu_0(\lambda_0^{-1}, Q)\end{aligned}$$

and so

$$\lambda^{-2}\mu_0(\lambda^{-1}, Q) \ge \lambda_0^{-2}\mu_0(\lambda_0^{-2}, Q) - \lambda^{-2}\mu_0(\lambda_0^{-1}, Q).$$

This implies

$$\liminf_{\lambda\to\infty}\{\lambda^{-2}\mu_0(\lambda^{-1}, Q)\} \ge \sup_{\lambda>0}\{\lambda^{-2}\mu_0(\lambda^{-1}, Q)\},$$

and hence (6.3.14) in the case $n = 2$ on noting that

$$\limsup_{\lambda\to\infty}\{\lambda^{-2}\mu_0(\lambda^{-1}, Q)\} \le \sup_{\lambda>0}\{\lambda^{-2}\mu_0(\lambda^{-1}, Q)\}$$

and setting $\lambda = \varepsilon^{-1}$. □

Corollary 6.3.9. *If Q is an open cube in $\mathbb{R}^n$ then as $\varepsilon \to 0$,*

$$\mu_0(\varepsilon, Q), \nu_0(\varepsilon, Q), \nu_M(\varepsilon, Q) \approx \varepsilon^{-n} \tag{6.3.15}$$

Proof. This follows from Theorems 6.3.7, 6.3.8 and, by Lemma 6.3.1,

$$\mu_0(\varepsilon, Q) \le \nu_0(\varepsilon, Q) \le \nu_M(\varepsilon, Q) + 1.$$

□

Remark 6.3.10. Similar proofs yield the same results as those in Theorems 6.3.7, 6.3.8 and Corollary 6.3.9 when the cubes Q are replaced by equilateral triangles and their n-dimensional analogues.

Remark 6.3.11. In the case $p = 2$ we have from Lemma 6.3.2 that

$$\nu_M(\varepsilon, Q) = \sharp\{m : \lambda_m(-\Delta_{N,Q}) \leq \varepsilon^{-2}\},$$

where the $\lambda_m(-\Delta_{N,Q})$ are the positive eigenvalues of the Neumann Laplacian on Q arranged in increasing order of magnitude and repeated according to multiplicity. If Q has side length ℓ, these eigenvalues are

$$(\frac{\pi}{\ell})^2 \Sigma_{j=1}^n m_j^2$$

where $m_j \in \mathbb{N}$. Thus

$$\nu_M(\varepsilon, Q) = \sharp\{(m_1, \cdots, m_n) \in \mathbb{N}^n : \Sigma_{j=1}^n m_j^2 \leq \varepsilon^{-2}\}.$$

This yields the result (see [46], Theorem XI.2.6)

$$\varepsilon^n \nu_M(\varepsilon, Q) = (2\pi)^{-n} \omega_n |Q| + O(\varepsilon), \tag{6.3.16}$$

as $\varepsilon \to 0$, where ω_n is the volume of the unit ball in $\mathbb{R}^n$. The same argument applies to the Dirichlet Laplacian to give

$$\varepsilon^n \nu_0(\varepsilon, Q) = \varepsilon^n \mu_0(\varepsilon, Q) = (2\pi)^{-n} \omega_n |Q| + O(\varepsilon) \tag{6.3.17}$$

as $\varepsilon \to 0$.

Example 6.3.12. To illustrate the use of the results of Section 6.3 we apply them to the rooms and passages domain of Example 5.6.4.

We write

$$\Omega = \bigcup_{j=1}^{\infty} Q_j,$$

where, in the notation of Example 5.2.4, Q_{2j-1} is the (square) room R_{2j-1} and Q_{2j} is the (rectangular) passage P_{2j}. Set

$$\Omega_0 := \bigcup_{j=1}^{2R} Q_j, \quad \Omega_R := \bigcup_{j=2R+1}^{\infty} Q_j.$$

We proved in Theorem 5.7.5 that the Poincaré inequality

$$\|f - f_\Omega | L_p(\Omega)\| \leq D(\Omega) \|\nabla f | L_p(\Omega)\|$$

holds for all $f \in W^1_p(\Omega)$ if and only if $\alpha \leq p+1$ in (5.6.14). The Poincaré inequality is similarly valid for Ω_R when $\alpha \leq p+1$ and, in view of Corollary 5.5.9. we have

$$D(\Omega_R) \lesssim k(\Omega_R) + N_R, \tag{6.3.18}$$

where

$$k(\Omega_R) := \sup_{\Omega_R} \rho \circ \tau(x) \lesssim h_{2R+1} \lesssim h^{-R},$$

and

$$N_R := \sup_{R \leq k < \infty} N(R,k)$$

where (see Example 5.7.4)

$$N(R,k) \approx \begin{cases} h^{-(p+1-\alpha)k} & \text{if} \quad \alpha > p-1, \\ h^{-2k-(p-1-\alpha)R} & \text{if} \quad \alpha < p-1, \\ h^{-2k}(k-R)^{p-1} & \text{if} \quad \alpha = p-1. \end{cases}$$

We suppose hereafter that $\alpha < p+1$, which means that the embeddings E, E_M are compact. In this case we have

$$N_R \lesssim h^{-(p+1-\alpha)}R$$

and consequently

$$D(\Omega_R) \lesssim h^{-\theta R}, \quad \theta := \min\{1, p+1-\alpha\}. \tag{6.3.19}$$

It follows that given $\varepsilon > 0$,

$$\|f - f_{\Omega_R}|L_p(\Omega_R)\| \leq \varepsilon \|\nabla f|L_p(\Omega_R)\|$$

for all $f \in W^1_p(\Omega_R)$ if, for some positive constant $C, h^{-\theta R} < \varepsilon/C$ and hence

$$R > \frac{\ln(1/C\varepsilon)}{\theta \ln h}. \tag{6.3.20}$$

On taking

$$R = [\frac{\ln(1/C\varepsilon)}{\theta \ln h}] + 1,$$

where $[\cdot]$ denotes the integer part, (6.3.20) implies that

$$\nu_M(\varepsilon, \Omega_R) = 0. \tag{6.3.21}$$

From Lemma 6.3.3, we have that

$$\begin{aligned} \nu_M(\varepsilon,\Omega) &\leq \{\sum_{j=1}^{R} [\nu_M(\varepsilon, Q_{2j-1}) + 1]\} + \{\sum_{j=1}^{R} [\nu_M(\varepsilon, Q_{2j}) + 1]\} + \{\nu(\varepsilon, \Omega_R) + 1\} \\ &= I_1 + I_2 + I_3, \end{aligned} \tag{6.3.22}$$

say. Let $l_1(\varepsilon, j), l_2(\varepsilon, j)$, $(l_1(\varepsilon, j) \leq l_2(\varepsilon, j))$, denote the side lengths of $Q_j, 1 \leq j \leq 2R$. Suppose that

$$1 \leq \varepsilon^{-1}\delta l_1(\varepsilon, 2j-1) = \varepsilon^{-1}\delta l_2(\varepsilon, 2j-1) \tag{6.3.23}$$

and set

$$M = [\varepsilon^{-1}\delta l_1(\varepsilon, 2j-1)]$$

where $\delta \in (0,1)$. Then $M \geq 1$ and

$$\varepsilon^{-1}\delta l_1(\varepsilon, 2j-1) = M + \phi, \quad \phi \in [0,1).$$

Divide $\varepsilon^{-1}Q_{2j-1}$ into M^2 squares of side $\mu = \delta^{-1}$ and an L-shaped strip S made up of $2M$ rectangles of sides $\phi\mu$ and μ, and a square of side $\phi\mu \times \phi\mu$. Further, subdivide each of the rectangles in S into $k(\mu)$ rectangles T say, which are sufficiently small that $\nu_M(1, T) = 0$; note that $k(\mu) = O(\mu^2)$. It follows from Lemmas 6.3.3 and 6.3.6 that with U the square $[0,1] \times [0,1]$,

$$\begin{aligned}
\nu_M(\varepsilon, Q_{2j-1}) &= \nu_M(1, \varepsilon^{-1}Q_{2j-1}) \\
&\leq M^2\{\nu_M(1, \mu U) + 1\} + (2M+1)k(\mu) \\
&\leq \varepsilon^{-2}\delta^2|Q_{2j-1}|[\nu_M(1, \mu U) + 1] + \{2\varepsilon^{-1}\delta l_1(\varepsilon, 2j-1) + 1\}k(\mu) \\
&\leq \varepsilon^{-2}|Q_{2j-1}|\delta^2[\nu_M(\delta, U) + 1] + K\varepsilon^{-1}\delta^{-1}l_1(\varepsilon, 2j-1).
\end{aligned} \tag{6.3.24}$$

Next, suppose that

$$\varepsilon^{-1}\delta l_1(\varepsilon, 2j-1) = \varepsilon^{-1}\delta l_2(\varepsilon, 2j-1) < 1. \tag{6.3.25}$$

We now divide $\varepsilon^{-1}Q_{2j-1}$ into $O(\mu^2)$ rectangles T for which $\nu_M(1, T) = 0$. Then

$$\begin{aligned}
\nu_M(\varepsilon, Q_{2j-1}) &= \nu_M(1, \varepsilon^{-1}Q_{2j-1}) \\
&\leq K\delta^{-2}.
\end{aligned} \tag{6.3.26}$$

We deal with the rectangles Q_{2j} in I_2 in a similar way. Suppose first that

$$1 \leq \varepsilon^{-1}\delta l_1(\varepsilon, 2j) < \varepsilon^{-1}\delta l_2(\varepsilon, 2j) \tag{6.3.27}$$

and set

$$M_k := [\varepsilon^{-1}\delta l_k(\varepsilon, 2j)], \quad k = 1, 2.$$

Then

$$\varepsilon^{-1}\delta l_k(\varepsilon, 2j) = M_k + \phi_k, \quad \phi_k \in [0,1).$$

Divide $\varepsilon^{-1}Q_{2j}$ into M_1M_2 squares of side $\mu = \delta^{-1}$ leaving an L-shaped strip S made up of M_1 rectangles of side $\mu \times \phi_2\mu$, M_2 rectangles of side $\phi_1\mu \times \mu$ and one rectangle of side $\phi_1\mu \times \phi_2\mu$. Again divide the rectangles in S into $k(\mu^2)$ rectangles T which are such that $\nu_M(1, T) = 0$. Then

$$\begin{aligned}\nu_M(\varepsilon, Q_{2j}) =& \nu_M(1, \varepsilon^{-1}Q_{2j}) \\ \le& M_1M_2\{\nu_M(1, \mu U) + 1\} \\ &+ (M_1 + M_2 + 1)k(\mu) \\ \le& \varepsilon^{-2}\delta^2|Q_{2j}|\big[\nu_M(1, \mu U) + 1\big] \\ &+ \big\{\varepsilon^{-1}\delta[l_1(\varepsilon, 2j) + l_2(\varepsilon, 2j)] + 1\big\}k(\mu) \\ \le& \varepsilon^{-2}|Q_{2j}|\delta^2\big[\nu_M(\delta, U) + 1\big] \\ &+ K\varepsilon^{-1}\delta^{-1}\big(l_1(\varepsilon, 2j) + l_2(\varepsilon, 2j)\big). \end{aligned} \tag{6.3.28}$$

If

$$\varepsilon^{-1}\delta l_1(\varepsilon, 2j) < 1 \le \varepsilon^{-1}\delta l_2(\varepsilon, 2j) \tag{6.3.29}$$

we divide $\varepsilon^{-1}Q_{2j}$ into M_2 rectangles whose sides are no greater than μ and then divide these rectangles into $O(\mu^2)$ rectangles T which are such that $\nu(1, T) = 0$. Then

$$\begin{aligned}\nu_M(\varepsilon, Q_{2j}) &= \nu_2(1, \varepsilon^{-1}Q_{2j}) \\ &\le K(M_2 + 1)\mu^2 \\ &\le K\varepsilon^{-1}\delta^{-1}l_2(\varepsilon, 2j).\end{aligned}$$

Finally, if

$$\varepsilon^{-1}\delta l_2(\varepsilon, 2j) < 1 \tag{6.3.30}$$

we get as in (6.3.26)

$$\nu_M(\varepsilon, Q_{2j}) + 1 \le K\delta^{-2}. \tag{6.3.31}$$

Let

$$\begin{aligned}\mathcal{R}_1 &:= \big\{j \in \{1, \cdots, R\} : 1 \le \varepsilon^{-1}\delta l_1(\varepsilon, 2j - 1)\big\}, \\ \mathcal{R}_2 &:= \big\{j \in \{1, \cdots, R\} : 1 > \varepsilon^{-1}\delta l_1(\varepsilon, 2j - 1)\big\}, \\ \mathcal{P}_1 &:= \big\{j \in \{1, \cdots, R\} : 1 \le \varepsilon^{-1}\delta l_1(\varepsilon, 2j)\big\}, \\ \mathcal{P}_2 &:= \big\{j \in \{1, \cdots, R\} : \varepsilon^{-1}\delta l_1(\varepsilon, 2j) < 1 \le \varepsilon^{-1}\delta l_2(\varepsilon, 2j)\big\}, \\ \mathcal{P}_3 &:= \big\{j \in \{1, \cdots, R\} : \varepsilon^{-1}\delta l_2(\varepsilon, 2j) < 1\big\}.\end{aligned} \tag{6.3.32}$$

Then in (6.3.22),

$$\begin{aligned} I_1 + I_2 \le& \Big(\sum_{j\in\mathcal{R}_1} + \sum_{j\in\mathcal{R}_2}\Big)\{\nu_M(\varepsilon, Q_{2j-1}) + 1\} \\ &+ \Big(\sum_{j\in\mathcal{P}_1} + \sum_{j\in\mathcal{P}_2} + \sum_{j\in\mathcal{P}_3}\Big)\{\nu_M(\varepsilon, Q_{2j}) + 1\} \\ \le& \sum_{j\in\mathcal{R}_1}\{\varepsilon^{-2}|Q_{2j-1}|(\delta^2[\nu_M(\delta, U) + 1]) + K\varepsilon^{-1}\delta^{-1}l_1(\varepsilon, 2j - 1)\} \\ &+ \sum_{j\in\mathcal{R}_2} K\delta^{-2} + \sum_{j\in\mathcal{P}_1}\{\varepsilon^{-2}|Q_{2j}|(\delta^2[\nu_M(\delta, U) + 1]) \end{aligned}$$

$$+ K\varepsilon^{-1}\delta^{-1}\big[l_1(\varepsilon, 2j) + l_2(\varepsilon, 2j)\big]\Big\}$$
$$+ \sum_{j\in\mathcal{P}_2} K\varepsilon^{-1}\delta^{-1}l_2(\varepsilon, 2j) + \sum_{j\in\mathcal{P}_3} K\delta^{-2}$$
$$\leq \varepsilon^{-2}\delta^2\big[\nu_M(\delta, U) + 1\big]|\Omega_0| + K\varepsilon^{-1}\delta^{-1}|\partial\Omega_0|$$
$$+ K\delta^{-2}\sharp\{j \in \{1, \cdots, 2R\} : l_1(\varepsilon, j) < \varepsilon\delta^{-1}\}$$
$$\leq \varepsilon^{-2}\delta^2\big[\nu_M(\delta, U) + 1\big]|\Omega_0| + K\varepsilon^{-1}\delta^{-1}|\partial\Omega_0|$$
$$+ K\delta^{-2}\ln(1/\varepsilon). \tag{6.3.33}$$

It follows from (6.3.21) and (6.3.22) that

$$\varepsilon^2\nu_M(\varepsilon, \Omega) \leq \delta^2\big[\nu_M(\delta, U) + 1)\big]|\Omega_0| + K\varepsilon\delta^{-1}|\partial\Omega_0|$$
$$+ K\varepsilon^2\delta^{-2}\ln(1/\varepsilon). \tag{6.3.34}$$

This gives, in particular,

$$\limsup_{\varepsilon\to 0} \varepsilon^2\nu_M(\varepsilon, \Omega) \leq |\Omega| \lim_{\delta\to 0} \delta^2\nu_M(\delta, U). \tag{6.3.35}$$

When $p = n = 2$, we have from (6.3.16)

$$\nu_M(\delta, U) = \frac{1}{4\pi}\delta^{-2} + O(\delta^{-1}). \tag{6.3.36}$$

Thus, on substituting this in (6.3.34),

$$\varepsilon^2\nu_M(\varepsilon, \Omega) \leq \frac{1}{4\pi}|\Omega| + K\varepsilon\delta^{-1}|\partial\Omega|$$
$$+ K\varepsilon^2\delta^{-2}\ln(1/\varepsilon) + O(\delta).$$

On choosing $\delta = \varepsilon^\sigma, \sigma < 1$, we arrive at

$$\varepsilon^2\nu_M(\varepsilon, \Omega) \leq \frac{1}{4\pi}|\Omega| + K\varepsilon^{1-\sigma}|\partial\Omega|$$
$$+ K\varepsilon^{2(1-\sigma)}\ln(1/\varepsilon) + O(\varepsilon^\sigma). \tag{6.3.37}$$

We derive a lower bound for $\varepsilon^2\mu_0(\varepsilon, \Omega)$ in a similar way. We choose R as before to satisfy (6.3.20) and apply Lemma 6.3.4 to give

$$\mu_0(\varepsilon, \Omega_0) \geq \sum_{j=1}^{R} \mu_0(\varepsilon, Q_{2j-1}) + \sum_{j=1}^{R} \mu_0(\varepsilon, Q_{2j}). \tag{6.3.38}$$

If

$$1 \leq \varepsilon^{-1}\delta l_1(\varepsilon, 2j-1) = \varepsilon^{-1}\delta l_2(\varepsilon, 2j-1) \tag{6.3.39}$$

we derive by the same procedure of subdividing $\varepsilon^{-1}Q_{2j-1}$ as before, with $M = \big[\varepsilon^{-1}\delta l_1(\varepsilon, 2j-1)\big]$,

$$\mu_0(\varepsilon, Q_{2j-1}) = \mu_0(1, \varepsilon^{-1}Q_{2j-1}) \geq M^2\mu_0(1, \mu U) = M^2\mu_0(\delta, U).$$

Similarly, with $M_k = [\varepsilon^{-1}\delta l_k(\varepsilon, 2j)], k = 1, 2$, we have when

$$1 \leq \varepsilon^{-1}\delta l_1(\varepsilon, 2j) < \varepsilon^{-1}\delta l_2 1(\varepsilon, 2j) \tag{6.3.40}$$

that

$$\mu_0(\varepsilon, Q_{2j}) \geq M_1 M_2 \mu_0(\delta, U). \tag{6.3.41}$$

The other cases are treated as before to give in all, since $M_k > \varepsilon^{-1}\delta l_k(\varepsilon, 2j) - 1$,

$$\mu_0(\varepsilon, \Omega_0) \geq \sum_{j=1}^{R} \{\varepsilon^{-2}\delta^2 l_1^2(\varepsilon, 2j-1) - 2\varepsilon^{-1}\delta l_1(\varepsilon, 2j-1) + 1\}\mu_0(\delta, U)$$
$$+ \sum_{j=1}^{R} \{\varepsilon^{-2}\delta^2 l_1(\varepsilon, 2j) l_2(\varepsilon, 2j) - \varepsilon^{-1}\delta[l_1(\varepsilon, 2j) + l_2(\varepsilon, 2j)] + 1\}\mu_0(\mu, U)$$

and so

$$\varepsilon^2\mu_0(\varepsilon, \Omega_0) \geq \delta^2\mu_0(\delta, U)|\Omega_0| - K\varepsilon\delta|\partial\Omega_0|\mu_0(\delta, U). \tag{6.3.42}$$

Also

$$\begin{aligned} |\Omega_0| &= |\Omega| - \Big(\sum_{j=R+1}^{\infty} |Q_{2j-1}| + \sum_{j=R+1}^{\infty} |Q_{2j}|\Big) \\ &\geq |\Omega| - O\Big(\sum_{j=R+1}^{\infty} h^{-2j}\Big) = |\Omega| - O(h^{-2R}) \\ &= |\Omega| - O(\varepsilon^{2/\theta}), \end{aligned}$$

where $\theta = \min\{1, p + 1 - \alpha\}$ and

$$|\partial\Omega| - |\partial\Omega_0| \approx \sum_{j=R+1}^{\infty} h^{-j} \approx \varepsilon^{1/\theta}.$$

Therefore

$$\begin{aligned} \varepsilon^2\mu_0(\varepsilon, \Omega) &\geq \varepsilon^2\mu_0(\varepsilon, \Omega_0) \\ &\geq [\delta^2\mu_0(\delta, U)]\big(|\Omega| - O(\varepsilon^{2/\theta})\big) \\ &\quad - K\varepsilon\delta|\partial\Omega|\mu_0(\delta, U) \\ &\geq [\delta^2\mu_0(\delta, U)]\{|\Omega| - O(\varepsilon^{2/\theta}) - O(\varepsilon\delta^{-1})\}. \end{aligned} \tag{6.3.43}$$

This gives, in particular,

$$\liminf_{\varepsilon\to 0} \varepsilon^2\mu_0(\epsilon, \Omega) \geq |\Omega| \lim_{\delta\to 0} \delta^2\mu_0(\delta, U) \tag{6.3.44}$$

which complements (6.3.35).

On setting $m = \nu_M(\varepsilon, \Omega)$ and noting that

$$a_{m+1}(E_M(\Omega)) < \varepsilon, \quad a_m(E_M(\Omega)) \geq \varepsilon,$$

we deduce from (6.3.35), (6.3.44) and Lemma 6.3.1 that

$$\limsup_{m\to\infty} m a_m^2(E_M(\Omega)) \leq |\Omega| \lim_{\delta\to 0} \delta^2 \nu_M(\delta, U), \tag{6.3.45}$$

$$\liminf_{m\to\infty} m a_m^2(E_M(\Omega)) \geq |\Omega| \lim_{\delta\to 0} \delta^2 \mu_0(\delta, U), \tag{6.3.46}$$

and similarly with $E_M(\Omega)$ replaced by $E_0(\Omega)$.

When $p = n = 2$ we have from (6.3.17) that

$$\mu_0(\delta, U) = \frac{1}{4\pi}\delta^{-2} + O(\delta^{-1}) \tag{6.3.47}$$

as $\delta \to 0$. On choosing $\delta = \varepsilon^\theta, \theta < 1$, we get from (6.3.43)

$$\varepsilon^2 \mu_0(\varepsilon, \Omega) \geq \frac{1}{4\pi}|\Omega| - O(\varepsilon^{1-\theta} + \varepsilon^\theta). \tag{6.3.48}$$

Hence, from (6.3.37) and (6.3.48), with $\sigma = \theta < 1$, we have when $p = n = 2$ that

$$\varepsilon^2 \mu_0(\varepsilon, \Omega) = \varepsilon^2 \nu_0(\varepsilon, \Omega) = \varepsilon^2 \nu_M(\varepsilon, \Omega) = \frac{1}{4\pi}|\Omega| + O\left(\varepsilon^{1-\theta} + \varepsilon^\theta\right), \tag{6.3.49}$$

where $\theta = p + 1 - \alpha,\ p < \alpha < p + 1$. In particular,

$$a_m(E_0(\Omega)), a_m(E_M(\Omega)), a_m(E_{\mathbb{C}}(\Omega)) \sim \frac{1}{4\pi}|\Omega| m^{-1/2} \tag{6.3.50}$$

as $m \to \infty$.

6.4 Further asymptotic estimates for a GRD Ω

In this section we establish upper and lower bounds for $\nu_{\mathbb{C}}(\varepsilon, \Omega)$ and $\mu_0(\varepsilon, \Omega)$ in terms of the analogous quantities for its generalised ridge when Ω is a GRD. These results complement those in Section 6.3 in that they provide a way of estimating the contribution of the boundary region of a general domain Ω which is left over after a union of rectangles is removed, as long as this boundary region is made up of GRDs. The union of rectangles can of course be handled by the technique of Section 6.3. We shall illustrate this application in Example 6.4.4.

The analysis depends on the isometry T and approximate inverse M of Section 5.4. The maps

$$T_1 : [F] \mapsto [TF] : L^1_{p,\mathbb{C}}(\Gamma, \mu) \to W^1_{p,\mathbb{C}}(\Omega),$$

$$\begin{aligned} T_0 &: [F] \mapsto [TF] : L_{p,\mathbb{C}}(\Gamma,\mu) \to L_{p,\mathbb{C}}(\Omega), \\ M_1 &: [f] \mapsto [Mf] : W^1_{p,\mathbb{C}}(\Omega) \to L^1_{p,\mathbb{C}}(\Gamma,\mu), \\ M_0 &: [f] \mapsto [Mf] : L_{p,\mathbb{C}}(\Omega) \to L_{p,\mathbb{C}}(\Gamma,\mu) \end{aligned} \tag{6.4.1}$$

are well defined and we have the commuting diagram

$$\begin{array}{ccccc} W^1_{p,\mathbb{C}}(\Omega) & \xrightarrow{M_1} & L^1_{p,\mathbb{C}}(\Gamma,\mu) & \xrightarrow{T_1} & W^1_{p,\mathbb{C}}(\Omega) \\ {\scriptstyle E_{\mathbb{C}}}\downarrow & & {\scriptstyle I_{\mathbb{C}}}\downarrow & & \downarrow{\scriptstyle E_{\mathbb{C}}} \\ L_{p,\mathbb{C}}(\Omega) & \xrightarrow{M_0} & L_{p,\mathbb{C}}(\Gamma,\mu) & \xrightarrow{T_0} & L_{p,\mathbb{C}}(\Omega) \end{array} \tag{6.4.2}$$

where $E_{\mathbb{C}} = E_{\mathbb{C}}(\Omega)$, and $I_{\mathbb{C}} = I_{\mathbb{C}}(\Gamma)$ is the embedding map defined in (6.2.22). To simplify notation, we suppress the dependence of the maps on Ω and Γ unless there is a possibility of confusion.

Theorem 6.4.1. *Let Ω be a GRD with generalised ridge $u(\Gamma)$. Then there exists a constant K, depending on Ω, such that*

$$a_m(E_{\mathbb{C}}) \le K\{k(\Omega) + a_m(I_{\mathbb{C}})\} \tag{6.4.3}$$

where $k(\Omega) = \sup_{\Omega}(\rho \circ \tau)(x)$. Therefore, given $\varepsilon > 0$,

$$\nu_{\mathbb{C}}(K(k(\Omega) + \varepsilon), \Omega) \le \nu_{\mathbb{C}}(\varepsilon, \Gamma) \tag{6.4.4}$$

in the notation of Lemmas 6.2.3 and 6.2.4

Proof. We first observe that in (6.4.2),

$$M_0 E_{\mathbb{C}} = I_{\mathbb{C}} M_1, \tag{6.4.5}$$

$$E_{\mathbb{C}} T_1 = T_0 I_{\mathbb{C}}. \tag{6.4.6}$$

The identity (6.4.5) gives

$$\begin{aligned} E_{\mathbb{C}} &= E_{\mathbb{C}} - T_0 M_0 E_{\mathbb{C}} + T_0 M_0 E_{\mathbb{C}} \\ &= E_{\mathbb{C}} - T_0 M_0 E_{\mathbb{C}} + T_0 I_{\mathbb{C}} M_1, \end{aligned}$$

and on using the properties of approximation numbers listed in Section 1.3, we conclude that

$$\begin{aligned} a_m(E_{\mathbb{C}}) &\le \|E_{\mathbb{C}} - T_0 M_0 E_{\mathbb{C}}\| + a_m(T_0 I_{\mathbb{C}} M_1) \\ &\le \|E_{\mathbb{C}} - T_0 M_0 E_{\mathbb{C}}\| + \|T_0\| a_m(I_{\mathbb{C}}) \|M_1\|. \end{aligned}$$

From (5.4.17) and the proof of part (1) in Example 5.5.2,

$$\|E_{\mathbb{C}} - T_0 M_0 E_{\mathbb{C}}\| \le \frac{(\alpha+2)^{n+1}}{n\delta_1} \{1 + 2^n \|\mathbb{M}|L_p(\mathbb{R}^n) \to L_p(\mathbb{R}^n)\|\} k(\Omega) \tag{6.4.7}$$

and from (5.4.22),

$$\|M_1\| \le \beta(\alpha+1)^n \|\mathbb{M}|L_p(\mathbb{R}^n) \to L_p(\mathbb{R}^n)\|. \tag{6.4.8}$$

Since T_0 is an isometry, (6.4.3) follows and (6.4.4) is an immediate consequence. □

Remark 6.4.2. It follows from (6.4.7) and (6.4.8) that the constant K in Theorem 6.4.1 depends only on n and the constants α, β and δ_1 in Definition 5.2.1.

We now show that estimates for the approximation numbers of $I_{\mathbb{C}}$ are given by those for the Hardy operator $H_c : L_p(\Gamma, \mu) \to L_p(\Gamma, \mu)$ defined by

$$H_c F(t) := \int_c^t F'(t) d\mu(t), \quad c, t \in \Gamma,$$

or equivalently, as shown in the proof of Corollary 5.5.8, the operator $T_c : L_p(\Gamma) \to L_p(\Gamma)$ defined by

$$T_c f(t) := v(t) \int_c^t u(t) f(t) dt \tag{6.4.9}$$

where

$$v(t) := \Big(\frac{d\mu}{dt}\Big)^{1/p}, u(t) := \Big(\frac{dt}{d\mu}\Big)^{1/p}. \tag{6.4.10}$$

Theorem 6.4.3. *For all $m \in \mathbb{N}$,*

$$\frac{1}{2} a_{m+1}(T_c) \le a_m(I_{\mathbb{C}}) \le a_m(T_c) \tag{6.4.11}$$

and hence for any $\varepsilon > 0$,

$$\nu(2\varepsilon, T_c) \le \nu_{\mathbb{C}}(\varepsilon, \Gamma) \le \nu(\varepsilon, T_c), \tag{6.4.12}$$

where

$$\nu(\varepsilon, T_c) := \max\big\{m \in \mathbb{N} : a_m(T_c) \ge \varepsilon\big\}.$$

Proof. For $F \in L_p^1(\Gamma, \mu)$ we have

$$\begin{aligned} F(t) - F(c) &= \int_c^t F'(s) ds \\ &= \int_c^t F'(s) v(s) u(s) ds \end{aligned}$$

since $v(s)u(s) = 1$ by (6.4.10). Hence

$$F(t) - F(c) = (R_c DF)(t)$$

where

$$D : [F] \to F'v : L^1_{p,\mathbb{C}}(\Gamma, \mu) \to L_p(\Gamma)$$

and

$$R_c F(t) := \int_c^t F(s) u(s) ds; \quad R_c : L_p(\Gamma) \to L_p(\Gamma, \mu).$$

Note that D is surjective, and is isometric since

$$\|DF\|_{p,\Gamma} = \|F'|L_p(\Gamma,\mu)\| = \|F|L^1_{p,\mathbb{C}}(\Gamma,\mu)\|.$$

If V_Γ denotes the canonical map $L_p(\Gamma) \to L_{p,\mathbb{C}}(\Gamma,\mu)$, we therefore have

$$I_{\mathbb{C}} = V_\Gamma R_c D \tag{6.4.13}$$
$$= V_\Gamma v^{-1} T_c D, \tag{6.4.14}$$

where v^{-1} denotes multiplication by $1/v$; observe that the map $v^{-1} : F \mapsto v^{-1}F$ is an isometry between $L_p(\Gamma)$ and $L_p(\Gamma,\mu)$. We conclude that (see Section 1.3)

$$\begin{aligned} a_m(I_{\mathbb{C}}) &\le \|V_\Gamma\|\|v^{-1}\|\|D\|a_m(T_c) \\ &\le a_m(T_c), \end{aligned}$$

since v^{-1} and D are isometries and $\|V_\Gamma\| \le 1$.

To derive the first inequality in (6.4.11) we introduce the one-dimensional operator

$$Af := \frac{1}{\mu(\Gamma)}\int_\Gamma f d\mu.$$

Denote the kernel of A by N and the restriction of V_Γ to N by V_N. Then V_N is injective since $V_N f = V_N g$ implies that $[f-g]=0$ and hence $f-g=c$, a constant. But this means that $c = A(f-g) = 0$ and hence $f = g$. Furthermore, V_N has range $L_{p,\mathbb{C}}(\Gamma)$, for $[f] = [f - Af]$ and $f - Af \in N$. Thus

$$V_N(I-A)f = [f] = V_\Gamma f, \tag{6.4.15}$$

where I is the identity on $L_p(\Gamma)$. On substituting this in (6.4.13) we get

$$\begin{aligned} a_m(I_{\mathbb{C}}) =& a_m(V_N(I-A)R_cD) \\ =& \inf\{\|V_N(I-A)R_cD - P|L^1_{p,\mathbb{C}}(\Gamma,\mu) \to L_{p,\mathbb{C}}(\Gamma,\mu)\| \\ & : \text{rank } P < m\} \\ =& \inf\{\|V_N(R_c - AR_c - V_N^{-1}PD^{-1})|L_p(\Gamma,\mu) \to L_{p,\mathbb{C}}(\Gamma,\mu)\| \\ & : \text{rank } P < m\} \end{aligned}$$

since D is an isometry of $L^1_{p,\mathbb{C}}(\Gamma,\mu)$ onto $L_p(\Gamma)$. From (6.4.15), since $\|Af|L_p(\Gamma,\mu)\| \le \|f|L_p(\Gamma,\mu)\|$, we have

$$\|V_N^{-1}|L_{p,\mathbb{C}}(\Gamma,\mu) \to L_p(\Gamma,\mu)\| \le 2$$

and so $AR_c + V_N^{-1}PD : L_p(\Gamma) \to L_p(\Gamma,\mu)$ is bounded and of rank no greater than rank $P+1 < m+1$. It follows that

$$\begin{aligned} a_m(I_{\mathbb{C}}) &\ge \|V_N^{-1}\|^{-1}\inf\{\|R_c - Q|L_p(\Gamma) \to L_p(\Gamma,\mu)\| : \text{rank } Q < m+1\} \\ &\ge \frac{1}{2}a_{m+1}(R_c) = \frac{1}{2}a_{m+1}(T_c). \end{aligned}$$

□

Example 6.4.4. We now apply the above results to determine estimates for the quantities $\nu_M(\varepsilon, \Omega), \mu_0(\varepsilon, \Omega)$ when Ω is the horn-shaped domain of Theorem 5.7.3(3) in $\mathbb{R}^2$, namely

$$\Omega = \{x = (x_1, x_2) : |x_2| < \Phi(x_1), 0 < x_1 < \infty\},$$

where

$$\Phi(t) = e^{-t^\theta}, \quad \theta > 2. \tag{6.4.16}$$

Recall that in this case, E and E_M are compact; in fact $\theta > 1$ is enough to ensure this, but taking $\theta > 2$ simplifies the analysis to follow.

For $R > 0$, set $\Omega_R := \{x \in \Omega : x_1 > R\}$. Then we have the Poincaré inequality

$$\|f - f_{\Omega_R}\|_{p,\Omega_R} \leq D(\Omega_R)\|\nabla f\|_{p,\Omega_R},$$

where by Corollary 5.5.9 there exists a constant K depending on Ω but not on R such that

$$D(\Omega_R) \leq K\{k(\Omega_R) + N_R\},$$

with

$$k(\Omega_R) = \sup_{\Omega_R} \rho \circ \tau,$$

$$N_R = \sup_{R \leq X < \infty} \mu(X, \infty)^{1/p} \Big(\int_R^X |\frac{dt}{d\mu}|^{p'} d\mu \Big)^{1/p'}.$$

In this example we have (see (5.2.15))

$$\frac{d\mu}{dt} = \Phi(t) = e^{-t^\theta}.$$

From the proof of Theorem 5.7.3(3), with $r = s = q = p$ and $N_R = \sup_{R \leq X < \infty} B_R(X)$, we have

$$N_R \lesssim \sup_{R \leq X < \infty} X^{(1-\theta)/p} R^{(1-\theta)/p'} = R^{1-\theta}$$

and so

$$\sup\{R : N_R \leq \varepsilon\} \gtrsim \varepsilon^{-1/(\theta-1)}. \tag{6.4.17}$$

The radius $\rho(t)$ of the ball B_t in Definition 5.2.1 is naturally taken to be the distance from the ridge point $u(t)$ to the boundary of Ω. If the normal from $u(t) = (t, 0)$ meets $\partial\Omega$ at $(s, \Phi(s))$, then

$$s - t = -\Phi(s)\Phi'(s) = \theta s^{\theta-1}\Phi(s)^2.$$

Hence

$$\ln [\Phi(t)/\Phi(s)] = \theta \int_t^s r^{\theta-1} dr$$

$$\begin{aligned} &\le \theta s^{\theta-1}(s-t) \\ &\le \theta^2 s^{2(\theta-1)}\Phi(s)^2 \\ &\lesssim 1. \end{aligned}$$

Consequently $\Phi(t) \approx \Phi(s)$,

$$\rho(t) \approx \Phi(t) = e^{-t^\theta} \tag{6.4.18}$$

and so

$$\sup\{t : \rho(t) \ge \varepsilon\} \approx \left[\ln\left(1/\varepsilon\right)\right]^{1/\theta}. \tag{6.4.19}$$

We therefore infer that for some constant K_0, $D(\Omega_R) < \varepsilon$ if $R > K_0\varepsilon^{-1/(\theta-1)}$. Let $l_\varepsilon = \varepsilon^{1-\sigma}$, where $\sigma \in (0,1)$, and set

$$k_\varepsilon := 1 + \left[\frac{K_0\varepsilon^{-1/(\theta-1)}}{l_\varepsilon}\right] \approx \varepsilon^{\sigma-\theta/(\theta-1)}, \tag{6.4.20}$$

$$R_\varepsilon := k_\varepsilon l_\varepsilon \approx \varepsilon^{-1/(\theta-1)}, \tag{6.4.21}$$

where $[\cdot]$ denotes the integer part. Then $D(R_\varepsilon) < \varepsilon$ and hence

$$\nu_M(\varepsilon, \Omega_{R_\varepsilon}) = 0. \tag{6.4.22}$$

We now partition Ω as follows:

$$\Omega = \bigcup_{j=1}^{k_\varepsilon} \{Q_j \cup T_j^+ \cup T_j^-\} \cup \Omega_{R_\varepsilon} \cup \mathcal{N}, \tag{6.4.23}$$

where

$$\begin{aligned} Q_j &= ((j-1)l_\varepsilon, jl_\varepsilon) \times \big(-\Phi(jl_\varepsilon), \Phi(jl_\varepsilon)\big), \\ T_j^+ &= \{x = (x_1, x_2) : (j-1)l_\varepsilon < x_1 < jl_\varepsilon, \Phi(jl_\varepsilon) < x_2 < \Phi(x_1)\}, \\ T_j^- &= \{x = (x_1, x_2) : (j-1)l_\varepsilon < x_1 < jl_\varepsilon, -\Phi(x_1) < x_2 < -\Phi(jl_\varepsilon)\}. \end{aligned}$$

The sets T_j^+, T_j^- are clearly GRDs with $\Gamma = ((j-1)\varepsilon, j\varepsilon), \tau$ the vertical projection and $\frac{d\mu}{dt} = |\Phi(t) - \Phi(j\varepsilon)|$. It is readily seen that the values of the constants α, β and δ_1 in Definition 5.2.1 can be chosen to be independent of j and ε; they compare with the corresponding constants for Ω.

We use Lemma 6.3.3 to determine an upper bound for $\nu_M(\varepsilon, \Omega)$. This gives from (6.4.23),

$$\begin{aligned} \nu_M(\varepsilon, \Omega) \le &\sum_{j=1}^{k_\varepsilon} \left[\nu_M(\varepsilon, Q_j) + 1\right] \\ &+ \sum_{j=1}^{k_\varepsilon} \left\{\left[\nu_M(\varepsilon, T_j^+) + 1\right] + \left[\nu_M(\varepsilon, T_j^-) + 1\right]\right\} \end{aligned}$$

$$+ \left[\nu_M(\varepsilon, \Omega_{R_\varepsilon}) + 1\right]$$
$$= I_1 + I_2 + I_3 \tag{6.4.24}$$

say. We already know from (6.4.22) that $I_3 = 1$.

We use the same technique as in Example 6.3.12 to estimate I_1. For the shorter of the side lengths $l_i(\varepsilon, j), i = 1, 2$ of Q_j we now have

$$l_1(\varepsilon, j) = \begin{cases} l_\varepsilon & \text{if} \quad 2e^{-(jl_\varepsilon)^\theta} \geq l_\varepsilon, \\ 2e^{-(jl_\varepsilon)^\theta} & \text{if} \quad 2e^{-(jl_\varepsilon)^\theta} < l_\varepsilon. \end{cases}$$

Hence, setting $n_\varepsilon := \left[l_\varepsilon^{-1}\left(\ln(2l_\varepsilon^{-1})\right)^{1/\theta}\right]$, we have

$$l_1(\varepsilon, j) = \begin{cases} l_\varepsilon & \text{if} \quad j \leq n_\varepsilon, \\ 2e^{-(jl_\varepsilon)^\theta} & \text{if} \quad j > n_\varepsilon. \end{cases}$$

We repeat the argument used for (6.3.33) to obtain, with the notation of (6.3.32) and $\Omega_0 := \bigcup_{j=1}^{k_\varepsilon} Q_j$,

$$\begin{aligned} I_1 &\leq \varepsilon^{-2}\delta^2\left[\nu_M(\delta, U) + 1\right]|\Omega_0| \\ &+ K\varepsilon^{-1}\delta^{-1} \sum_{j\in\mathcal{P}_1} \left[l_1(\varepsilon, j) + l_2(\varepsilon, j)\right] + K\varepsilon^{-1}\delta^{-1} \sum_{j\in\mathcal{P}_2} l_2(\varepsilon, j) \\ &+ K\delta^{-2} \sum_{j\in\mathcal{P}_3} 1. \end{aligned}$$

Since

$$\begin{aligned} \sum_{j=1}^{k_\varepsilon} l_2(\varepsilon, j) &= \Big(\sum_{j=1}^{n_\varepsilon} + \sum_{j=n_\varepsilon+1}^{k_\varepsilon}\Big) l_2(\varepsilon, j) \\ &= \sum_{j=1}^{n_\varepsilon} 2e^{-(jl_\varepsilon)^\theta} + \sum_{j=n_\varepsilon+1}^{k_\varepsilon} l_\varepsilon \\ &\leq 2l_\varepsilon^{-1} \int_0^\infty e^{-x^\theta} dx + k_\varepsilon l_\varepsilon \\ &\lesssim l_\varepsilon^{-1} + \varepsilon^{-1/(\theta-1)} \\ &\approx \varepsilon^{-1+\sigma} \end{aligned}$$

if $\sigma < \left(\frac{\theta-2}{\theta-1}\right)$ and

$$\sum_{j\in\mathcal{P}_3} 1 \leq k_\varepsilon \approx \varepsilon^{\sigma-\theta/(\theta-1)},$$

we obtain

$$\varepsilon^2 I_1 \leq \delta^2\left[\nu_M(\delta, U) + 1\right]|\Omega_0|$$

$$+ K\delta^{-1}\varepsilon^{\sigma} + K\delta^{-2}\varepsilon^{\sigma+\left(\frac{\theta-2}{\theta-1}\right)}. \tag{6.4.25}$$

Let T stand for T_j^+ say, and set $a = (j-1)l_\varepsilon, b = jl_\varepsilon$. We have already noted that T is a GRD with the choice

$$\frac{d\mu}{dt} = \Phi(t) - \Phi(b), \quad \Gamma = (a,b)$$
$$\tau(x_1, x_2) = x_1;$$

the map u in Definition 5.2.1 can be any Lipschitz map which takes Γ into T. We now repeat the above procedure and divide the generalised ridge Γ of T into $m_\varepsilon = [l_\varepsilon/\kappa\varepsilon]$ intervals of length $\tilde{\varepsilon} = \kappa\varepsilon$, and one of length less than $\tilde{\varepsilon}$, where κ is a constant to be chosen later. We write

$$T = \bigcup_{i=1}^{m_\varepsilon} \{R_i \cup S_i\} \cup S_{m_\varepsilon+1} \cup \mathcal{N}, \tag{6.4.26}$$

where

$$R_i = \big(a + (i-1)\tilde{\varepsilon}, a + i\tilde{\varepsilon}\big) \times \big(\Phi(b), \Phi(a + i\tilde{\varepsilon})\big),$$
$$S_i = \big\{x = (x_1, x_2) : a + (i-1)\tilde{\varepsilon} < x_1 < a + i\tilde{\varepsilon}, \Phi(a + i\tilde{\varepsilon}) < x_2 < \Phi(x_1)\big\},$$

for $i = 1, 2, \cdots, m_\varepsilon$ and

$$S_{m_\varepsilon+1} = \big\{x = (x_1, x_2) : a + m_\varepsilon\tilde{\varepsilon} < x_1 < b, \Phi(b) < x_2 < \Phi(x_1)\big\}.$$

From Lemma 6.3.3 and (6.4.26),

$$\nu_M(\varepsilon, T) \le \sum_{i=1}^{m_\varepsilon} \big[\nu_M(\varepsilon, R_\varepsilon) + 1\big] + \sum_{i=1}^{m_\varepsilon+1} \big[\nu_M(\varepsilon, S_\varepsilon) + 1\big]$$
$$= I_4 + I_5 \tag{6.4.27}$$

say. As for I_1 above, we have

$$\varepsilon^2 I_4 \le \delta^2 \big[\nu_M(\delta, U) + 1\big]|T|$$
$$+ K\varepsilon\delta^{-1} \sum_{i=1}^{m_\varepsilon} \big[l_1(\varepsilon, i) + l_2(\varepsilon, i)\big] + K\varepsilon^2\delta^{-2}m_\varepsilon.$$

Since $\Phi(a + i\tilde{\varepsilon}) - \Phi(b) = O(b - a) = O(l_\varepsilon)$, we have

$$\sum_{i=1}^{m_\varepsilon} \big[\tilde{\varepsilon} + \Phi(a + i\tilde{\varepsilon}) - \Phi(b)\big] \le K\big\{\varepsilon m_\varepsilon + l_\varepsilon m_\varepsilon\big\}$$
$$\le K\varepsilon^{1-2\sigma},$$

and hence

$$\varepsilon^2 I_4 \leq \delta^2\big[\nu_M(\delta,U)+1\big]|T| \\ + K\delta^{-1}\varepsilon^{2-2\sigma} + K\delta^{-2}\varepsilon^{2-\sigma}. \tag{6.4.28}$$

Let $c = a+(i-1)\tilde{\varepsilon}, d = a+i\tilde{\varepsilon}$, for $i = 1,2,\cdots,m_\varepsilon$ and $d = b$ when $i = m_\varepsilon+1$. Then $S = S_i$ is a GRD with

$$\frac{d\mu}{dt} = \Phi(t) - \Phi(d), \quad \Gamma = (c,d),$$
$$\tau(x_1,x_2) = x_1.$$

Set

$$u(t) = \Big(\frac{dt}{d\mu}\Big)^{1/p}, \quad v(t) = \Big(\frac{d\mu}{dt}\Big)^{1/p}$$

and

$$T_c f(t) = v(t)\int_c^t u(s)f(s)ds.$$

Then from (6.4.3) and (6.4.11)

$$a_1\big(E_{\mathbb{C}}(S)\big) \lesssim k(S) + a_1(T_c). \tag{6.4.29}$$

As in the remark after (6.4.23), the constant implied in (6.4.29) is independent of j and ε.

Let $t,s \in [0,R_\varepsilon]$. Then, from (6.4.21),

$$\Big|\ln\Big[\frac{\Phi(t)}{\Phi(s)}\Big]\Big| = \theta\Big|\int_t^s r^{\theta-1}dr\Big| \\ \leq \theta R_\varepsilon^{\theta-1}|t-s| \\ \lesssim \varepsilon^{-1}|t-s|.$$

Consequently Φ has bounded oscillation over any interval of length $O(\varepsilon)$ in $[0,R_\varepsilon]$. From Theorem 2.2.1 we have that

$$\begin{aligned}\|T_c\|^p &\leq 4^p \sup_{c<z<d}\Big(\int_c^z\big[\Phi(t)-\Phi(d)\big]^{-p'/p}dt\Big)^{p/p'}\Big(\int_z^d\big[\Phi(t)-\Phi(d)\big]dt\Big) \\ &\lesssim \sup_{c<z<d}\Big(\int_c^z\big[\Phi(z)-\Phi(d)\big]^{-p'/p}dt\Big)^{p/p'}\Big(\int_z^d\big[\Phi(z)-\Phi(d)\big]dt\Big) \\ &\approx \sup_{c<z<d}(z-c)^{p-1}(d-z) \\ &\approx (d-c)^p = \tilde{\varepsilon}^p.\end{aligned} \tag{6.4.30}$$

Also

$$k(S) \approx \Phi(c) - \Phi(d) \\ = O(c-d) = O(\tilde{\varepsilon}).$$

Therefore, for sufficiently small κ, it follows from (6.4.29) and (6.2.16) that

$$a_1(E_M(S)) < \varepsilon.$$

Hence, $\nu_M(\varepsilon, S) = 0$ and in (6.4.27)

$$I_5 = m_\varepsilon + 1 \approx \varepsilon^{-\sigma}. \tag{6.4.31}$$

From (6.4.28) and (6.4.31) we conclude that

$$\begin{aligned} \varepsilon^2 \nu_M(\varepsilon, T) \leq \delta^2 \big[\nu_M(\delta, U) + 1\big] |T| \\ + K\big(\delta^{-1}\varepsilon^{2-2\sigma} + \delta^{-2}\varepsilon^{2-\sigma}\big). \end{aligned} \tag{6.4.32}$$

The sets T_j^- can be treated in the same way to give from (6.4.24), (6.4.25), (6.4.27), and (6.4.32)

$$\begin{aligned} \varepsilon^2 \nu_M(\varepsilon, \Omega) &\leq \delta^2 \big[\nu_M(\delta, U) + 1\big] |\Omega| \\ &+ K\big(\delta^{-1}\varepsilon^{\sigma} + \delta^{-2}\varepsilon^{\sigma + \left(\frac{\theta-2}{\theta-1}\right)}\big) \\ &+ K\big(\delta^{-1}\varepsilon^{2-2\sigma} + \delta^{-2}\varepsilon^{2-\sigma} + \varepsilon^2\big)k_\varepsilon + \varepsilon^2. \end{aligned}$$

The choice

$$\sigma = \frac{1}{2}\Big(\frac{\theta - 2}{\theta - 1}\Big), \quad \varepsilon^\sigma < \delta < 1 \tag{6.4.33}$$

yields

$$\varepsilon^2 \nu_M(\varepsilon, \Omega) \leq \delta^2 \big[\nu_M(\delta, U) + 1\big] |\Omega| + K\delta^{-1}\varepsilon^{\frac{1}{2}\left(\frac{\theta-2}{\theta-1}\right)}. \tag{6.4.34}$$

When $p = n = 2$, (6.3.16) gives

$$\varepsilon^2 \nu_M(\varepsilon, \Omega) \leq \frac{1}{4\pi}|\Omega| + K\big\{\delta + \delta^{-1}\varepsilon^{\frac{1}{2}\left(\frac{\theta-2}{\theta-1}\right)}\big\}.$$

On choosing $\delta = \varepsilon^{\frac{1}{4}\left(\frac{\theta-2}{\theta-1}\right)}$ we obtain

$$\varepsilon^2 \nu_M(\varepsilon, \Omega) \leq \frac{1}{4\pi}|\Omega| + K\big(\varepsilon^{\frac{1}{4}\left(\frac{\theta-2}{\theta-1}\right)}\big). \tag{6.4.35}$$

To determine a lower bound for $\mu_0(\varepsilon, \Omega)$ we again divide $(0, R_\varepsilon)$ into k_ε intervals of equal length $l_\varepsilon = \varepsilon^{1-\sigma}$, where $0 < \sigma < \frac{\theta-2}{\theta-1}$ and suppose that $\varepsilon^\sigma < \delta < 1$. Let

$$\begin{aligned} M_1 &:= \big[\varepsilon^{-1}\delta l_\varepsilon\big], \\ M_2^{(j)} &:= \big[\varepsilon^{-1}\delta h(\varepsilon, j)\big], \quad h(\varepsilon, j) := 2e^{-(jl_\varepsilon)^\theta}. \end{aligned} \tag{6.4.36}$$

Then,

$$\begin{aligned} M_1 M_2^{(j)} &\geq \big(\varepsilon^{-1}\delta l_\varepsilon - 1\big)\big(\varepsilon^{-1}\delta h(\varepsilon, j) - 1\big) \\ &= \varepsilon^{-2}\delta^2 |Q_j| - \big[\varepsilon^{-1}\delta\big(l_\varepsilon + h(\varepsilon, j)\big) - 1\big] \\ &= \varepsilon^{-2}\delta^2 |Q_j| \big(1 - \frac{\varepsilon^\sigma}{\delta}\big) - \varepsilon^{-\sigma}\delta + 1. \end{aligned}$$

On repeating the argument applied to (6.3.38), we derive

$$\begin{aligned} \varepsilon^2 \mu_0(\varepsilon, \Omega) &\geq \sum_{j=1}^{k_\varepsilon} M_1 M_2^{(j)} \mu_0(\delta, U) \\ &\geq \delta^2 \mu_0(\delta, U)\big(1 - \frac{\varepsilon^\sigma}{\delta}\big) \sum_{j=1}^{k_\varepsilon} |Q_j| - O\big(\varepsilon^{2-\sigma}\delta^{-1}k_\varepsilon\big) \\ &\geq \delta^2 \mu_0(\delta, U)\big(1 - \frac{\varepsilon^\sigma}{\delta}\big)|\Omega_0| - O\big(\delta^{-1}\varepsilon^{\left(\frac{\theta-2}{\theta-1}\right)}\big) \end{aligned} \tag{6.4.37}$$

by (6.4.20); note that in the penultimate inequality, we have used the result $\mu_0(\delta, U) \approx \delta^{-2}$, established in Corollary 6.3.9.

From the proof of Theorem 5.7.3 (3) we have

$$\begin{aligned} |\Omega_{R_\varepsilon}| &= \int_{R_\varepsilon}^\infty e^{-t^\theta} dt \\ &\lesssim R_\varepsilon^{1-\theta} e^{-R_\varepsilon^\theta} \lesssim \varepsilon \end{aligned}$$

by (6.4.21). Also, for the sets T_j^+, T_j^- in the above analysis for the upper bound,

$$\begin{aligned} \sum_{j=1}^{k_\varepsilon} |T_j^+| &= \sum_{j=1}^{k_\varepsilon} \int_{(j-1)l_\varepsilon}^{jl_\varepsilon} \{e^{-t^\theta} - e^{-(j\varepsilon)^\theta}\} dt \\ &\lesssim \sum_{j=1}^{k_\varepsilon} l_\varepsilon^2 \\ &\approx \varepsilon^{\left(\frac{\theta-2}{\theta-1}\right)-\sigma}. \end{aligned}$$

The inequality (6.4.37) therefore gives

$$\begin{aligned} \varepsilon^2 \mu_0(\varepsilon, \Omega) \geq{} & \delta^2 \mu_0(\delta, U)\big(1 - \delta^{-1}\varepsilon^\sigma\big)\big\{|\Omega| - O\big(e^{\left[\left(\frac{\theta-2}{\theta-1}\right)-\sigma\right]}\big) \\ & - O\big(\delta^{-1}\varepsilon^{\left(\frac{\theta-2}{\theta-1}\right)}\big)\big\} \end{aligned} \tag{6.4.38}$$

when

$$0 < \sigma < \big(\frac{\theta - 2}{\theta - 1}\big), \quad \varepsilon^\sigma < \delta < 1. \tag{6.4.39}$$

The asymptotic bounds (6.3.45), (6.3.46) for $a_m(E_M(\Omega))$ and $a_m(E_0(\Omega))$ hold also in this example.

When $p = n = 2$, (6.4.38) and (6.3.17) lead to

$$\varepsilon^2 \mu_0(\varepsilon, \Omega) \geq \frac{1}{4\pi}|\Omega| - O\left(\delta + \delta^{-1}\varepsilon^{\sigma} + \varepsilon^{\left[\left(\frac{\theta-2}{\theta-1}\right)-\sigma\right]}\right)$$

which, on choosing $\delta = \varepsilon^{\sigma/2}, \sigma = \frac{1}{2}\left(\frac{\theta-2}{\theta-1}\right)$, becomes

$$\varepsilon^2 \mu_0(\varepsilon, \Omega) \geq \frac{1}{4\pi}|\Omega| - O\left(\varepsilon^{\frac{1}{4}\left(\frac{\theta-2}{\theta-1}\right)}\right). \tag{6.4.40}$$

Combining (6.4.35) and (6.4.40) yields

$$\begin{aligned} \varepsilon^2 \nu_M(\varepsilon, \Omega) &= \varepsilon^2 \nu_{\mathbb{C}}(\varepsilon, \Omega) = \varepsilon^2 \mu_0(\varepsilon, \Omega) \\ &= \frac{1}{4\pi}|\Omega| + O\left(\varepsilon^{\frac{1}{4}\left(\frac{\theta-2}{\theta-1}\right)}\right). \end{aligned} \tag{6.4.41}$$

In particular,

$$a_m(E_0(\Omega)), a_m(E_M(\Omega)), a_m(E_{\mathbb{C}}(\Omega)) \sim \frac{1}{4\pi}|\Omega| m^{-1/2} \tag{6.4.42}$$

as $m \to \infty$.

6.5 Notes

6.3. Except in the case $p = 2$ for any n and $n = 1$ for any p, it is not known whether or not the limits in Theorems 6.3.7 and 6.3.8 are equal: in the aforementioned exceptional cases the limits are equal by Remark 6.3.11 and Theorem 2.4.4. A positive answer to this question, together with estimates for the error terms in (6.3.12) and (6.3.14) would enable the techniques demonstrated in Example 6.3.12 and Section 6.4 to yield asymptotic results with errors for general values of p, just as is shown in the case $p = 2$ by using Remark 6.3.11 and choosing δ in terms of ε.

References

1. Achache, M., Some problems associated with linear differential equations, M.Sc. thesis, University of Wales, 1988.
2. Adams, R.A., *Sobolev spaces*, Academic Press, New York, 1975.
3. Adams, D.R. and Hedberg, L.I., *Function spaces and potential theory*, Springer-Verlag, Berlin, 1996.
4. Aleksandrov, S., Janson, S., Peller, V.V. and Rochberg, R., An interesting class of operators with unusual Schatten-von Neumann behaviour, preprint.
5. Alvino, A., Trombetti, G. and Lions, P.-L., On optimization problems with prescribed rearrangements, Nonlinear Anal. **13** (1989), 185-220.
6. Amick, C.J., Some remarks on Rellich's theorem and the Poincaré inequality, J. London Math. Soc. **18** (1978), 81-93.
7. Ariño, M. and Muckenhoupt, B., Maximal functions on classical Lorentz spaces and Hardy's inequality with weights for non-increasing functions, Trans. American Math. Soc. **320** (1990), 727-735.
8. Aronszajn, N. and Smith, K.T., Theory of Bessel potentials, Part I, Ann. Inst. Fourier **11** (1961), 385-475.
9. Aubin, T., Problèmes isopérimetriques et espaces de Sobolev, J. Diff. Geom. **11** (1976), 573-598.
10. Baernstein, A., A unified approach to symmetrization, in Partial Differential Equations of Elliptic Type (A. Alvino, E. Fabes and G. Talenti, eds.), Symposia Math. **35**, Cambridge University Press, 1994.
11. Bastero, J., Milman, M. and Ruiz Blaser, F., A note on $L(\infty, q)$ spaces and Sobolev embeddings, Indiana Univ. Math. J. **52** (2003), 1215-1230.
12. Beesack, P.R. and Heinig, H.P., Hardy's inequalities with indices less than 1, Proc. American Math. Soc. **83** (1981), 532-536.
13. Bennett, C. and Rudnick, K., On Lorentz-Zygmund spaces, Dissertationes Mathematicae **175** (1980), 1-72.
14. Bennett, C., De Vore, R. and Sharpley, C., Weak -L^∞ and BMO, Annals of Math. **113** (1981), 601-611.
15. Bennett, C. and Sharpley, R., *Interpolation of operators*, Academic Press, New York, 1988.
16. Bennewitz, C. and Saitō, Y., An embedding norm and the Linqvist trigonometric functions, Electronic J. of Diff. Eq. **86** (2002),1-6.

17. Bennewitz, C. and Saitō, Y., Operator norms of Sobolev embedding operators on an interval, preprint.
18. Bingham, N.H., Goldie, C.M. and Teugels, J.L., *Regular Variation,* Cambridge Univ. Press, Cambridge, 1987.
19. Bliss, G.A., An integral inequality, J. London Math. Soc. **5** (1930), 40-46.
20. Bojarski, B., Remarks on Sobolev imbedding inequalities, in: Proc. Conf. Complex Anal. (Joensu, 1987), pp. 52-68, Lecture Notes in Math. **1351**, Springer-Verlag, Berlin, 1988.
21. Bradley, J. S., Hardy inequality with mixed norms, Canadian Math. Bull. **21** (1978), 405-408.
22. Brézis, H. and Wainger, S., A note on limiting cases of Sobolev embeddings, Comm. Partial Diff. Equations **5** (1980), 773-789.
23. Brothers, J. and Ziemer, W., Minimal rearrangements of Sobolev functions, J. Reine Angew. Math. **384** (1988), 153-179.
24. Brown, R.C., Edmunds, D.E. and Rákosník, J., Remarks on inequalities of Poincaré type, Czech. Math. J. **45** (1995), 351-377.
25. Buckley, S.M. and Koskela, P., Sobolev-Poincaré implies John, Math. Res. Letters **2** (1995), 577-593.
26. Caetano, A.M., About the approximation numbers in function spaces, J. Approx. Theory, **94** (1998), 383-395.
27. Canavati, J.A. and Galaz-Fontes, F., Compactness of imbeddings between Banach spaces and applications to Sobolev spaces, J. London Math. Soc. **41** (1990), 511-525.
28. Carl, B., Entropy numbers, $s-$numbers and eigenvalue problems, J. Functional Anal. **41** (1981), 290-306.
29. Carl, B. and Stephani, I., *Entropy, compactness and approximation of operators,* Cambridge Univ. Press, Cambridge, 1990.
30. Carl, B. and Triebel, H., Inequalities between eigenvalues, entropy numbers, and related quantities of compact operators in Banach spaces, Math. Ann. **251** (1980), 129-133.
31. Chisholm, R.S. and Everitt, W.N., On bounded integral operators in the space of square-integrable functions, Proc. Roy. Soc. Edinburgh A **69** (1971), 199-204.
32. Cianchi, A., A sharp embedding theorem for Orlicz-Sobolev spaces, Indiana Univ. Math. J. **45** (1996), 39-65.
33. Cianchi, A., Symmetrization and second-order Sobolev inequalities, Preprint, Università degli Studi di Firenze, 2001.
34. Cianchi, A. and Fusco, N., Functions of bounded variation and rearrangements, Preprint, Università degli Studi di Firenze, 2001.
35. Cianchi, A. and Pick, L., Sobolev embeddings into BMO, VMO and L^∞, Arkiv. Math. **36** (1998), 317-340.
36. Copson, E.T., Note on series of positive terms, J. London Math. Soc. **2** (1927), 9-12.
37. Copson, E.T., Note on series of positive terms, J. London Math. Soc. **3** (1928), 49-51.
38. Courant, R. and Hilbert, D, *Methoden der mathematischen Physik II,* Springer-Verlag, Berlin, 1937
39. Cruz-Uribe, D. and Krbec, M., Localisation and extrapolation in Lorentz-Orlicz spaces, to appear.

40. Cwikel, M. and Pustylnik, E., Sobolev type embeddings in the limiting case, J. Fourier Anal. Appl. **4** (1998), 433-446.
41. Davies, E.B., Some norm bounds and quadratic form inequalities for Schrödinger operators (II), J. Operator Theory **12** (1984), 177-196.
42. Davies, E.B., The Hardy constant, Quart. J. Math. Oxford **46** (1995), 417-431.
43. Drábek, P. and Manásevich, R., On the solution to some p-Laplacian nonhomogeneous eigenvalue problems in closed form, Diff. and Int. Equs. **12** (1999), 723-740.
44. Dunford, N. and Schwartz, J.T., *Linear operators I*, Interscience, New York and London, 1958.
45. Edmunds, D.E. and Evans, W.D., Spectral theory and embeddings of Sobolev spaces, Quart. J. Math. Oxford **30** (1979), 431-453.
46. Edmunds, D.E. and Evans, W.D., *Spectral theory and differential operators*, Oxford University Press, Oxford, 1987.
47. Edmunds, D.E., Evans, W.D. and Harris, D.J., Approximation numbers of certain Volterra integral operators, J. London Math. Soc. (2) **37** (1988), 471-489.
48. Edmunds, D.E., Evans, W.D. and Harris, D.J., Two-sided estimates of the approximation numbers of certain Volterra integral operators, Studia Math. **124** (1997), 59-80.
49. Edmunds, D.E., Gurka, P. and Opic, B., Double exponential integrability of convolution operators in generalised Lorentz-Zygmund spaces, Indiana Univ. Math. J. **44** (1995), 19-43.
50. Edmunds, D.E., Gurka, P. and Opic, B., Double exponential integrability, Bessel potentials and embedding theorems, Studia Math. **115** (1995), 151-181.
51. Edmunds, D.E., Gurka, P. and Opic, B., Sharpness of embeddings in logarithmic Bessel potential spaces, Proc. Roy. Soc. Edinburgh **126A** (1996), 995-1009.
52. Edmunds, D.E., Gurka, P. and Opic, B., On embeddings of logarithmic Bessel potential spaces, J. Functional Anal. **146** (1997), 116-150.
53. Edmunds, D.E., Gurka, P. and Opic, B., Norms of embeddings of logarithmic Bessel potential spaces, Proc. Amer. Math. Soc. **126** (1998), 2417-2425.
54. Edmunds, D.E., Gurka, P. and Opic, B., Optimality of embeddings of logarithmic Bessel potential spaces, Quart. J. Math. Oxford **51** (2000), 185-209.
55. Edmunds, D. E., Gurka, P. and Pick, L., Compactness of Hardy-type integral operators in weighted Banach function spaces, Studia Math. **109** (1994), 73-90.
56. Edmunds, D.E. and Haroske, D., Spaces of Lipschitz type, embeddings and entropy numbers, Dissertationes Math. **380** (1999), 1-43.
57. Edmunds, D.E. and Haroske, D., Embeddings in spaces of Lipschitz type, entropy and approximation numbers, and applications, J. Approx. Theory **104** (2000), 226-271.
58. Edmunds, D.E. and Hurri-Syrjänen, R., Weighted Poincaré inequalities and Minkowski content, Proc. Roy. Soc. Edinburgh **125A** (1995), 817-825.
59. Edmunds, D.E. and Hurri-Syrjänen, R., Remarks on the Hardy inequality, J. Inequalities and Appl. **1** (1997), 125-137.
60. Edmunds, D.E. and Hurri-Syrjänen, R., On Hardy-type inequalities, Math. Nachr. **194** (1998), 23-33.
61. Edmunds, D.E. and Hurri-Syrjänen, R., Sobolev inequalities of exponential type, Israel J. Math. **123** (2001), 61-92.
62. Edmunds, D.E. and Hurri-Syrjänen,R., Rellich's theorem in irregular domains, Houston J. Math., to appear.

63. Edmunds, D.E., Kerman, R. and Lang, J., Remainder estimates for the approximation numbers of weighted Hardy operators acting on L^2, J. d'Anal. Math. **85** (2001), 225-243.
64. Edmunds, D.E., Kerman, R. and Pick, L., Optimal Sobolev imbeddings involving rearrangement-invariant quasinorms, J. Functional Anal. **170** (2000), 307-355.
65. Edmunds, D.E., Kokilashvili, V. and Meskhi, A., *Bounded and compact integral operators*, Kluwer Academic Publishers, Dordrecht, 2002.
66. Edmunds, D.E. and Krbec, M., On decomposition in exponential Orlicz spaces, Math. Nachr. **213** (2000), 77-88.
67. Edmunds, D.E. and Krbec, M., Decomposition and Moser's lemma, Revista Mat. Complutense Madrid **15** (2002), 57-74.
68. Edmunds, D. E. and Lang, J., Behaviour of the approximation numbers of a Sobolev embedding in the one-dimensional case, J. Functional Anal. **206** (2004), 149-166.
69. Edmunds, D.E. and Opic, B., Weighted Poincaré and Friedrichs inequalities, J. London Math. Soc. **47** (1993), 79-96.
70. Edmunds, D.E. and Opic, B., Boundedness of fractional maximal operators between classical and weak-type Lorentz spaces, Dissertationes Mathematicae **410** (2002), 1-53.
71. Edmunds, D.E. and Opic, B., On equivalent quasi-norms on Lorentz spaces, Proc. American Math. Soc. **131** (2003), 745-754.
72. Edmunds, D.E., Opic, B. and Pick, L., Poincaré and Friedrichs inequalities in abstract Sobolev spaces, Math. Proc. Camb. Phil. Soc. **113** (1993), 355-379.
73. Edmunds, D.E., Opic, B. and Rákosník, J., Poincaré and Friedrichs inequalities in abstract Sobolev spaces II, Math. Proc. Camb. Phil. Soc. **115** (1994), 159-173.
74. Edmunds, D.E. and Triebel, H., *Function spaces, entropy numbers, differential operators,* Cambridge Univ. Press, Cambridge, 1996.
75. Edmunds, D.E. and Triebel, H., Eigenfunctions of isotropic fractal drums, Operator Theory Adv. Appl. **110** (1999), 81-102.
76. Edmunds, D.E. and Triebel, H., Sharp Sobolev embeddings and related Hardy inequalities: the critical case, Math. Nachr. **207** (1999), 79-92.
77. Eggleston, H.G., *Convexity,* Cambridge Univ. Press, Cambridge, 1966.
78. Enflo, P., A counterexample to the approximation problem in Banach spaces, Acta Math. **130** (1973), 309-317.
79. Evans, L.C. and Gariepy, R.F., *Measure theory and fine properties of functions,* CRC Press, Boca Raton, Florida, 1992.
80. Evans, W.D. and Harris, D.J., Sobolev embeddings for generalized ridged domains, Proc. Lond. Math. Soc. **54** (1987), 141-175.
81. Evans, W.D. and Harris, D.J., On the approximation numbers of Sobolev embeddings for irregular domains, Quart. J. Math. Oxford **40** (1989), 13-42.
82. Evans, W.D. and Harris, D.J., Fractals, trees and the Neumann Laplacian, Math. Ann. **296** (1993), 493-512.
83. Evans, W.D., Harris, D.J. and Lang, J., Two-sided estimates for the approximation numbers of Hardy-type operators in L^1 and L^∞, Studia Math. **130** (1998), 171-192.
84. Evans, W.D., Harris, D.J. and Lang, J., The approximation numbers of Hardy-type operators on trees, Proc. London Math. Soc. **83** (2001), 390-418.

85. Evans, W.D., Harris, D.J. and Pick, L., Weighted Hardy and Poincaré inequalities on trees, J. Lond. Math. Soc. **52** (1995), 121-136.
86. Evans, W.D., Harris, D.J. and Pick, L., Ridged domains, embedding theorems and Poincaré inequalities, Math. Nachr. **221** (2001), 41-74.
87. Evans, W.D., Harris, D.J. and Saitō, Y. On the approximation numbers of Sobolev embeddings on singular domains and trees, Quart. J. Math. Oxford, to appear.
88. Evans, W.D., Opic, B. and Pick, L., Interpolation of operators on scales of generalized Lorentz-Zygmund spaces, Math. Nachr. **182** (1996), 127-181.
89. Evans, W.D., Opic, B. and Pick, L., Real interpolation with logarithmic functors, J. Inequalities and Appl. **7** (2002), 187-269.
90. Falconer, K.J., *The geometry of fractal sets*, Cambridge University Press, Cambridge, 1985.
91. Farkas, W., Function spaces of generalised smoothness and pseudodifferential operators associated to a negative definite function, Habilitationsschrift, Univ. München, 2002.
92. Farkas, W. and Leopold, H.-G., Characterisations of function spaces of generalised smoothness, Preprint, Jena, 2001.
93. Federer, H., *Geometric measure theory*, Springer-Verlag, Berlin, 1969.
94. Fraenkel, L.E., On regularity of the boundary in the theory of Sobolev spaces, Proc. London Math. Soc. (3) **39** (1979), 385-427.
95. Fremlin, D.H., Skeletons and central sets, Proc. London Math. Soc. **74** (1997), 701-720.
96. Fusco, N., Lions, P.-L. and Sbordone, C., Sobolev embedding theorems in borderline cases, Proc. Amer. Math. Soc. **124** (1996), 561-565.
97. Gehring, F.W. and Martio, O., Lipschitz classes and quasiconformal mappings, Ann. Acad. Sci. Fenn. Ser. A I Math. **10** (1985), 203-219.
98. Gehring, F.W. and Osgood, B.G., Uniform domains and quasi-hyperbolic metric, J. d'Analyse Math. **36** (1979), 50-74.
99. Gilbarg, D. and Trudinger, N.S., *Elliptic partial differential equations of second order*, Springer-Verlag, Berlin-Heidelberg-New York, 1977.
100. Giusti, E., *Minimal surfaces and functions of bounded variation,* Birkhäuser, Boston, 1984.
101. Gogatishvili, A., Neves, J.S. and Opic, B., Optimality of embeddings of Bessel-potential type spaces into Lorentz-Karamata spaces, to appear.
102. Gogatishvili, A., Neves, J.S. and Opic, B., Optimality of embeddings of Bessel-potential type spaces into Lipschitz-Lorentz spaces, to appear.
103. Gogatishvili, A., Opic, B. and Trebels, W., Limiting reiteration for real interpolation with slowly varying functions, to appear.
104. Goldman, M.L., A description of the traces of some function spaces, Trudy Mat. Inst. Steklov **150** (1979), 99-127. English transl.: Proc. Steklov Inst. Math. 1981, no. 4 (150).
105. Goldman, M.L., A method of coverings for describing general spaces of Besov type, Trudy Mat. Inst. Steklov **156** (1980), 47-81. English transl.: Proc. Steklov Inst. Math. 1983, no. 2 (156).
106. Goldman, M.L., Imbedding theorems for anisotropic Nikol'skii-Besov spaces with moduli of continuity of general type, Trudy Mat. Inst. Steklov **170** (1984), 86-104. English transl.: Proc. Steklov Inst. Math. 1987, no. 1 (170).

107. Goldman, M.L., Embedding constructive and structural Lipschitz spaces in symmetric spaces, Trudy Mat. Inst. Steklov **173** (1986), 90-112. English transl.: Proc. Steklov Inst. Math. 1987, no. 4 (173).
108. Goldman, M.L., Traces of functions with restrictions on the spectrum, Trudy Mat. Inst. Steklov **187** (1989), 69-77. English transl.: Proc. Steklov Inst. Math. 1990, no. 3 (187).
109. Goldman, M.L., A criterion for the embedding of different metrics for isotropic Besov spaces with arbitrary moduli of continuity, Trudy Mat. Inst. Steklov **201** (1992), 186-218. English transl.: Proc. Steklov Inst. Math. 1994, no. 2 (201).
110. Gol'dstein, V. and Gurov, L., Applications of change of variables operators for exact embedding theorems, Int. Equ. Oper.Th. **19** (1994), 1-24
111. Grosse-Erdmann, K.-G., *The Blocking Technique, Weighted Mean Operators and Hardy's Inequality*, Lecture Notes in Mathematics 1679, Springer-Verlag, Berlin, 1998.
112. Gurka, P. and Opic, B., Global limiting embeddings of logarithmic Bessel potential spaces, Math. Ineq. Appl. **1** (1998), 565-584.
113. Gurka, P. and Opic, B., Sharp embeddings of Besov spaces with logarithmic smoothness, to appear.
114. Hajłasz, P., Sobolev spaces on an arbitrary metric space, Potential Anal. **5** (1996), 403-415.
115. Hajłasz, P., Sobolev inequalities, truncation method, and John domains, Papers on Analysis, Vol. dedicated to Olli Martio, Report Univ. Jyväskylä **83**, (2001), pp.109-126.
116. Hajłasz, P. and Koskela, P., Sobolev met Poincaré, Mem. American Math. Soc. **145** (2000).
117. Hansson, K., Imbedding theorems of Sobolev type in potential theory, Math. Scand. **45** (1979), 77-102.
118. Hardy, G. H., Note on a theorem of Hilbert, Math. Z. **6** (1920), 314-317.
119. Hardy, G.H., Littlewood, J.E. and Pólya, G., *Inequalities*, Second Edition, Cambridge University Press, 1952.
120. Haroske, D., On more general Lipschitz spaces, Z. Anal. Anwendungen **19** (2000), 781-800.
121. Haroske, D., Envelopes in function spaces-a first approach, to appear.
122. Hedberg, L.I., On certain convolution inequalities, Proc. Amer. Math. Soc. **36** (1972), 505-510.
123. Heinonen, J., *Lectures on analysis on metric spaces*, Springer-Verlag, Berlin, 2001.
124. Herron, D.A., Metric boundary conditions for plane domains, Proc. 13th. Nevanlinna Colloquium.
125. Herron, D.A. and Vuorinen, M., Positive harmonic functions in uniform and admissible domains, Analysis **8** (1988), 187-206.
126. Hewitt, E. and Stromberg, K., *Real and abstract analysis*, Springer-Verlag, New York, 1965.
127. Hörmander, L., *The Analysis of Linear Partial Differential Equations I*, Springer-Verlag, Berlin, 1983.
128. Hörmander, L., *Notions of Convexity*, Birkhäuser, Boston, 1994.
129. Hurri, R., Poincaré domains in $\mathbb{R}^n$, Ann. Acad. Sci. Fenn. Ser. A I Math. Dissertationes **71** (1988), 1-42.
130. Jawerth, B. and Milman, M., Extrapolation theory with applications, Mem. Amer. Math. Soc. **440** (1991).

131. Jerison, D.S. and Kenig, C.E., Boundary behavior of harmonic functions in non-tangentially accessible domains, Adv. Math. **46** (1982), 80-147.
132. John, F., Rotation and strain, Comm. Pure Appl. Math. **4** (1961), 391-414.
133. Jones, P.W., Quasiconformal mappings and extendability of functions in Sobolev spaces, Acta Math. **147** (1981), 71-88.
134. Kalyabin, G.A., Characterization of spaces of generalized Liouville differentiation, Mat. Sb. Nov. Ser. **104** (1977), 42-48.
135. Kalyabin, G.A., Description of functions in classes of Besov-Lizorkin-Triebel type, Trudy Mat. Inst. Steklov **156** (1980), 82-109. English transl.: Proc. Steklov Inst. Math. 1983, no. 2 (156).
136. Kawohl, B., *Rearrangements and convexity of level sets in PDE*, Lecture Notes in Math. **1150**, Springer-Verlag, Berlin, 1985.
137. Kinnunen, J., The Hardy-Littlewood maximal function of a Sobolev function, Israel J. Math. **100** (1997), 117-124.
138. Kinnunen, J. and Lindqvist, P., The derivative of the maximal function, J. reine angew. Math. **503** (1998), 161-167.
139. Kinnunen, J. and Martio, O., Hardy's inequality for Sobolev functions, Math. Res. Lett. **4** (1997), 489-500.
140. Kinnunen, J. and Saksman, E., Regularity of the fractional maximal function, University of Helsinki Mathematics Report, Preprint 308 (2001).
141. Kolyada, V.I., Rearrangements of functions, and embedding theorems, Russian Math. Surveys **44** (1989), 73-117.
142. Kolyada, V.I., Rearrangements of functions and embedding of anisotropic spaces of Sobolev type, East J. on Approximations **4** (1998), 111-198.
143. König, H., *Eigenvalue distribution of compact operators*, Birkhäuser, Basel, 1986.
144. Krasnosel'skii, M.A. and Rutickii, Ya.B., *Convex functions and Orlicz spaces*, Noordhoff, Groningen, 1961.
145. Krbec, M., Opic, B., Pick, L. and Ràkosnìk, J., Some recent results on Hardy-type operators in weighted function spaces and related topics, *Function spaces, differential operators and non-linear analysis*, Vol. **133**, Proc. Spring School, Prague, 1993, Teubner-Text, 1993, pp. 158-182.
146. Krbec, M. and Schmeisser, H.-J., Imbeddings of Brézis-Wainger type. The case of missing derivatives, Proc. Roy. Soc. Edinburgh **131A**, (2001), 1-34.
147. Kufner, A., *Weighted Sobolev spaces,* Wiley, Chichester, 1985.
148. Kufner, A., John, O. and Fučik, S., *Function Spaces,* Academia, Prague, 1977.
149. Lai, Q. and Pick, L., The Hardy operator, L_∞ and BMO, J. London Math. Soc. **48** (1993), 167-177.
150. Landau, E., A note on a theorem concerning series of positive terms, J. London Math. Soc. **1** (1926), 38-39.
151. Lang, J., Improved estimates for the approximation numbers of Hardy-type operators, J. Approx. Theory, to appear.
152. Lang, J. and Nekvinda, A., A difference between the continuous and the absolutely continuous norms in Banach function spaces, Czech. Math. J. **47** (1997), 221-232.
153. Lapidus, M.L., Fractal drum, inverse spectral problems for elliptic operators and a partial resolution of the Weyl-Berry conjecture, Trans. Amer. Math. Soc. **325** (1991), 465-529.
154. Leopold, H.-G., Embeddings and entropy numbers in Besov spaces of generalized smoothness, Proc. Conf. Function Spaces V, Poznan, 1998.

155. Levin, V.I., On a class of integral inequalities, Recueil Math. **4** (46) (1938), 309-331.
156. Lewis, J.L., Uniformly fat sets, Trans. Amer. Math. Soc. **308** (1988), 177-196.
157. Lifschits, M.A. and Linde, W., Approximation and entropy numbers of Volterra operators with application to Brownian motion, AMS Memoirs, **157**, no. 745 (2002).
158. Lizorkin, P.I., Spaces of generalized smoothness, Appendix to the Russian edition of [217].
159. Lomakina, E. and Stepanov, V., On the compactness and approximation numbers of Hardy-type integral operators in Lorentz spaces, J. London Math. Soc.(2) **53** (1996), 369-382.
160. Lomakina, E. and Stepanov, V., On the Hardy-type integral operator in Banach function spaces, Preprint, 1997 .
161. Lomakina, E. and Stepanov, V., On the asymptotic behaviour of the approximation numbers and estimates of Schatten-von Neumann norms of the Hardy-type integral operator, Preprint, 2000.
162. Luxemburg, W.A.J., Banach function spaces, Thesis, Technische Hogeschool te Delft, 1955.
163. Manakov, V.M., On the best constant in weighted inequalities for Riemann-Liouville inegrals, Bull. London Math. Soc. **24** (1992), 442-448.
164. Malý, J. and Pick, L., An elementary proof of sharp Sobolev embeddings, Proc. Amer. Math. Soc. **130** (2002), 555-563.
165. Marcus, M., Mizel, V.J. and Pinchover, Y., On the best constant for Hardy's inequality in $\mathbb{R}^n$, Trans. Amer. Math. Soc. **350** (1998), 3237-3255.
166. Martio, O., John domains, bilipschitz balls and Poincaré inequality, Roumaine Math. Pures Appl. **33** (1988), 107-112.
167. Martio, O. and Sarvas, J., Injectivity theorems in plane and space, Ann. Acad. Sci. Fenn. Ser. A I Math. **4** (1978/1979), 383-401.
168. Martio, O. and Väisälä, J., Global L^p-integrability of the derivatives of a quasiconformal mapping, Complex Variables, Theory and Appl. **9** (1988), 309-319.
169. Martio, O. and Vuorinen, M., Whitney cubes, $p-$capacity and Minkowski content, Exposition. Math. **5** (1987), 17-40.
170. Matskewich, T. and Sobolevskii, P.E., The best possible constant in a generalized Hardy's inequality for convex domains in $\mathbb{R}^n$, in Proc. Conf. on Elliptic and Parabolic P.D.E.'s and Applications, Capri, 1994.
171. Maz'ya, V.G., *Sobolev spaces,* Springer-Verlag, Berlin, 1985.
172. Maz'ya, V.G. and Poborchi, S.V., *Differentiable functions on 'bad' domains*, World Scientific Press, Singapore, 1997.
173. McShane, E.J., Extension of range of functions, Bull. Amer. Math. Soc. **40** (1934), 837-842.
174. Meyers, N.G. and Serrin, J., $H = W$, Proc. Nat. Acad. Sci. USA **51** (1964), 1055-1056.
175. Milman, M., *Extrapolation and optimal decompositions, with applications to analysis*, Lecture Notes in Math. **1580,** Springer-Verlag, Berlin, 1994.
176. Moura, S., Function spaces of generalised smoothness, Dissert. Math. **398** (2001), 1-88.
177. Muckenhoupt, B., Hardy's inequality with weights, Studia Math. **44** (1972), 31-38.

178. Naimark, K. and Solomyak, M., Eigenvalue estimates for the weighted Laplacian on metric trees, Proc. London Math. Soc. **80** (2000), 690-724.
179. Netrusov, Yu.V., Embedding theorems for Lizorkin-Triebel spaces , Notes of Scientific Seminar LOMI **159** (1987), 103-112 (Russian).
180. Netrusov, Yu.V., Embedding theorems for traces of Besov spaces and Lizorkin-Triebel spaces, Dokl. Akad. Nauk SSSR **298** (1988), 1326-1330. English transl. : Soviet Math. Dokl. 1988, no. 1 (37).
181. Netrusov, Yu.V., Metric estimates of the capacities of sets in Besov spaces, Trudy Mat. Inst. Steklov **190** (1989), 159-185. English transl.: Proc. Steklov Inst. Math. 1992, no. 1 (190).
182. Neves, J.S., Fractional Sobolev-type spaces and embeddings, D. Phil. thesis, University of Sussex, 2001.
183. Neves, J.S., On decompositions in generalised Lorentz-Zygmund spaces, Boll. Un. Mat. Ital. B, **4**-B (8) (2001), 239-267.
184. Neves, J.S., Extrapolation results on general Besov-Hölder-Lipschitz spaces, Math. Nachr. **230** (2001), 117-141.
185. Neves, J.S., Lorentz-Karamata spaces, Bessel and Riesz potentials and embeddings, Dissertationes Mathematicae **405** (2002).
186. Neves, J.S., Bessel-Lorentz-Karamata spaces and embeddings: the super-limiting case, to appear.
187. Newman, J. and Solomyak, M., Two-sided estimates of singular values of a class of integral operators on the semi-axis, Integral Equations and Operator Theory **20** (1994), 335-349.
188. Opic, B., Embeddings of Bessel potential and Sobolev type spaces, Colloq. Depart. Anal. Mat., CURSO 1999-2000, Sección 1, Número 48, Universidad Complutense de Madrid, 2001, pp. 100-118.
189. Opic, B., New characterizations of Lorentz spaces, Proc. Roy. Soc. Edinburgh **133A** (2003), 439-448.
190. Opic, B., On equivalent quasi-norms on Lorentz spaces, in *Function Spaces, Differential Operators and Non-linear Analysis*, The Hans Triebel Anniversary Volume, Birkhauser, Basel-Boston-Berlin, 2003, 415-426.
191. Opic, B. and Kufner, A., *Hardy-type inequalities*, Pitman Research Notes in Math. Series **219**, Longman Sci. and Tech., Harlow, 1990.
192. Opic, B. and Pick, L., On generalized Lorentz-Zygmund spaces, Math. Ineq. and Appl. **2** (1999), 391-467.
193. Opic, B. and Trebels, W., Bessel potentials with logarithmic components and Sobolev-type embeddings, Anal. Math. **26** (2000), 299-319.
194. Opic, B. and Trebels, W., Sharp embeddings of Bessel potential spaces with logarithmic smoothness, Math. Proc. Camb. Phil. Soc. **134** (2003), 347-384.
195. Peetre, J., Espaces d'interpolation et théorème de Soboleff, Ann. Inst. Fourier **16** (1966), 279-317.
196. Pick, L., Optimal Sobolev embeddings, Nonlinear analysis, function spaces and applications **6** (M. Krbec and A. Kufner, eds.), Math. Inst. Czech Acad. Sci. Prague, 1999, pp. 156-199.
197. Pick, L., Supremum operators and optimal Sobolev inequalities, Function spaces, differential operators and nonlinear analysis (V. Mustonen and J. Rákosník, eds.), Math. Inst. Czech Acad. Sci. Prague, 2000, pp. 207-219.
198. Pietsch, A., *Operators ideals*, North-Holland, Amsterdam, 1980.
199. Pietsch, A., *Eigenvalues and s−numbers*, Cambridge Univ. Press, Cambridge, 1987.

200. Pohozaev, S.I., On the eigenfunctions of the equation $\Delta u + \lambda f(u) = 0$, Soviet Math. Dokl. **6** (1965), 1408-1411.
201. Pólya, G. and Szegö, G., *Isoperimetric inequalities in mathematical physics*, Princeton Univ. Press, Princeton, 1951.
202. Rao, M.M. and Ren, B.D., *Theory of Orlicz spaces*, Marcel Dekker, New York, 1991.
203. Read, T.T., Generalisations of Hardy's inequality, preprint.
204. Reshetnyak, Yu.G., Integral representations of differentiable functions in domains with nonsmooth boundary (Russian), Sib. Mat. Sb. **21** (1980), 108-116.
205. Sawyer, E. T., Weighted Lebesgue and Lorentz norm inequalities for the Hardy operator, Trans. American Math. Soc. **281** (1984), 329-337.
206. Schmidt, E., Über die Ungleichung, welche die Integrale über eine Potenz einer Funktion und über eine andere Potenz ihrer Ableitung verbindet, Math. Ann. **117** (1940), 301-326.
207. Smith, W. and Stegenga, D., Hölder domains and Poincaré domains, Trans. American Math. Soc. **319** (1990), 67-100.
208. Smith, W. and Stegenga, D., Exponential integrabilty of the quasi-hyperbolic metric on Hölder domains, Ann. Acad. Sci. Fenn. Ser. A I Math. **16** (1991), 345-360.
209. Smith, W. and Stegenga, D., Sobolev imbeddings and integrability of harmonic functions on Hölder domains, Potential Theory (Nagoya, 1990) (1992), 303-313, de Gruyter, Berlin.
210. Solomyak, M.Z., A remark on the Hardy inequalities, Integr. Equat. Oper. Th. **19** (1994), 120-124.
211. Stein, E.M., *Singular integrals and differentiability properties of functions*, Princeton University Press, Princeton, 1970.
212. Stepanov, V., Weighted norm inequalities for integral operators and related topics, *in Non-linear Analysis, Function Spaces and Applications 5*, 139-176, Olympia Press, Prague, 1994.
213. Strichartz, R.S., A note on Trudinger's extension of Sobolev's inequality, Indiana Univ. Math. J. **21** (1972), 841-842.
214. Talenti, G., Best constant in Sobolev inequality, Ann. Mat. Pura Appl. **110** (1976), 353-372.
215. Talenti, G., An inequality between u^* and $|grad\ u|^*$, *General inequalities* **6** (Oberwolfach, 1990), Internat. Ser. Numer. Math., Vol. **103** (W.Walter, ed.), Birkhäuser, Basel, 1992, pp. 175-182.
216. Talenti, G., Inequalities in rearrangement invariant function spaces, *Nonlinear analysis, function spaces and applications,* Vol. **5**, Proc. Spring School, Prague, 1994, Prometheus Publishing House, 1994, pp. 177-230.
217. Torchinsky, A., *Real variable methods in harmonic analysis*, Academic Press, San Diego, 1986.
218. Triebel, H., *Interpolation theory, function spaces, differential operators*, North-Holland, Amsterdam, 1978.
219. Triebel, H., Approximation numbers and entropy numbers of embeddings of fractional Besov-Sobolev spaces in Orlicz spaces, Proc. London Math. Soc. **66** (1993), 589-618.
220. Triebel, H., Sharp Sobolev embeddings and related Hardy inequalities: the sub-critical case, Math. Nachr. **208** (1999), 167-178.
221. Triebel, H., *The structure of functions*, Birkhäuser Verlag, Basel, 2001.

222. Trudinger, N.S., On embeddings into Orlicz spaces and some applications, J. Math. Mech. **17** (1967), 473-484.
223. Väisälä, J., Uniform domains, Tohoku Math. J. **40** (1988), 101-118.
224. Väisälä, J., Quasiconformal maps of cylindrical domains, Acta Math. **162** (1989), 201-225.
225. Väisälä, J., Exhaustions of John domains, Ann. Acad. Sci. Fenn. Ser. A I Math. **19** (1994), 47-57.
226. Wannebo, A., Hardy inequalities, Proc. American Math. Soc. **109** (1990), 85-95.
227. Wannebo, A., Hardy inequalities and embeddings in domains generalising $C^{0,\alpha}$ domains, Proc. American Math. Soc. **122** (1994), 1181-1190.
228. Wojtaszczyk, P., *Banach spaces for analysts*, Cambridge Univ. Press, Cambridge, 1991.
229. Yano, S., Notes on Fourier analysis (XXIX): An extrapolation theorem, J. Math. Soc. Japan **3** (1951), 296-305.
230. Yudovich, V.I., Some estimates connected with integral operators and with solutions of elliptic equations, Soviet Math. Doklady **2** (1961), 746-749.
231. Ziemer, W., *Weakly differentiable functions*, Springer-Verlag, Berlin, 1989.
232. Zygmund, A., *Trignometric Series*, Volume II, Cambridge Univ. Press, Cambridge, 2nd edition, 1959.

Author Index

Subject Index

Notation Index

Springer Monographs in Mathematics

This series publishes advanced monographs giving well-written presentations of the "state-of-the-art" in fields of mathematical research that have acquired the maturity needed for such a treatment. They are sufficiently self-contained to be accessible to more than just the intimate specialists of the subject, and sufficiently comprehensive to remain valuable references for many years. Besides the current state of knowledge in its field, an SMM volume should also describe its relevance to and interaction with neighbouring fields of mathematics, and give pointers to future directions of research.

Abhyankar, S.S. **Resolution of Singularities of Embedded Algebraic Surfaces** 2nd enlarged ed. 1998
Andrievskii, V.V.; Blatt, H.-P. **Discrepancy of Signed Measures and Polynomial Approximation** 2002
Ara, P.; Mathieu, M. **Local Multipliers of C*-Algebras** 2003
Armitage, D.H.; Gardiner, S.J. **Classical Potential Theory** 2001
Arnold, L. **Random Dynamical Systems** corr. 2nd printing 2003 (1st ed. 1998)
Aubin, T. **Some Nonlinear Problems in Riemannian Geometry** 1998
Auslender, A.; Teboulle M. **Asymptotic Cones and Functions in Optimization and Variational Inequalities** 2003
Bang-Jensen, J.; Gutin, G. **Digraphs** 2001
Baues, H.-J. **Combinatorial Foundation of Homology and Homotopy** 1999
Brown, K.S. **Buildings** 3rd printing 2000 (1st ed. 1998)
Cherry, W.; Ye, Z. **Nevanlinna's Theory of Value Distribution** 2001
Ching, W.K. **Iterative Methods for Queuing and Manufacturing Systems** 2001
Crabb, M.C.; James, I.M. **Fibrewise Homotopy Theory** 1998
Dineen, S. **Complex Analysis on Infinite Dimensional Spaces** 1999
Elstrodt, J.; Grunewald, F. Mennicke, J. **Groups Acting on Hyperbolic Space** 1998
Edmunds, D.E.; Evans, W.D. **Hardy Operators, Function Spaces and Embeddings** 2004
Fadell, E.R.; Husseini, S.Y. **Geometry and Topology of Configuration Spaces** 2001
Fedorov, Y.N.; Kozlov, V.V. **A Memoir on Integrable Systems** 2001
Flenner, H.; O'Carroll, L. Vogel, W. **Joins and Intersections** 1999
Gelfand, S.I.; Manin, Y.I. **Methods of Homological Algebra** 2nd ed. 2003
Griess, R.L.Jr. **Twelve Sporadic Groups** 1998
Gras, G. **Class Field Theory** 2003
Ivrii, V. **Microlocal Analysis and Precise Spectral Asymptotics** 1998
Jech, T. **Set Theory** (3rd revised edition 2002)
Jorgenson, J.; Lang, S. **Spherical Inversion on SLn (R)** 2001
Kanamori, A.; **The Higher Infinite** (2nd edition 2003)
Khoshnevisan, D. **Multiparameter Processes** 2002
Koch, H. **Galois Theory of *p*-Extensions** 2002
Kozlov, V.; Maz'ya, V. **Differential Equations with Operator Coefficients** 1999
Landsman, N.P. **Mathematical Topics between Classical & Quantum Mechanics** 1998
Lebedev, L.P.; Vorovich, I.I. **Functional Analysis in Mechanics** 2002
Lemmermeyer, F. **Reciprocity Laws: From Euler to Eisenstein** 2000
Malle, G.; Matzat, B.H. **Inverse Galois Theory** 1999
Mardesic, S. **Strong Shape and Homology** 2000
Margulis, G.A. **On Some Aspects of the Theory of Anosov Systems** 2004
Murdock, J. **Normal Forms and Unfoldings for Local Dynamical Systems** 2002
Narkiewicz, W. **The Development of Prime Number Theory** 2000
Parker, C.; Rowley, P. **Symplectic Amalgams** 2002
Peller, V. (Ed.) **Hankel Operators and Their Applications** 2003
Prestel, A.; Delzell, C.N. **Positive Polynomials** 2001
Puig, L. **Blocks of Finite Groups** 2002
Ranicki, A. **High-dimensional Knot Theory** 1998
Ribenboim, P. **The Theory of Classical Valuations** 1999
Rowe, E.G.P. **Geometrical Physics in Minkowski Spacetime** 2001

Rudyak, Y.B. **On Thom Spectra, Orientability and Cobordism** 1998
Ryan, R.A. **Introduction to Tensor Products of Banach Spaces** 2002
Saranen, J.; Vainikko, G. **Periodic Integral and Pseudodifferential Equations with Numerical Approximation** 2002
Schneider, P. **Nonarchimedean Functional Analysis** 2002
Serre, J-P. **Complex Semisimple Lie Algebras** 2001 (reprint of first ed. 1987)
Serre, J-P. **Galois Cohomology** corr. 2nd printing 2002 (1st ed. 1997)
Serre, J-P. **Local Algebra** 2000
Serre, J-P. **Trees** corr. 2nd printing 2003 (1st ed. 1980)
Smirnov, E. **Hausdorff Spectra in Functional Analysis** 2002
Springer, T.A. Veldkamp, F.D. **Octonions, Jordan Algebras, and Exceptional Groups** 2000
Sznitman, A.-S. **Brownian Motion, Obstacles and Random Media** 1998
Taira, K. **Semigroups, Boundary Value Problems and Markov Processes** 2003
Tits, J.; Weiss, R.M. **Moufang Polygons** 2002
Uchiyama, A. **Hardy Spaces on the Euclidean Space** 2001
Üstünel, A.-S.; Zakai, M. **Transformation of Measure on Wiener Space** 2000
Yang, Y. **Solitons in Field Theory and Nonlinear Analysis** 2001

Printed by Publishers' Graphics LLC